Molecular Plasmonics

Molecular Plasmonics

Theory and Applications

Volodymyr I. Chegel
Andrii M. Lopatynskyi

JENNY STANFORD
PUBLISHING

Published by

Jenny Stanford Publishing Pte. Ltd.
Level 34, Centennial Tower
3 Temasek Avenue
Singapore 039190

Email: editorial@jennystanford.com
Web: www.jennystanford.com

British Library Cataloguing-in-Publication Data
A catalogue record for this book is available from the British Library.

Molecular Plasmonics: Theory and Applications

ISBN 978-981-4800-65-5 (Hardcover)
ISBN 978-0-429-29511-9 (eBook)

Contents

Preface

Our laboratory has been involved in surface plasmon resonance (SPR) research, particularly on nanochips, for more than 30 years and has developed one of the world's first localized SPR spectrometers using plasmonic nanochips. We received an invitation from Jenny Stanford Publishing to write a book on SPR, and that is when we conceived the idea of a book on nanochips, which are arrays of highly conductive noble metal nanostructures on dielectric substrates. After a discussion at the publisher's editorial office, two questions gripped our minds, "Since plasmonic nanochips form a novel and important direction of research, how can we discuss them discretely from molecular plasmonics in our book?" and "How will this presentation be different from the existing publications on molecular plasmonics?" This led us to decide that we will present information on topics that have not been sufficiently covered in existing publications on SPR.

We found it necessary to emphasize on topics that have been relatively bypassed in existing publications, such as SPR studies of conformational transformations, electrochemical SPR, metamaterials exploiting SPR, and, of course, the theme of plasmonic nanochips. So these topics found their first and foremost place in this book. They have been complemented with topics, such as results of studies in the fields of bio- and chemosensors and surface enhancement using plasmonics methods, that have already been covered well in other publications, but without which the book would be incomplete. The chapter on theoretical description of the physical mechanisms of plasmonics completes the concept of this book. The information has been presented in the form of short completed studies, which would be helpful for researchers looking forward to master this new field of research. In fact, molecular plasmonics is a classic example of an interdisciplinary scientific field in which the correctness of achieved results must be confirmed by as many complementary methods as possible, and completed studies published in reputable publications are an important resource for it. Therefore, this book includes studies that were conducted by the authors either in their laboratory or in some of the best scientific laboratories of the world together with their fellow researchers.

We are sincerely grateful to scientists at the Hebrew University of Jerusalem (laboratory of Prof. Itamar Willner), Israel; the University of Michigan (laboratory of Prof. L. Jay Guo), USA; the National Institute for Materials Science (laboratory of Prof. Katsuhiko Ariga), Japan; and the Taras Shevchenko National University of Kyiv (department of Prof. Valeri Lozovski), Ukraine, and to the colleagues at the National Academy of Sciences of Ukraine (department of Prof. Yuri Shirshov), where research on molecular plasmonics is receiving good support with respect to resources and is being done in an established manner.

We hope that this book will be useful for students who are beginning their professional careers as well as experienced researchers, including those who are engaged in interdisciplinary research in the ever-growing field of molecular plasmonics.

Volodymyr I. Chegel

Andrii M. Lopatynskyi

Summer 2020

Chapter 1

Molecular Plasmonics

1.1 Introduction

Unstoppable scientific progress in the development and preservation of human civilization is the solution to many of the challenges facing humankind in the third millennium. Humanity is constantly opening new opportunities for itself, but it also faces new challenges, among which are the challenges in the fight against new diseases, in preserving the environment, and in confronting terrorist threats. According to this, research in biochemistry, chemosensory, and materials science is of considerable scientific and practical interest. Surface plasmon resonance (SPR) takes a special place among the modern scientific methods, which combine the possibilities for research in the aforementioned directions. Surface oscillations of free electrons in high-conductive materials (surface plasmons) upon their resonance excitation with light create a sensitive electromagnetic field that penetrates the adjacent medium and can be used as an active supersensitive probe to changes in its refractive index. The non-destructive nature of the surface plasmon field allows investigating biomolecules in their natural state without the use of different types of labels in real time. The SPR method is not limited to biomolecular studies, which are mostly nanoscale. The ability to measure the physical parameters of a substance in a real-time mode is also necessary in the world of macromolecules,

Molecular Plasmonics: Theory and Applications
Volodymyr I. Chegel and Andrii M. Lopatynskyi

ISBN 978-981-4800-65-5 (Hardcover), 978-0-429-29511-9 (eBook)
www.jennystanford.com

for example, while studying conformational transformations in polymers during the polymerization process or under the influence of external factors. In this case, an important additional advantage of the SPR method is the possibility of macromolecular studies using electrochemical approaches. The process kinetics is one of the major factors in the study of materials that exhibit redox properties, and electrochemical SPR provides an opportunity for real-time study of redox transformations in materials with electrorefractive, electrochromic, and conductive functions.

Research is being conducted with the use of SPR to enhance optical transitions—in particular, in the directions of surface-enhanced fluorescence and surface-enhanced infrared spectroscopy. Among the number of urgent problems of the SPR method, one can distinguish the need for a detailed explanation of the nature of the optical response in multilayer structures with pronounced heterogeneity, the study of processes at the media boundary under the external factors influence, and the determination of mechanisms for the interaction of biomolecules of various types and shapes with an electromagnetic field on the surface of a solid.

In contrast to the plasma wave that occurs at the surface plasmon resonance, localized surface plasmon is a collective oscillation of conduction electrons excited by the electromagnetic field of an incident light, which is confined in three-dimensional space. When the size of the metal particles decreases to the nanometer range, their optical properties change sharply with the appearance of localized surface plasmon resonance (LSPR) and their behavior differs significantly from the bulk material. At the same time, a significant dependence of the optical parameters of nanoparticles on their size and shape arises. That is why it is necessary to study the interaction mechanisms of nanosized metal particles with molecules and to explain the optical response of sensor structures on their basis, which operate due to the LSPR phenomenon. Because of the small size of individual biomolecules, in order to achieve an effective interaction between a molecule and a nanoparticle, the LSPR method requires precise control of the localized surface plasmon electromagnetic field spatial profile for the placement of molecules within this field. The use of LSPR method in nanomedicine imposes additional conditions on the shape, size, LSPR wavelength, and surface functionalization of metallic nanoparticles when used for targeted delivery, visualization, and plasmonic excitation inside the human body. At present, LSPR has already proven itself as a promising scientific method, and with the growth of production capabilities of nanomaterials, its role is rapidly increasing, and research into its use and search for new applications are becoming more and more relevant.

As a result of the scientific community's efforts in SPR and LSPR research, during the last decades a new promising scientific trend has been formed: plasmonics. This book presents generalized research results of its authors using the SPR and LSPR methods related to the studies of a variety of molecules and molecular complexes, as well as their interaction with external factors and objects of influence. This new direction of plasmonics was called the molecular plasmonics.

1.2 Overview of Current Research Progress in Molecular Plasmonics

The development of plasmonics has led to the appearance of advanced methods and analysis tools for applications in molecular research and to the formation of a scientific field known as molecular plasmonics. Nanosized electromagnetic fields of surface plasmons can interact with polymeric macromolecules, cells, and biomolecules of different sizes via optical, thermal, and mechanical influences. The control of the interaction between these objects and surface plasmons allowed the development of approaches for the effective detection, analysis, capture, transport, and manipulation of these objects. For example, since the electromagnetic field concentrated near the particles is sensitive to molecular interactions, metal nanoparticles can function as sensors for understanding biological processes of the molecular level [1]. Due to the increased light scattering and absorption cross section, metal nanoparticles are extremely sensitive markers for immunological analysis and molecular spectroscopy [2]. The high intensity and large gradient of electromagnetic (EM) fields of LSPR in the near field of a particle lead to the appearance of significant optical forces, which are the basis for the plasmonic tweezers used for the study of individual molecules [3].

Due to the small size of single molecules, to achieve effective interaction between the molecule and the surface plasmon, the molecular plasmonics requires precise control of the plasmons' spatial profile and the molecules' position within the limits of the metallic nanoparticles near field. The use of metal nanoparticles in vivo imposes additional conditions related to their shape, size, wavelength, and surface functionalization for directed delivery, visualization, and plasmonic excitation inside the human body. Progress in the nanomanufacturing, measurement methods, tools and calculations provided the opportunity to accurately simulate and control the profile of the near and far fields of LSPR in a wide range of wavelengths. This approach made possible the necessary time, spatial, and spectral control of the interactions between the plasmon and the molecule.

1.2.1 Sensors Based on SPR and LSPR Phenomena

Surface plasmon resonance is exceptionally sensitive to changes in the refractive index of the medium near the surface of the metal film or nanoparticle due to the nanosized localization and amplification of the EM field. The measurements of angular SPR spectra or LSPR spectra in the light extinction or scattering modes register changes in the local refractive index due to the presence of molecules. In particular, the SPR angular position or the LSPR peak wavelength can be correlated with molecular adsorption, desorption, or conformational changes that cause changes in the refractive index. The high sensitivity of the nanoparticle EM near field in combination with the advantages of light (non-destructive action, high speed, and direction) makes the SPR and LSPR (plasmonic) sensors promising for the study of biological molecules and reactions.

An important stage for the creation of functional SPR sensors is the development of physical criteria for determining the complex refractive index and geometric parameters of the layered molecular structures and recording changes in these parameters. It was shown in Ref. [4] that the processes of surface plasmon polariton resonance excitation in the Kretschmann configuration sensors are well described by a one-dimensional model of a multilayer system based on the effective optical constants matrix and layer thicknesses, which takes into account the presence of transition layers and the imperfectness of the geometric surface. Due to this, an approach was developed that allows the use of the SPR angular spectrum shape to determine by mathematical analysis at least three parameters (effective refractive index and absorption coefficient, thickness) of the investigated structure.

An important area of SPR research is the development of new sensing techniques, in particular, to expand the capabilities and improve the limit of detection of sensor devices. For example, Lavine et al. [5] deposited molecularly imprinted lightly crosslinked *N*-(*N*-propyl)acrylamide granules on a glass plate covered with a thin gold layer (SPR chip) by the preparation of thin organic layers by centrifugation (spincoating), and the swelling of these granules was used to measure the concentration of theophylline. Gabai et al. [6] performed measurements of glucose concentration using copolymer films of "boronic acid/acrylamide" on the SPR chip surface made by electropolymerization.

It is known that the SPR chip angular reflectivity spectrum significantly depends on the electromagnetic interaction between the surface plasmon polariton that is excited on the metal surface and the localized surface plasmon of the metal nanoparticles located on this surface, which is expressed in the minimum angle shift of the SPR curve and the increase in reflection [7, 8]. It has been shown that this physical effect can be used to enhance the molecular biorecognition events [9–11] and biocatalytic transformations [12, 13] with the use of gold or silver nanoparticles as labels.

For the creation of novel sensor devices, electrochemical SPR (ESPR) spectroscopy is a promising method, which is based on the combination of the SPR measurements with the electrochemical control of molecular processes occurring on the metal–electrolyte boundary. In particular, the ESPR method was used to detect cofactor molecules [14], glucose [6, 15], and hydrogen peroxide [16]. The applicability of the method for the investigation of the influence of electric field on the molecular systems properties has been shown. In Ref. [17], the Stark effect in molecular adsorbates at different light wavelengths was studied using SPR spectroscopy. The influence of the applied electrostatic field on the processes of hybridization and DNA denaturation was investigated with SPR in Ref. [18]. By measuring reflectivity, the dependence of the degree of soybean peroxidase immobilization on the gold substrate on the magnitude of the applied electric field [19] was recorded.

As in the case of SPR, at the initial stage of the LSPR sensor development, it is necessary to use mathematical modeling in order to evaluate the potential sensitivity of the sensor and optimize its parameters. Lee and El-Sayed [20] studied the metallic nanorods LSPR spectrum sensitivity dependence on changes in the refractive index of environment, which depend on the size, shape, and metal type of nanorods. Yan et al. [21] have shown that the position of the LSPR absorption peak, its half-width, and the intensity increase nonlinearly with the increase in the shell thickness of the "gold core–dielectric shell" nanostructures. Xu and Käll [22] developed theoretical approaches to take into account the particle–particle and substrate–particle interactions in the model of LSPR sensors based on gold nanoparticles. Westcott et al. [23] studied the spectral properties of LSPR extinction of the "dielectric core–metal shell" nanostructures. Haes et al. [24] showed that an increase in the ratio of the silver nanostructure geometric sizes with the shape of a cut tetrahedron provides greater shifts in the LSPR peak in the light extinction spectrum when the dielectric coating is formed on its surface. Murray et al. [25] found that gold nanorods provide greater sensitivity to changes in the local refractive index of the environment than nanoparticles with disk shapes. Malinsky et al. [26] theoretically studied the sensitivity of the silver nanoparticles LSPR extinction peak position to the changes in the refractive index of the environment in the model based on the Mie theory.

The LSPR properties of multilayer nanoparticles of different geometry and composition are actively studied, and these results can be used to create highly sensitive LSPR sensors. Khlebtsov et al. demonstrated that the sensitivity to a biomolecular coating of quartz nanospheres coated with a layer of gold may be higher compared to the sensitivity of spherical gold nanoparticles with the same volume [27]. In Refs. [28, 29], the same authors proposed a multilayer model for gold and silver nanoparticles, which allows describing the interaction between biomolecules immobilized on nanoparticles and analyte molecules in a solution. Wu et al. [30] showed significant sensitivity of three-component nanostructures of SiO_2–Ag–Au and SiO_2–Au–Ag types to the dielectric properties of the environment.

The practical implementation of sensors based on the LSPR phenomenon was preceded by experimental studies of the optical properties of metallic nanostructures. In 1995, Kreibig and Vollmer [31] demonstrated that the optical density of the immobilized monolayer of gold colloidal nanoparticles depends on the refractive index of the surrounding liquid medium. Several experimental works were published, which considered the influence of the parameters (for example, shape, size, and interparticle distance) of nanostructured systems on the properties of light extinction and optical dichroism [32–36]. Schatz et al. [37] and Van Duyne et al. [38, 39] showed that ordered monolayers of silver or gold nanostructures on the surface of mica or glass can be fabricated using nanosphere lithography, which allows registering the biomolecular interaction. Chumanov et al. [40] showed the possibility of creating stable monolayers of silver nanoparticles obtained from colloidal solutions on solid or flexible substrates by virtue of a transition polymer layer. Such arrays of silver nanostructures exhibit extremely narrow peaks in the extinction spectra, which may be promising for the creation of highly sensitive biosensors.

Yonzon et al. studied the binding of Concanavalin A to mannose-functionalized nanoparticles in real time [41]. Haes et al. used LSPR sensor to detect ligands (amyloid derivatives capable of diffusion into biological tissues) at a concentration of 100 fM [42]. Alivisatos et al. developed a kind of plasmon molecular ruler, which measures the modulation of the LSPR spectrum depending on changes in the electromagnetic interaction caused by changing the distance between a pair of metal nanoparticles, to detect the hybridization of DNA oligonucleotides to single-stranded DNA [43]. Recently, 3D plasmonic molecular rulers have been developed based on bundled nanoparticles (plasmonic oligomers); 3D rulers allow getting the full spatial configuration of biological processes and their dynamic development. Atwater et al. developed elastic plasmonic materials [44]. The integration of split ring resonators into polydimethylsiloxane allowed, through the mechanical deformation of the polymer, changing the strength of the electromagnetic interaction between the resonators, which makes it possible to regulate the response of the metamaterial.

Since molecular resonances lead to spectrally selective optical absorption by molecules and using an electronic coupling between their molecular resonance and nanoparticles LSPR, mechanisms of significant changes in the LSPR spectra were discovered upon overlapping of the aforementioned resonances [45]. Such increased sensitivity to molecular absorption opens a way for the creation of highly sensitive resonance biosensors. Wiederrecht et al. reported hybridization in J-aggregate–metallic nanosphere complexes [46]. Halas et al. studied wavelength-dependent behavior of hybrid nanostructures formed by Au nanoshells and J-aggregates [47]. Au nanoshells made it possible to easily adjust the LSPR wavelength in a wide spectral range around the absorption peak of the J-aggregates. Wang et al. used Au nanorods to study the resonance interaction with H-aggregates at different LSPR wavelength positions of nanorods [48]. To solve the issues related to the fixed LSPR position in metallic nanoparticles, Zheng et al. have developed tunable plasmonic systems with variable light incidence angle changing the LSPR spectrum [49].

The LSPR phenomenon in high-conductive nanostructures can also be used to localize and amplify the electromagnetic field in order to provide conditions for increasing the efficiency of optical transitions in molecular systems. For example, if a fluorophore molecule is placed near the nanostructured surface of a high-conductive metal, then under certain conditions one can observe an increase in the intensity of its emission compared with the case of the absence of nanostructures [50–52]. Sensors constructed on this principle give an opportunity to increase the sensitivity of fluorescence measurements (for example, to capture a signal even from individual molecules [53, 54]), which determines their potential for applications in the biochemistry and medicine fields. However, enhancement of the dyes' fluorescence on silver and gold nanostructures substantially depends on the conditions of the resonant energy transfer of plasmon oscillations from the nanostructured metal surface to the dye molecule located near this surface [55–57]. Therefore, studies were actively carried out on the fluorescence enhancement affected by the shape and size of the metallic nanostructures themselves [58], the distance between the fluorophore molecules and the plasmon-generating surface [50, 55, 59], and also the characteristics of the molecule, such as its quantum yield and excitation lifetime [55, 56, 60]. Sorokin et al. showed that the fluorescence in the J-aggregates of cyanine dyes was enhanced when colloidal silver nanoparticles (two times) and gold nanostructure arrays (eight times) coated with polyelectrolyte layers were used to amplify the signal [61, 62]. In both cases, the optimal total thickness of the polymer layers was 16 nm. The simulation of the emission of fluorescent molecule layer with a thickness of 5 nm on a spherical gold nanoparticle with a diameter of 80 nm showed that the optimal distance between the fluorophore and the metal surface when the emission enhancement is observed is about 20 nm; in other cases, smaller amplification or quenching was observed [60]. In Ref. [63], in an experimental study of dyes emission on silver and gold nanostructures, it was noted that for monitoring fluorescence amplification, the distance between the dye and the metal surface should be 24–25 nm, and the fluorescence quenching is observed at a distance of 15 nm. Theoretical calculations [60] showed that the greatest increase in fluorescence is possible when the dipole moments of the molecules are directed normally to the plane of the nanostructured surface and at an optimal distance between the molecule and the metal surface. The magnitude of the molecular fluorescence intensity enhancement that can be achieved is from several to tens of times, according to literature data [53, 64].

To summarize, improvement in sensitivity and selectivity of plasmonic sensors is achieved through progress in a number of aspects: modeling and manufacturing of metallic nanostructures; surface functionalization; understanding of the interactions between the surface plasmon and the molecule.

1.2.2 Research in Material Science Field

One of the promising applications of molecular plasmonics is the non-destructive study of the properties of thin organic and inorganic films. Thus, the ESPR method was used for the study of electropolymerization processes [65] and the study of redox properties of polymers [66]. Damos et al. [67] performed electropolymerization of ultrathin films of methylene blue and investigated them using an ESPR method in real time. In Ref. [14], ESPR method was demonstrated for photonic transduction of the redox properties of an inorganic three-dimensional polymer Prussian blue. For the three redox states of this substance, different spectra of the SPR were observed; considering that the redox state does not affect the thickness of the film, these differences were explained by the change in the refractive index of the polymer. Thus, in this experiment, the electrochemical information, which is typical for the three redox states, has been converted into optical information by means of SPR, indicating the perspective of memory devices development for photonics with three stable states. Reference [14] demonstrates the fundamental possibility of using complex multi-stage redox transformations to create stable and reproducible photonic systems with multiple switching.

Due to its high sensitivity and surface nature, SPR method makes it easy to excite molecules and track their relaxation [68]. Therefore, it is ideally suited for studying conformational changes in gel-like polymeric materials by registering changes in the refractive index during the measurement of the object being studied in real time. In Ref. [6], the immobilization of an acrylamidophenylboronic acid–acrylamide copolymer by an ESPR is considered with the subsequent study of cyclic glucose-induced swelling of a polymer. The structure of the hydrogel (thickness, liquid saturation) and the kinetics of glucose-induced swelling and shrinking were studied. Investigation of these processes in acrylamidophenylboronic acid–acrylamide copolymers opens prospects for their use as matrices in sensors for glucose or in glucose-activated drug release systems.

An important application of SPR-based methods is the study of various properties of nanomaterials and nanostructured systems. In Ref. [69], it was found that charging a gold nanoparticle by electrons or removing electrons from it leads to significant shifts in the LSPR band. These spectral shifts were explained by changes in plasma frequency caused by the growth of charge density, which is the result of the metal nanoparticles electrolytic charging [70]. For example, the transformation of gold nanoparticles to the electron-depleted state by changing the voltage from −0.16 V to 0.82 V (relative to the silver quasi-reference electrode) induces a shift in the LSPR spectrum toward lower energies [69]. In the study in Ref. [71], with the help of SPR spectroscopy and electrochemical measurements, photoelectrochemical charging of gold nanoparticles attached to the gold surface through the auxiliary monolayer of cystamine was demonstrated by using the light excitation of CdS quantum dots bound to gold nanoparticles.

It is worth noting that SPR phenomenon and methods on its basis play a significant role in the creation and research of nanostructured materials with unique optical properties, so-called metamaterials, which in recent years have been actively studied due to promising applications in the fields of laser optics, optoelectronics, and chemical and biological sensors [72–75]. An important contribution to the production of such metamaterials can be the use of plasmonic nanostructures, such as gold and silver nanoparticles, nanorods, nanodisks, nanorectangles, and nanoprisms [76–80]. Of particular interest are metamaterials, which the reversible variation of physical properties is possible in, for example, materials that exhibit a dynamic change in the LSPR parameters [81–83].

1.2.3 Promising Research Directions

1.2.3.1 Plasmonic nanoscopy and visualization

Light has significant advantages for visualization, including remote and non-destructive character and short response time. However, the complexity of optical imaging of nanoscale objects is associated with diffraction limitation. Advances in nanotechnology have helped to develop techniques that have overcome this limitation. For example, near-field scanning optical microscopy (plasmonic nanoscopy) allowed nano-dimensional biological and medical imaging with high spatial and time resolution. Estrada and Gratton [84] used the high resolution, typical for plasmonic nanoscopy, to study the fluorescence lifetime of a single molecule depending on the distance to the laser-illuminated gold nanoparticle, which also serves as a nanoscopic probe. This study helped to highlight the physical mechanisms associated with quenching and enhancement of the dye fluorescence during LSPR excitation in nanoparticles. Using plasmonic nanoscopy, Estrada and Gratton obtained 3D images of biological fibers such as collagen and actin filaments with high resolution, moving a separate Au nanoparticle along the fibers in the near-infrared femtosecond pulses exposure mode and measuring its trajectory. Thus, metal nanoparticles with a special surface functionalization can serve as probes for 3D in vivo molecular imaging. "Chemical vision," which combines nanoscale imaging and molecular recognition, is one of the most important new directions in plasmonic nanoscopy.

1.2.3.2 Applications based on thermal effects

With optical near-field effects, metal nanoparticles can quickly convert the energy of the absorbed photon into heat through optical absorption as a result of LSPR. Among the various metallic nanoparticles proposed for photothermal therapy, gold nanoshells, nanorods, and nanocontainers have been most studied due to their wide LSPR range that extends into the near-infrared spectral region, where absorption by living tissues is minimal. Elliott et al. determined the quantitative characteristics of the nanoshells' interaction with laser radiation to study the influence of the nanoshells' concentration and the laser power on the photothermal effect [85]. Stern et al. evaluated the effect of nanoshell concentration in mice on the treatment of prostate cancer [86]. When the authors directed the infrared laser light through the skin of mice to the tumor, resonance absorption of energy by nanoshells raised the local temperature of malignant formations from 37°C to 45°C, killing cancer cells and leaving the surrounding healthy tissue intact. El-Sayed et al. used gold nanoparticles covered with antibodies for targeted delivery and photothermal therapy of epithelial carcinoma [87].

For therapies that require chemical drugs or genes, rather than direct heat treatment, plasmon-enhanced photothermal effects can also be used to develop nanocarriers that will enable optically controlled delivery of drug/oligonucleotide molecules. Particularly, Huang et al. developed aptamer/DNA-gold nanoparticle nanocarrier for directed drug delivery [88]. When irradiating the nanostructure by light with a wavelength corresponding to LSPR, the shell may become heated to a temperature that destabilizes the bond between the nanoparticle and the molecule and leads to the release of the drug that allows the treatment of cancer cells in defined spatiotemporal intervals.

1.2.3.3 Mechanical applications

The ability to capture, retain, and control molecules or biomolecules with nanosized precision is important for the analysis and understanding of biochemical processes. Among the works in this direction, it is necessary to highlight studies on the development and use of plasmonic tweezers. In this approach, a laser beam focused on a plasmonic nanoparticle is used to hold and control the biomolecules in the near-field gradient zone of the nanoparticle at low laser radiation intensities. Miao and Lin have demonstrated that the LSPR near field enhanced by an array of self-assembled gold nanoparticles can be used to hold objects up to micron size at low laser intensities [89]. Electromagnetic interactions between adjacent plasmonic nanostructures can be used to achieve better control of the EM fields and to increase the plasmonic tweezers productivity. For example, in a closely spaced dimer of plasmonic nanoparticles, stable holding of small particles and molecules was observed. Grigorenko et al. reported the retention of molecules with a pair of gold nanoparticles in the standard configuration of optical tweezers [3].

1.3 Conclusion

Surface plasmon excitation in metallic films and nanoparticles, besides the widely used biosensing and chemical sensing, allows investigating the physical properties of traditional materials and metamaterials, creating new materials and operating with the amplification of local electromagnetic fields. These advantages of plasmonic systems offer new approaches to emerging problems and improvements to well-known processes. Future development of these applications is entirely dependent on research in the field of nanophysics and new technologies for the creation of nanomaterials. In the case of biomedical applications, the key requirements are the strict control of the size and dispersion of the nanoparticles shape, as well as their functionalization.

Molecular plasmonics that investigates the interactions between molecules and surface plasmons of metallic nanostructures offers great opportunities for the detection, visualization, control, delivery, and heating of biological molecules, and provides a number of powerful tools for biological and medical research. Plasmon-enhanced thermal effects are fundamental for photothermal therapy and light-activated drug-delivery systems that can provide new instruments for disease control. Based on nanosized LSPR localization and enhancement of EM fields, metallic nanoparticles can be used to measure biological events, for fluorescence regulation, and in research on the single molecule level.

References

1. Kairdolf, B. A., Qian, X., and Nie, S. (2017). Bioconjugated nanoparticles for biosensing, in vivo imaging, and medical diagnostics, *Anal. Chem.*, **89**, pp. 1015–1031.
2. Corrie, S. R. and Plebanski, M. (2018). The emerging role of nanomaterials in immunological sensing: A brief review, *Mol. Immunol.*, **98**, pp. 28–35.
3. Grigorenko, A. N., Roberts, N. W., Dickinson, M. R. et al. (2008). Nanometric optical tweezers based on nanostructured substrates, *Nat. Photon.*, **2**, pp. 365–370.
4. Chegel, V. I. (2002). An influence of dielectric characteristics of ambient and external factors on parameters of physical and biological sensors based on surface plasmon resonance, PhD Thesis, Institute of Semiconductors Physics of National Academy of Science of Ukraine, Kyiv, Ukraine (in Ukrainian).
5. Lavine, B. K., Westover, D. J., Kaval, N. et al. (2007). Swellable molecularly imprinted poly*N*-(*N*-propyl) acrylamide particles for detection of emerging organic contaminants using surface plasmon resonance spectroscopy, *Talanta*, **72**, pp. 1042–1048.
6. Gabai, R., Sallacan, N., Chegel, V. et al. (2001). Characterization of the swelling of acrylamidophenylboronic acid–acrylamide hydrogels upon interaction with glucose by Faradaic impedance spectroscopy, chronopotentiometry, quartz-crystal microbalance (QCM) and surface plasmon resonance (SPR) experiments, *J. Phys. Chem. B.*, **105**, pp. 8196–8202.
7. Holland, W. R. and Hall, D. G. (1983). Surface-plasmon dispersion relation: Shifts induced by the interaction with localized plasma resonances, *Phys. Rev. B.*, **27**, pp. 7765–7768.
8. Lyon, L. A., Musick, M. D., Smith, P. C. et al. (1999). Surface plasmon resonance of colloidal Au-modified gold films, *Sens. Actuators B Chem.*, **54**, pp. 118–124.
9. Lyon, L. A., Musick, M. D., and Natan, M. J. (1998). Colloidal Au-enhanced surface plasmon resonance immunosensing, *Anal. Chem.*, **70**, pp. 5177–5183.
10. He, L., Musick, M. D., Nicewarner, S. R. et al. (2000). Colloidal Au-enhanced surface plasmon resonance for ultrasensitive detection of DNA hybridization, *J. Am. Chem. Soc.*, **122**, pp. 9071–9077.

11. Fathi, F., Rashidi, M. R., and Omidi, Y. (2018). Ultra-sensitive detection by metal nanoparticles-mediated enhanced SPR biosensors, *Talanta*, **192**, pp. 118–127.

12. Lioubashevski, O., Chegel, V. I., Patolsky, F. et al. (2004). Enzyme-catalyzed bio-pumping of electrons into Au-nanoparticles: A surface plasmon resonance and electrochemical study, *J. Am. Chem. Soc.*, **126**, pp. 7133–7143.

13. Antiochia, R., Bollella, P., Favero, G. et al. (2016). Nanotechnology-based surface plasmon resonance affinity biosensors for in vitro diagnostics, *Int. J. Anal. Chem.*, **2016**, 2981931.

14. Raitman, O. A., Katz, E., Willner, I. et al. (2001). Photonic transduction of a three-state electronic memory and of electrochemical sensing of NADH using surface plasmon resonance spectroscopy, *Angew. Chem. Int. Ed.*, **40**, pp. 3649–3652.

15. Tian, L., Qiu, J., Zhou, Y.-C. et al. (2010). Application of polypyrrole/GOx film to glucose biosensor based on electrochemical-surface plasmon resonance technique, *Microchim. Acta*, **169**, pp. 269–275.

16. Kang, X., Cheng, G., and Dong, S. (2001). A novel electrochemical SPR biosensor, *Electrochem. Commun.*, **3**, pp. 489–493.

17. Wang, S., Boussaad, S., Wong, S. et al. (2000). High-sensitivity Stark spectroscopy obtained by surface plasmon resonance measurement, *Anal. Chem.*, **72**, pp. 4003–4008.

18. Heaton, R. J., Peterson, A. W., and Georgiadis, R. M. (2001). Electrostatic surface plasmon resonance: Direct electric field-induced hybridization and denaturation in monolayer nucleic acid films and label-free discrimination of base mismatches, *Proc. Natl. Acad. Sci. U. S. A.*, **98**, pp. 3701–3704.

19. Morrow, R., McKenzie, D. R., Bilek, M. M. M. et al. (2007). Electric field effects on adsorption/desorption of proteins and colloidal particles on a gold film observed using surface plasmon resonance, *Phys. B Phys. Cond. Matt.*, **394**, pp. 203–207.

20. Lee, K.-S. and El-Sayed, M. A. (2006). Gold and silver nanoparticles in sensing and imaging: Sensitivity of plasmon response to size, shape, and metal composition, *J. Phys. Chem. B*, **110**, pp. 19220–19225.

21. Yan, S., Wang, Y., Wen, T. et al. (2006). A study on the optical absorption properties of dielectric-mediated gold nanoshells, *Physica E*, **33**, pp. 139–143.

22. Xu, H. and Käll, M. (2002). Modeling the optical response of nanoparticle-based surface plasmon resonance sensors, *Sens. Actuators B Chem.*, **87**, pp. 244–249.

23. Westcott, S. L., Jackson, J. B., Radloff, C. et al. (2002). Relative contributions to the plasmon line shape of metal nanoshells, *Phys. Rev. B*, **66**, 155431.

24. Haes, A. J., Zou, S., Schatz, G. C. et al. (2004). A nanoscale optical biosensor: The long range distance dependence of the localized surface plasmon resonance of noble metal nanoparticles, *J. Phys. Chem. B*, **108**, pp. 109–116.

25. Murray, W. A., Auguie, B., and Barnes, W. L. (2009). Sensitivity of localized surface plasmon resonances to bulk and local changes in the optical environment, *J. Phys. Chem. C*, **113**, pp. 5120–5125.

26. Malinsky, M. D., Kelly, K. L., Schatz, G. C. et al. (2001). Chain length dependence and sensing capabilities of the localized surface plasmon resonance of silver nanoparticles chemically modified with alkanethiol self-assembled monolayers, *J. Am. Chem. Soc.*, **123**, pp. 1471–1482.

27. Khlebtsov, B. N. and Khlebtsov, N. G. (2007). Biosensing potential of silica/gold nanoshells: Sensitivity of plasmon resonance to the local dielectric environment, *J. Quant. Spectrosc. Radiat. Transf.*, **106**, pp. 154–169.

28. Khlebtsov, N. G., Bogatyrev, V. A., Khlebtsov, B. N. et al. (2003). A multilayer model for gold nanoparticle bioconjugates: Application to study of gelatin and human IgG adsorption using extinction and light scattering spectra and the dynamic light scattering method, *Colloid J.*, **65**, pp. 622–635.

29. Khlebtsov, N. G. (2008). Optical models for conjugates of gold and silver nanoparticles with biomacromolecules, *J. Quant. Spectrosc. Radiat. Transf.*, **89**, pp. 143–153.

30. Wu, D., Xu, X., and Liu, X. (2008). Tunable near-infrared optical properties of three-layered metal nanoshells, *J. Chem. Phys.*, **129**, 074711.

31. Kreibig, U. and Vollmer, M. (1995). *Optical Properties of Metal Clusters* (Springer-Verlag, Berlin).

32. Mock, J. J., Smith, D. R., and Schultz, S. (2003). Local refractive index dependence of plasmon resonance spectra from individual nanoparticles, *Nano Lett.*, **3**, pp. 485–491.

33. Mock, J. J., Barbic, M., Smith, D. R. et al. (2002). Shape effects in plasmon resonance of individual colloidal silver nanoparticles, *J. Chem. Phys.*, **116**, pp. 6755–6759.

34. Su, K. H., Wei, Q. H., Zhang, X. et al. (2003). Interparticle coupling effects on plasmon resonances of nanogold particles, *Nano Lett.*, **3**, pp. 1087–1090.

35. Canfield, B. K., Kujala, S., Kauranen, M. et al. (2005). Remarkable polarization sensitivity of gold nanoparticle arrays, *Appl. Phys. Lett.*, **86**, 183109.

36. Wei, J., Guo, Y., Li, J., et al. (2017). Optically active ultrafine Au–Ag alloy nanoparticles used for colorimetric chiral recognition and circular dichroism sensing of enantiomers, *Anal. Chem.*, **89**, pp. 9781–9787.

37. Malinsky, M. D., Kelly, K. L., Schatz, G. C. et al. (2001). Nanosphere lithography: Effect of substrate on the localized surface plasmon resonance spectrum of silver nanoparticles, *J. Phys. Chem. B*, **105**, pp. 2343–2350.

38. Haynes, C. L. and Van Duyne, R. P. (2003). Dichroic optical properties of extended nanostructures fabricated using angle-resolved nanosphere lithography, *Nano Lett.*, **3**, pp. 939–943.

39. Riboh, J. C., Haes, A. J., McFarland, A. D. et al. (2003). A nanoscale optical biosensor: Real-time immunoassay in physiological buffer enabled by improved nanoparticle adhesion, *J. Phys. Chem. B*, **107**, pp. 1772–1780.

40. Chumanov, G. and Malinich, S. (2006). Coupled planar silver nanoparticle arrays as refractive index sensors, *J. Opt. A Pure Appl. Opt.*, **8**, pp. 144–147.

41. Yonzon, C. R., Jeoung, E., Zou, S. L. et al. (2004). A comparative analysis of localized and propagating surface plasmon resonance sensors: The binding of Concanavalin A to a monosaccharide functionalized self-assembled monolayer, *J. Am. Chem. Soc.*, **126**, pp. 12669–12676.

42. Haes, A. J., Chang, L., Klein, W. L. et al. (2005). Detection of a biomarker for Alzheimer's disease from synthetic and clinical samples using a nanoscale optical biosensor, *J. Am. Chem. Soc.*, **127**, pp. 2264–2271.

43. Sonnichsen, C., Reinhard, B. M., Liphardt, J. et al. (2005). A molecular ruler based on plasmon coupling of single gold and silver nanoparticles, *Nat. Biotechnol.*, **23**, pp. 741–745.

44. Pryce, I. M., Kelaita, Y. A., Aydin, K. et al. (2011). Compliant metamaterials for resonantly enhanced infrared absorption spectroscopy and refractive index sensing, *ACS Nano*, **5**, pp. 8167–8174.

45. Haes, A. J., Zou, S., Zhao, J. et al. (2006). Localized surface plasmon resonance spectroscopy near molecular resonances, *J. Am. Chem. Soc.*, **128**, pp. 10905–10914.

46. Wiederrecht, G. P., Wurtz, G. A., and Hranisavljevic, J. (2004). Coherent coupling of molecular excitons to electronic polarizations of noble metal nanoparticles, *Nano Lett.*, **4**, pp. 2121–2125.

47. Fofang, N. T., Park, T. H., Neumann, O. et al. (2008). Plexcitonic nanoparticles: Plasmon-exciton coupling in nanoshell-J-aggregate complexes, *Nano Lett.*, **8**, pp. 3481–3487.

48. Ni, W., Ambjornsson, T., Apell, S. P. et al. (2009). Observing plasmonic–molecular resonance coupling on single gold nanorods, *Nano Lett.*, **10**, pp. 77–84.

49. Zheng, Y. B., Juluri, B. K., Jensen, L. L. et al. (2010). Dynamically tuning plasmon-exciton coupling in arrays of nanodisk-J-aggregate complexes, *Adv. Mater.*, **22**, pp. 3603–3607.

50. Chumanov, G., Sokolov, K., Gregory, B. W. et al. (1995). Colloidal metal films as a substrate for surface-enhanced spectroscopy, *J. Phys. Chem.*, **99**, pp. 9466–9471.

51. Tam, F., Goodrich, G. P., Johnson, B. R. et al. (2007). Plasmonic enhancement of molecular fluorescence, *Nano Lett.*, **7**, pp. 496–501.

52. Xu, S., Cao, Y., Zhou, J. et al. (2011). Plasmonic enhancement of fluorescence on silver nanoparticle films, *Nanotechnology*, **22**, 275715.

53. Kühn, S., Håkanson, U., Rogobete, L. et al. (2006). Enhancement of single-molecule fluorescence using a gold nanoparticle as an optical nanoantenna, *Phys. Rev. Lett.*, **97**, 017402.

54. Zhang, J., Fu, Y., Chowdhury, M. H. et al. (2007). Metal-enhanced single-molecule fluorescence on silver particle monomer and dimer: Coupling effect between metal particles, *Nano Lett.*, **7**, pp. 2101–2107.

55. Krutyakov, Y. A., Kudrinskiy, A. A., Olenin, A. Y. et al. (2008). Synthesis and properties of silver nanoparticles: Advances and prospects, *Russ. Chem. Rev.*, **77**, pp. 233–257.

56. Chen, Y., Munechika, K., and Ginger, D. S. (2007). Dependence of fluorescence intensity on the spectral overlap between fluorophores and plasmon resonant single silver nanoparticles, *Nano Lett.*, **7**, pp. 690–696.

57. Geddes, C. D., Parfenov, A., Roll, D. et al. (2003). Electrochemical and laser deposition of silver for use in metal-enhanced fluorescence, *Langmuir*, **19**, pp. 6236–6241.

58. Gartia, M. R., Hsiao, A., Sivaguru, M. et al. (2011). Enhanced 3D fluorescence live cell imaging on nanoplasmonic substrate, *Nanotechnology*, **22**, 365203.

59. Drozdowicz-Tomsia, K. and Goldys, E. M. (2011). Gold and silver nanowires for fluorescence enhancement, Chapter 14, In *Nanowires: Fundamental Research*, Hashim, A. (ed.), (InTech, Rijeka, Croatia) pp. 309–332.

60. Kruszewski, S., Wybranowski, T., Cyrankiewicz, M. et al. (2008). Enhancement of FITC fluorescence by silver colloids and silver island films, *Acta Phys. Polon.*, **113**, pp. 1599–1608.

61. Sorokin, A. V., Zabolotskii, A. A., Pereverzev, N. V. et al. (2015). Metal-enhanced fluorescence of pseudoisocyanine J-aggregates formed in layer-by-layer assembled films, *J. Phys. Chem. C*, **119**, pp. 2743–2751.

62. Sorokin, A. V., Zabolotskii, A. A., Pereverzev, N. V. et al. (2014). Plasmon controlled exciton fluorescence of molecular aggregates, *J. Phys. Chem. C*, **118**, pp. 7599–7605.

63. Dragan, A. I. and Geddes, C. D. (2011). Excitation volumetric effects (EVE) in metal-enhanced fluorescence, *Phys. Chem. Chem. Phys.*, **13**, pp. 3831–3838.

64. Bharadwaj, P. and Novotny, L. (2007). Spectral dependence of single molecule fluorescence enhancement, *Opt. Exp.*, **21**, pp. 14266–14274.

65. Kang, X., Jin, Y., Cheng, G. et al. (2002). In situ analysis of electropolymerization of aniline by combined electrochemistry and surface plasmon resonance, *Langmuir*, **18**. pp. 1713–1718.

66. Chegel, V., Raitman, O., Katz, E. et al. (2001). Photonic transduction of electrochemically-triggered redox-functions of polyaniline films using surface plasmon resonance spectroscopy, *Chem. Commun.*, pp. 883–884.

67. Damos, F. S., Luz, R. C. S., and Kubota, L. T. (2005). Study of poly (methylene blue) ultrathin films and its properties by electrochemical surface plasmon resonance, *J. Electroanal. Chem.*, **581**, pp. 231–240.

68. Akkilic, N., Mustafaev, M., and Chegel, V. (2008). Conformational dynamics of poly(acrylic acid)-bovine serum albumin polycomplexes at different pH conditions, *Macromol. Symp.*, **269**, pp. 138–144.

69. Templeton, A. C., Pietron, J. J., Murray, R. W. et al. (2000). Solvent refractive index and core charge influences on the surface plasmon absorbance of alkanethiolate monolayer-protected gold clusters, *J. Phys. Chem. B*, **104**, pp. 564–570.

70. Lioubashevski, O., Chegel, V. I., Patolsky, F. et al. (2004). Enzyme-catalyzed bio-pumping of electrons into Au-nanoparticles: A surface plasmon resonance and electrochemical study, *J. Am. Chem. Soc.*, **126**, pp. 7133–7143.

71. Zayats, M., Kharitonov, A. B., Pogorelova, S. P. et al. (2003). Probing photoelectrochemical processes in Au-CdS nanoparticle arrays by surface plasmon resonance: Application for the detection of acetylcholine esterase inhibitors, *J. Am. Chem. Soc.*, **125**, pp. 16006–16014.

72. Austin, D., Mullin, N., Bismuto, A. et al. (2010). Transmission properties of plasmonic metamaterial quantum cascade lasers, *IEEE Photonics Technol. Lett.*, **22**, pp. 1217–1219.

73. Ren, M., Jia, B., Ou, J.-Y. et al. (2011). Nanostructured plasmonic medium for terahertz bandwidth all-optical switching, *Adv. Mater.*, **23**, pp. 5540–5544.

74. Gu, Y., Li, Q., Xiao, J. et al. (2011). Plasmonic metamaterials for ultrasensitive refractive index sensing at near infrared, *J. Appl. Phys.*, **109**, 023104.

75. Sreekanth, K. V., Alapan, Y., ElKabbash, M. et al. (2016). Extreme sensitivity biosensing platform based on hyperbolic metamaterials, *Nat. Mater.*, **15**, pp. 621–627.

76. Zhang, X., Sun, B., Friend, R. et al. (2006). Metallic photonic crystals based on solution-processible gold nanoparticles, *Nano Lett.*, **6**, pp. 651–655.

77. Malynych, S. and Chumanov, G. (2003). Light-induced coherent interactions between silver nanoparticles in two-dimensional arrays, *J. Am. Chem. Soc.*, **125**, pp. 2896–2898.

78. Nicholls, L. H., Rodríguez-Fortuño, F. J., Nasir, M. E. et al. (2017). Ultrafast synthesis and switching of light polarization in nonlinear anisotropic metamaterials, *Nat. Photonics*, **11**, pp. 628–633.

79. Zheng, Y. B., Yang, Y.-W., Jensen, L. et al. (2009). Active molecular plasmonics: Controlling plasmon resonances with molecular switches, *Nano Lett.*, **9**, pp. 819–825.

80. Yang, H., Owiti, E., Li, S. et al. (2017). Ag nanoprism metamaterial based tunable plasmonic logic gate, *J. Phys. D Appl. Phys.*, **50**, 445103.

81. Iyer, K. S., Zdyrko, B., Malynych, S. et al. (2011). Reversible submergence of nanoparticles into ultrathin block copolymer films, *Soft Matter*, **7**, pp. 2538–2542.

82. Qian, Z. and Ginger, D. S. (2017). Reversibly reconfigurable colloidal plasmonic nanomaterials, *J. Am. Chem. Soc.*, **139**, pp. 5266–5276.

83. Liu, L., Gao, Z., Jiang, B. et al. (2018). Reversible assembly and dynamic plasmonic tuning of Ag nanoparticles enabled by limited ligand protection, *Nano Lett.*, **18**, pp. 5312–5318.

84. Estrada, L. C. and Gratton, E. (2011). 3D nanometer images of biological fibers by directed motion of gold nanoparticles, *Nano Lett.*, **11**, pp. 4656–4660.

85. Elliott, A. M., Stafford, R. J., Schwartz, J. et al. (2007). Laser-induced thermal response and characterization of nanoparticles for cancer

treatment using magnetic resonance thermal imaging, *Med. Phys.*, **34**, pp. 3102–3108.

86. Stern, J. M., Stanfield, J., Kabbani, W. et al. (2008). Selective prostate cancer thermal ablation with laser activated gold nanoshells, *J. Urol.*, **179**, pp. 748–753.

87. El-Sayed, I. H., Huang, X., and El-Sayed, M. A. (2006). Selective laser photo-thermal therapy of epithelial carcinoma using anti-EGFR antibody conjugated gold nanoparticles, *Cancer Lett.*, **239**, pp. 129–135.

88. Luo, Y. L., Shiao, Y. S., and Huang, Y. F. (2011). Release of photoactivatable drugs from plasmonic nanoparticles for targeted cancer therapy, *ACS Nano*, **5**, pp. 7796–7804.

89. Miao, X. and Lin, L. Y. (2007). Large dielectrophoresis force and torque induced by localized surface plasmon resonance of Au nanoparticle array, *Opt. Lett.*, **32**, pp. 295–297.

Chapter 2

Physics of the Phenomenon and Theoretical Background of Surface Plasmon Resonance Method

2.1 Introduction

The study of surface plasmons is of interest both from fundamental and applications points of view [1–4]. By altering the structure of a metal surface, the properties of surface plasmons can be tailored, which offers the potential for developing new types of photonic devices. Surface plasmons are studied for their use in sub-wavelength optics [5, 6], data storage, light generation [7, 8], microscopy and bio-photonics [9–11]. Therefore, surface plasmon excitation on surfaces covered by meso-particles, bio- and organic molecules, ultrathin organic and biopolymer films, and its applications are currently a highly investigated field [12–15]. The surface plasmon resonance (SPR) phenomenon is most widely used in modern sensors as a sensitive method for the study of physical properties of molecular layers or coverings at the surface of noble metal thin films and nanoparticles [16–19].

This chapter presents the theoretical background of SPR method from the point of view of its application primarily for sensing and related fields and results of several relevant theoretical studies. Specifically, Section 2.2 describes principal aspects of the SPR

Molecular Plasmonics: Theory and Applications
Volodymyr I. Chegel and Andrii M. Lopatynskyi

ISBN 978-981-4800-65-5 (Hardcover), 978-0-429-29511-9 (eBook)
www.jennystanford.com

phenomenon in thin noble metal films and its applications. These include studies on 3D quantification of molecular covers using SPR and influence of the shape of the particles covering the metal surface on the dispersion relations of surface plasmons. Section 2.3 includes theoretical background of the localized surface plasmon resonance (LSPR) phenomenon and presents results of studies related to the LSPR sensor sensitivity optimization. These include investigations of LSPR response depending on the geometrical parameters of the sensitive element, comparative analysis of different LSPR response measurement modes, and LSPR response analysis for light-absorbing molecular coatings.

2.2 SPR Phenomenon and Theoretical Background for Its Application in Sensing

2.2.1 General Interpretation of SPR Phenomenon and Most Common Theoretical Research Methods

Surface plasmon polariton (SPP) oscillations on the metal surface are a density charge wave that propagates along the metal–dielectric interface. It is known that SPP can be excited only when the dielectric permittivities ε_d and ε_m of the adjacent environments have opposite signs [20]. As a result, a surface plasmon wave cannot interact with the electromagnetic field of light that falls on a metal film, the carrier of plasmon oscillations, and the excitation of SPP can be achieved in the case of total internal reflection with the use of a prism or diffraction grating. At the same time, only *p*-polarized light participates in the excitation of the SPP.

The dispersion equations connect the frequency of the wave with its wave vector and indicate conditions when such waves can be excited. Using Eq. (2.1), it can be noticed that the wave vector of the SPP is defined as

$$k_{\text{spp}}(\omega_{\text{spp}}) = \frac{\omega_{\text{spp}}}{c}\sqrt{\frac{\varepsilon_d \varepsilon_m}{\varepsilon_d + \varepsilon_m}}, \qquad (2.1)$$

where ω_{spp} is the SPP frequency and c is the speed of light. In Fig. 2.1, dependence $\omega_{\text{spp}}(k_{\text{spp}})$ is shown. Of course, with the change in surface conditions, excitation conditions of SPPs are changing too.

So the dispersion equations depend on the surface conditions where the SPP is excited.

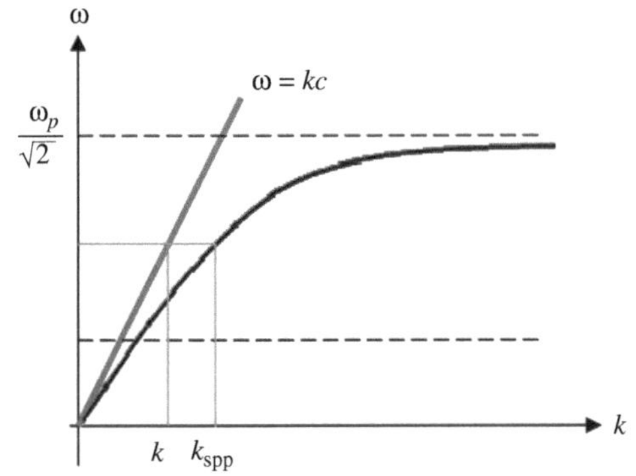

Figure 2.1 Surface plasmon polariton dispersion relation. Reprinted from Ref. [21], Copyright 2008, with permission from Elsevier.

Surface plasmons cannot be excited by light falling directly from a less optically dense medium, since the value of photon wave vectors $k = \frac{\omega}{c}\sqrt{\varepsilon_d}$ is not enough. One of the simplest ways for the coordination of the mentioned vectors (maintaining the law of conservation of wave vector component parallel to the metal–dielectric interface) is using the attenuated total reflection (ATR) method. For example, in the Kretschmann configuration [22] (Fig. 2.2), an optical prism with a high refractive index $\sqrt{\varepsilon_p} > \sqrt{\varepsilon_d}$ or a glass substrate coupled to the prism by an immersion liquid for optical contact is coated with a metal film. In the ATR method, light reflects with angles higher than the critical angle of the total internal reflection (θ_c). In this case, the total internal reflection is observed ($|R| \approx 1$). However, part of the light passes through the glass and excites SPP in the metal when the film is thin enough to let energy flow reach the metal–dielectric interface (Fig. 2.3).

Therefore, the SPR phenomena appears in case $k_x = k_{spp}$ (Fig. 2.2), which can be achieved with incidence angle changing in range $\theta_c < \theta < 90°$ and can be observed as a minimum in the angular dependence of reflected *p*-polarized light intensity $R(\theta)$ (further, SPR curve or spectrum). Rapid decrease in reflection curve, which can be observed upon angular sweep, represents absorption of light energy and appearance of resonance in a surface layer of electron plasma. The minimum reflectance intensity corresponds to the resonance angle θ_{spp}, which can be calculated using the equation

$$\sqrt{\varepsilon_p}\sin\theta_{spp} \approx \sqrt{\frac{\varepsilon'_m(\omega)\varepsilon_d}{\varepsilon_d+\varepsilon'_m(\omega)}}, \tag{2.2}$$

where only a real part of the complex dielectric function $\varepsilon_m(\omega)=\varepsilon'_m(\omega)+\varepsilon''_m(\omega)$ was used. Of course, surface plasmons cannot exist without damping in metal film and $\varepsilon''_m(\omega)\neq 0$, but in most cases, $|\varepsilon'_m(\omega)| >> |\varepsilon''_m(\omega)|$. In Fig. 2.4 one can observe the SPR curve for gold. The dielectric coating located at the metal surface, for example, as a layer of adsorbed molecules, results in an increase in surface plasmon wave vector value:

$$k^1_{spp} = k^0_{spp} + \Delta k_{spp}. \tag{2.3}$$

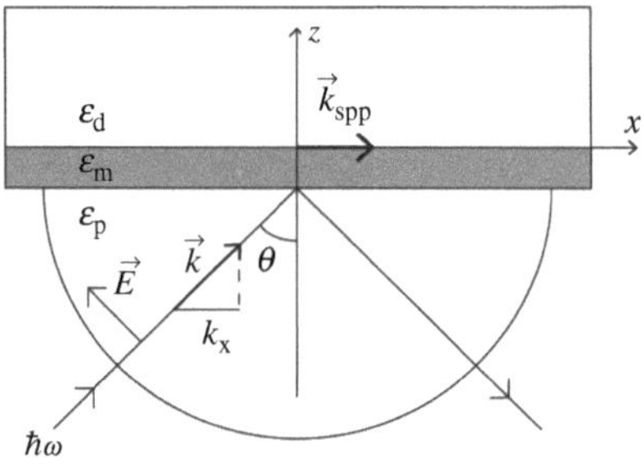

Figure 2.2 Kretschmann configuration of surface plasmon resonance excitation using the ATR method.

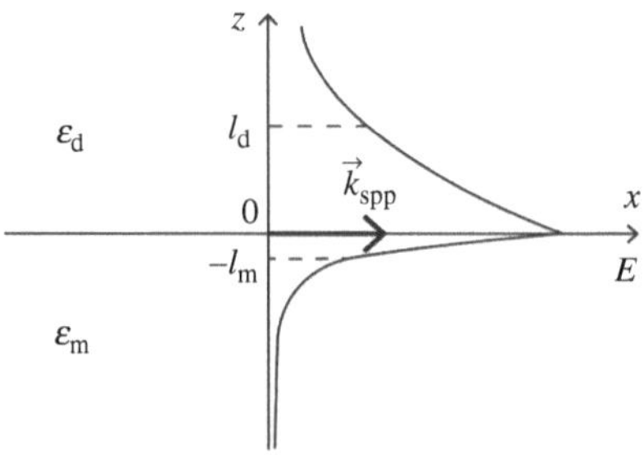

Figure 2.3 SPP wave vector and electric field distribution at the metal–dielectric interface.

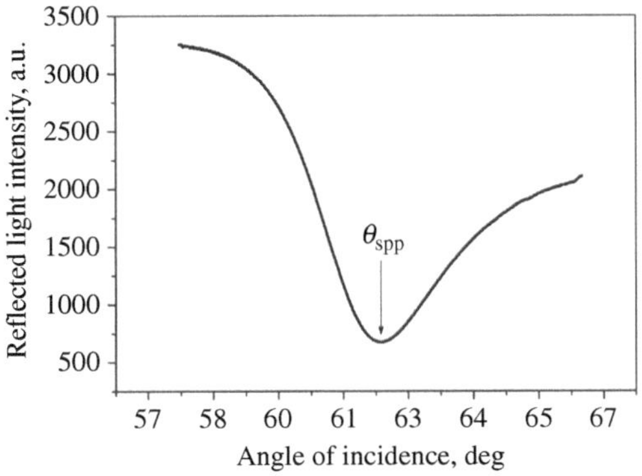

Figure 2.4 Dependence of reflected light intensity on the angle of incidence (SPR curve) for gold.

According to Eq. (2.2), this leads to a shift in the minimum position $d\theta$ of the SPR curve. Using Fresnel equations for $d\theta$ calculation, it becomes possible to determine the dielectric layer's optical thickness. Using Maxwell's equations, it is possible to describe the propagation of plane, monochromatic, linearly polarized electromagnetic field in a multilayer thin-film system. Experimentally obtained SPR curve depends on optical constants (refractive indices and absorption coefficients) of all phases, which interact with electromagnetic wave (material of prism, metal layer, surface-adsorbed substance, external medium, and other phases, which can be included in the system under study depending on measurement conditions) and also on geometrical thickness of all layers, including gold film and adsorbed layer. This dependence can be represented as [23]:

$$R_p(\theta) = -(u_{0p} - Y_p)/(u_{0p} + Y_p), \tag{2.4}$$

where $R_p(\theta)$ stands for the reflection coefficient of p-polarized electromagnetic wave, which is incident on the interface at angle θ, and Y_p is the total admittance of all reflective layers for a specific wavelength, which can be calculated with the following equation:

$$\begin{pmatrix} 1 \\ Y_p \end{pmatrix} = \left(\prod_{j=1}^{m+1} \left\| \begin{matrix} \cos\beta & (i/u_{jp})\sin\beta \\ iu_{jp}\sin\beta & \cos\beta \end{matrix} \right\| \right) \begin{pmatrix} 1 \\ u_{jp} \end{pmatrix}, \tag{2.5}$$

where m is the total number of layers for multilayer system (Fig. 2.5), excluding external medium; j is the number of considered layer; β_j is the phase thickness of the jth layer

$$\beta_j = 2 N_j \pi \frac{d}{\lambda} \cos\theta_j, \tag{2.6}$$

u_{jp} is an admittance of the jth layer (for p-polarized electromagnetic waves)

$$u_{jp} = N_j/\cos\theta_j = N_j^2/\sqrt{N_j^2 - N_0^2 \sin^2\theta_0}\ , \tag{2.7}$$

where $N_j = n_j - ik_j$ is the complex refractive index of investigated layer, θ_j is the angle of incidence in the jth layer, λ is the wavelength, d_j is the jth layer thickness, and θ_0 is the external angle of incidence. Admittances u_{0p} and u_{m+1p} describe the prism and the external medium, respectively.

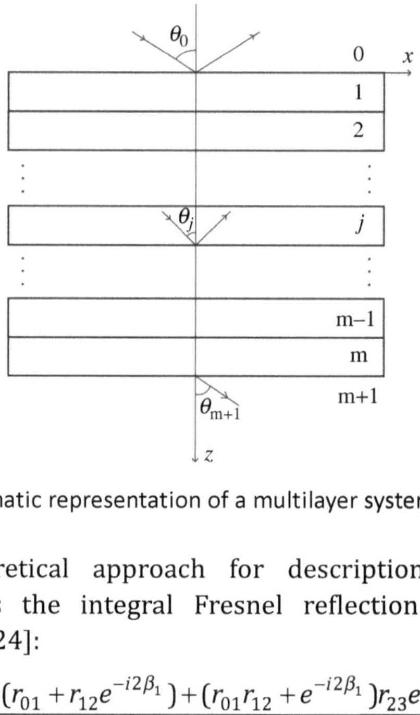

Figure 2.5 Schematic representation of a multilayer system model.

Other theoretical approach for description of multilayer structures uses the integral Fresnel reflection coefficient for p-polarization [24]:

$$R_p(\theta) = \frac{(r_{01} + r_{12}e^{-i2\beta_1}) + (r_{01}r_{12} + e^{-i2\beta_1})r_{23}e^{-i2\beta_2}}{(1 + r_{01}r_{12}e^{-i2\beta_1}) + (r_{12} + r_{01}e^{-i2\beta_1})r_{23}e^{-i2\beta_2}}, \tag{2.8}$$

where r_{01}, r_{12}, r_{23} are Fresnel reflection coefficients for corresponding interfaces and β_1, β_2 are the phase thicknesses of layers. Equation (2.8) allows calculations of three-layer structures. In the case of more layers, it is convenient to use calculations within the framework of scattering matrix formalism [25]. The scattering matrix S is a

2×2 matrix connecting the complex strengths of electric fields on the first and last interfaces of the multilayer structure:

$$\begin{bmatrix} E^{+}(z_{01}) \\ E^{-}(z_{01}) \end{bmatrix} = \begin{bmatrix} S_{11} & S_{12} \\ S_{21} & S_{22} \end{bmatrix} \cdot \begin{bmatrix} E^{+}(z_{m(m+1)}) \\ E^{-}(z_{m(m+1)}) \end{bmatrix}, \tag{2.9}$$

where the indexes "+" and "–" denote two components of the total field propagating in positive and negative directions relative to the z axis. The matrix S can be represented as a product of the interface matrices I and the layer matrices L, describing the influence of the individual layers and interfaces in the multilayer structure:

$$S = I_{01} L_1 I_{12} L_2 ... L_m I_{m(m+1)}. \tag{2.10}$$

Interface and layer matrices can be expressed as follows:

$$I_{j(j+1)} = (1/t_{j(j+1)}) \begin{bmatrix} 1 & r_{j(j+1)} \\ r_{j(j+1)} & 1 \end{bmatrix}, \quad L_j = \begin{bmatrix} e^{i\beta_j} & 0 \\ 0 & e^{-i\beta_j} \end{bmatrix}, \tag{2.11}$$

where $t_{j(j+1)}$ and $r_{j(j+1)}$ are the amplitude Fresnel transmission and reflection coefficients of p-polarized light for interface $j(j + 1)$, $\beta_j = 2\pi d_j \tilde{N}_j \cos\theta_j / \lambda$ is the phase thickness of the jth layer, d_j is the thickness of the jth layer, $\tilde{N}_j$ is the complex refractive index of the jth layer, and θ_j is the angle of incidence of light in the jth layer. Amplitude Fresnel transmission and reflection coefficients of p-polarized light for interface $j(j + 1)$ are defined by the following equations [24]:

$$\begin{aligned} t_{j(j+1)} &= \frac{2\tilde{N}_j \cos\theta_j}{\tilde{N}_{j+1} \cos\theta_j + \tilde{N}_j \cos\theta_{j+1}}, \\ r_{j(j+1)} &= \frac{\tilde{N}_{j+1} \cos\theta_j - \tilde{N}_j \cos\theta_{j+1}}{\tilde{N}_{j+1} \cos\theta_j + \tilde{N}_j \cos\theta_{j+1}}. \end{aligned} \tag{2.12}$$

By calculating the scattering matrix of a multilayer system, one can determine the intensity reflection coefficient of p-polarized light from a multilayer structure:

$$R = \left(\frac{S_{21}}{S_{11}} \right)^2, \tag{2.13}$$

where the indexes denote the corresponding elements of the matrix S. From the calculated $R(\theta_0)$ dependence for the investigated multilayer structure at $\theta_0 > \theta_c$, an angular position of the minimum θ_{spp} can be obtained, which corresponds to the SPR phenomenon.

Computer modeling showed that both approaches produce nearly the same results. Although Eq. (2.4) describes SPR phenomenon as a function dependent on the incidence angle θ of monochromatic light (which means λ is a constant), it can be easily changed to obtain the reflectance equation as a function of λ, where θ is constant. In practice for the creation of SPR sensors, both angular and wavelength spectra are used.

2.2.2 Theoretical Background of SPR Method According to Green's Function

2.2.2.1 Surface molecular layer susceptibility and reflection coefficient

As mentioned before, the main principle behind using the SPR method for studying the physical properties of coatings on solid surfaces is the measurement of the shifts in resonant angle if a molecular covering is present on the sensitive surface of a sensor [26]. The angle shift is, of course, dependent on molecular concentration as well as the type of the molecules. The physical models usually used for the description of this shift are based mainly on the concept of an additional layer on the surface of SPR transducer, which is characterized by the effective thickness and refractive index, analogous to the similar idea of ellipsometry of thin films [26–29]. Another approach is based on the idea of ultrathin film representation [30]. The main point of this approach is the representation of the molecular layer as an effective ultrathin homogeneous film characterized by any susceptibility, which was calculated with self-consistent equations for local field using molecular polarization and effective film thickness. Similar approaches do not allow one to obtain information about the concentration or individual dielectric properties of molecules at the surface. To describe the optical properties of molecular coverings at surfaces, one needs to take into account the individual properties of the adsorbed molecules, their interaction with the surface, and intermolecular (lateral) interactions. According to Bobbert and Vlieger [31], one solution to the problem of light reflection from a substrate covered with spherical particles can be obtained by defining the reflected electromagnetic wave as the sum of Fresnel's plane wave and the number of spherical waves, which are raised at the scattering on the spherical particles in accordance with the Mie theory. Another method of calculating the reflection coefficient for a surface covered by a molecular layer is based on Green's function [32]. This subsection covers an approach based on the concept of linear response for the monolayer of non-point-like protein molecules, which have the shape of oblate or prolate ellipsoids.

As it was mentioned in Subsection 2.2.1, the use of the light scattering matrix for a multilayer system is the most common approach for calculating the angular dependence of the reflected light intensity on the excitation of SPR in the Kretschmann configuration [33]. The determination of effective optical constants in this approach yields an approximate estimate of the thickness and complex refractive index of the molecular layer, while the use of such layer parameters as polarization and surface concentration of molecules in the application of Green's function significantly increases the informative value and reliability of the calculations. In this case, in order to calculate the reflection coefficient, it is necessary to know the effective susceptibility of the molecular layer. Let us consider a dilute thin layer of oblate or prolate ellipsoid organic molecules that are homogeneously distributed on the surface. According to the Lippmann–Schwinger equation, the field at an arbitrary point in the system obeys the equation [34]:

$$E_i(\vec{R},\omega) = E_i^{(0)}(\vec{R},\omega) - a\sum_{\alpha=1}^{Q}\int_{V_\alpha} d\vec{R}' G_{ij}(\vec{R},\vec{R}',\omega)\chi_{jl}(\omega)E_l(\vec{R}',\omega), \quad (2.14)$$

where $E_i^{(0)}(\vec{R},\omega)$ is the external long-range field, a is the coefficient defined by the system of units (for SI, $a = \omega^2/c^2\varepsilon_0$, where ω is the angular light velocity, c is the speed of light, ε_0 is the permittivity of vacuum), Q is the number of molecules on the surface, V_α is the molecular volume, $\chi_{jl}(\omega)$ is the molecular susceptibility, $G_{ij}(\vec{R},\vec{R}',\omega)$ is a photon propagator that describes the propagation of light of frequency ω from point $\vec{R}'$ to point $\vec{R}$ [35]. Summation is made over all positions that are occupied by molecules. Because molecular linear dimensions are much less than light wavelength and average distances between the molecules are considered larger than molecular linear dimension (submonolayer cover), one can make the next approximation:

$$\sum_{\alpha}\int_{V_\alpha} d\vec{R}' G_{ij}(\vec{R},\vec{R}',\omega)\chi_{jl}(\omega)E_l(\vec{R}')$$
$$\approx \sum_{\alpha} G_{ij}(\vec{r}-\vec{r}_\alpha, z, z_\alpha, \omega)\tilde{\chi}_{jl}(\omega)E_l(\vec{r}_\alpha, z_\alpha), \qquad (2.15)$$

where $\tilde{\chi}_{jl}(\omega)=V_\alpha \chi_{jl}(\omega)$; $\vec{r}$, $\vec{r}_\alpha$, z, z_α are vector and scalar coordinates of the observation point and the center of αth molecule. Here $\tilde{\chi}_{jl}(\omega)$ is the response on the local (total) field, which connects the polarization of the molecule and the local field via

$$P_j(\vec{r}_\alpha, z_\alpha, \omega)=\tilde{\chi}_{jl}(\omega)E_l(\vec{r}_\alpha, z_\alpha, \omega). \qquad (2.16)$$

The averaging over molecular coordinates if molecules are homogeneously distributed along the surface is performed using the equation:

$$\overline{\sum_{\alpha=1}^{Q} G_{ij}(\vec{r}-\vec{r}_\alpha, z, z_\alpha, \omega)\tilde{\chi}_{jl}(\omega)E_l(\vec{r}_\alpha, z_\alpha, \omega)}$$
$$= \frac{1}{S^{Q-1}}\int d\vec{r}_1 d\vec{r}_2 ... d\vec{r}_N \sum_{\alpha=1}^{Q}\int \frac{d\vec{k}}{(2\pi)^2} e^{-i\vec{k}(\vec{r}-\vec{r}_\alpha)} G_{ij}(\vec{k}, z, z_\alpha, \omega)$$
$$\tilde{\chi}_{jl}(\omega)\int \frac{d\vec{k}'}{(2\pi)^2} e^{-i\vec{k}'\vec{r}_\alpha} E_l(\vec{k}', z_\alpha, \omega) \qquad (2.17)$$
$$= N_S \int \frac{d\vec{k}}{(2\pi)^2} e^{-i\vec{k}\vec{r}} G_{ij}(\vec{k}, z, z_\alpha, \omega)\tilde{\chi}_{jl}(\omega)E_l(\vec{k}, z_\alpha, \omega),$$

where S is the area of the surface at which the Q molecules are situated; $N_S = Q/S$ is molecular surface concentration; and $\vec{k}$, $\vec{k}'$ are the wave vectors for different points of a field. Then, an equation of self-consistent field in the Weil representation can be written as

$$E_i(\vec{k}, z_\alpha, \omega)=E_i^{(0)}(\vec{k}, z_\alpha, \omega)$$
$$-N_S a G_{ij}(\vec{k}, z_\alpha, z_\alpha, \omega)\tilde{\chi}_{jl}(\omega)E_l(\vec{k}, z_\alpha, \omega). \qquad (2.18)$$

Making Fourier transformation in the plane of the surface, one obtains from Eq. (2.16)

$$E_l(\vec{k}, z_\alpha, \omega)=\left(\tilde{\chi}_{jl}(\omega)\right)^{-1} P_j(\vec{k}, z_\alpha, \omega). \qquad (2.19)$$

Then, Eq. (2.18) can be rewritten in the form:

$$\left(\tilde{\chi}_{ij}(\omega)\right)^{-1} P_j(\vec{k},z_\alpha,\omega)$$
$$= E_i^{(0)}(\vec{k},z_\alpha,\omega) - N_S a G_{ij}(\vec{k},z_\alpha,z_\alpha,\omega) P_j(\vec{k},z_\alpha,\omega). \quad (2.20)$$

The solution to this equation is

$$P_j(\vec{k},z_\alpha,\omega) = \left[\left(\tilde{\chi}_{ji}(\omega)\right)^{-1} + N_S a G_{ij}(\vec{k},z_\alpha,z_\alpha,\omega)\right]^{-1} E_i^{(0)}(\vec{k},z_\alpha,\omega). \quad (2.21)$$

Then, the effective susceptibility of submonolayer of the molecules at the surface, which connects the Fourier-transformants of layer polarization and external field, has the form

$$\tilde{X}_{ij}(\vec{k},z_\alpha,\omega) = \left[\left(\tilde{\chi}_{ij}(\omega)\right)^{-1} + N_S a G_{ji}(\vec{k},z_\alpha,z_\alpha,\omega)\right]^{-1}. \quad (2.22)$$

For SPR simulation, one needs to know the reflection coefficient of the molecular layer (see Fig. 2.6). For calculation of the reflection coefficient, let us consider the planar layered medium, the electrodynamical properties of which are characterized by photon propagator $G_{ji}(\vec{k},z,z',\omega)$. Let the light propagation from semi-space $z > 0$ to the same semi-space be described by the photon propagator $G_{ij}^{(+,+)}(\vec{k},z,z',\omega)$, the light propagation from semi-space $z < 0$ to semi-space $z > 0$ by the photon propagator $G_{ij}^{(+,-)}(\vec{k},z,z',\omega)$, and the light propagation from semi-space $z > 0$ to semi-space $z < 0$ by the photon propagator $G_{ij}^{(-,+)}(\vec{k},z,z',\omega)$. Then, an effective susceptibility of the molecular layer situated at the surface of semi-space $z < 0$ is defined by Eq. (2.22) with the photon propagator $G_{ij}^{(-,-)}(\vec{k},z,z',\omega)$. If the field $E_i^{(0)}(\vec{k},z,\omega)$ acts at the molecular layer, the field reflected by the layer will be written as

$$E_i^{(R)}(\vec{k},z,\omega) = N_S G_{ij}^{(+,-)}(\vec{k},z,z_\alpha,\omega) X_{jl}(\vec{k},\omega) E_l^{(0)}(\vec{k},z_\alpha,\omega) \quad (2.23)$$

Then, the reflection coefficient, which connects the amplitudes of reflected by the molecular layer and incident p-polarized fields $E_p^{(R)} = R_p E_p^{(0)}$, can be written in the form

$$\begin{aligned} R_p^{(M)}(\theta,\omega) &= G_{xj}^{(+,-)}(\vec{k},z,z_\alpha,\omega) N_S X_{jx}(\vec{k},\omega) \\ &+ G_{zj}^{(+,-)}(\vec{k},z,z_\alpha,\omega) N_S X_{jz}(\vec{k},\omega) \\ &+ \left[G_{xj}^{(+,-)}(\vec{k},z,z_\alpha,\omega) N_S X_{jz}(\vec{k},\omega)\right. \\ &+ \left.G_{zj}^{(+,-)}(\vec{k},z,z_\alpha,\omega) N_S X_{jx}(\vec{k},\omega)\right] \cos\theta \sin \end{aligned} \quad (2.24)$$

where θ is the incident angle. Because light is reflected by both the molecular layer and the surface, the total reflection coefficient should be written as the sum

$$R_p^{(T)}(\theta,\omega)=R_p^{(0)}(\theta,\omega)+R_p^{(M)}(\theta,\omega), \qquad (2.25)$$

where $R_p^{(0)}$ is the Fresnel reflection coefficient of the surface.

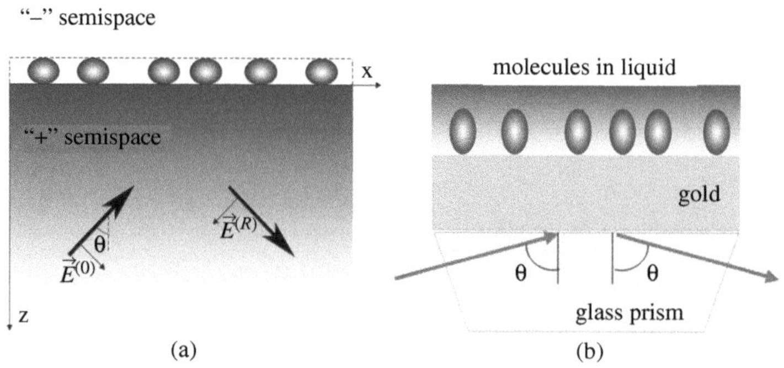

Figure 2.6 (a) Reflection of the light by molecular layer situated at a surface. (b) Schematic presentation of the system under investigation. Reprinted from Ref. [36], Copyright 2008, with permission from Elsevier.

2.2.2.2 Nanoparticles shape influence on the dispersion dependences of SPR

Green's function allows considering both molecular surface concentration and shape of molecules. Let us analyze the molecular coating of a metal surface where the molecules are represented as homogeneous particles with the shape of ellipsoids uniformly distributed along the plane of the metal surface. To calculate the effective susceptibility, one can use an approach described in Refs. [34, 37]. As a result, the effective susceptibility of the molecular layer can be obtained as

$$\mathrm{X}_{ij}(\mathbf{k},\omega)=\left[\chi_{ij}^{-1}(\omega)-N_S G_{ji}(\mathbf{k},l,l,\omega)\right]^{-1}, \qquad (2.26)$$

where $G_{ji}(\mathbf{k}, l, l, \omega)$ is the electrodynamic Green's function of the environment where the molecular layer is located, N_S is the concentration of surface particles, and $\chi_{ij}(\omega)$ is the single surface molecule susceptibility. Obviously, in this case, one should use Green's function for two semi-spaces with a flat boundary. Since we assume that molecules can be represented as homogeneous ellipsoidal particles, the polarizability of an ellipsoidal particle on a surface can be used for molecular susceptibility:

$$\tilde{\chi}_{ij}(\omega)=\begin{pmatrix} \chi_{||}(\omega) & 0 & 0 \\ 0 & \chi_{||}(\omega) & 0 \\ 0 & 0 & \chi_{\perp}(\omega) \end{pmatrix}, \qquad (2.27)$$

where $\chi_{\perp}(\omega)$ and $\chi_{||}(\omega)$ are normal and lateral parts of the linear response tensor, respectively.

To see the influence of the shape and concentration of dielectric nanoparticles on the dispersion dependence of SPPs, let us investigate the following system. Assume that nanoparticles are located on the metal surface, whose optical properties are described by a dielectric function $\varepsilon_m(\omega)=1-\omega_p^2/\omega^2$ with a plasma frequency $\omega_p = 2 \times 10^5$ cm^{-1}, which is a typical value for metals (for example, $\omega_p = 1.2 \times 10^5$ cm^{-1} for aluminum and $\omega_p = 4 \times 10^5$ cm^{-1} for gold). The geometric parameters of the nanoparticles were selected in the range of 1–10 nm in such a way that their volume remained unchanged and was equal to $V_p = 8.38 \times 10^{-21}$ cm^3, and the surface concentration of the monolayer coating was set to be equal to $N_S = 0.25 \times 10^{13}$ cm^{-2}. It is worth noting that exactly such sizes of nanostructures are common for biomolecules. The dielectric permittivity of the model nanoparticles was chosen equal to $\varepsilon_p = 5$.

In the case of *p*-type waves (Fig. 2.7), the presence of a layer of nanoparticles on the metal surface leads to the splitting of the dispersion curve of the typical SPP into four branches. It should be noted that this effect is common for both prolate and oblate nanoparticles. However, for spherical nanoparticles, only two dispersion curves can be observed. The obtained result can be explained by the fact that the appearance of the four modes is the result of the interaction of the *x*- and *z*-components of the electric field of the SPP with the longitudinal and lateral oscillations of ellipsoidal nanoparticles. In the case of spherical nanoparticles, the equality of polarizability values along the axes of OX and OZ leads to the appearance of only two dispersion curves. It should be noted that this fact is not universal for consideration of the interaction of nanoparticles with the surface of solid. In the case discussed in this subsection, the degree of interaction between the surface and the spherical nanoparticle does not cause significant differences between the various components of polarizability.

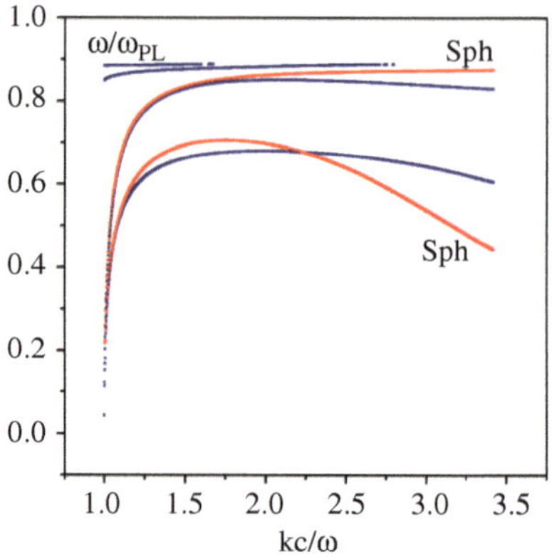

Figure 2.7 Dispersion curves for *p*-polarized surface wave upon interaction with monolayer of nanoparticles with oblate ellipsoid shape (blue lines) and spherical shape (red lines). The concentration of nanoparticles is $n = N_S$. Reprinted from Ref. [21], Copyright 2008, with permission from Elsevier.

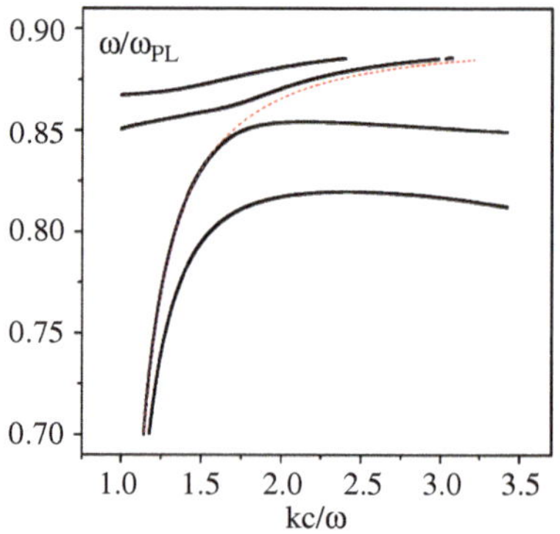

Figure 2.8 Dispersion curves for *p*-polarized surface wave upon interaction with monolayer of nanoparticles with oblate ellipsoid shapes. The concentration of nanoparticles is $n = 0.1N_S$. The dashed line represents the dispersion curve of the surface wave, which is typical for the free surface. Reprinted from Ref. [21], Copyright 2008, with permission from Elsevier.

Figure 2.8 shows the influence of lateral interactions, which occur due to the nanoparticles' concentration dependence, on the form of the dispersion curves of the *p*-polarized surface wave. Comparing these results with the ones mentioned before in Fig. 2.7, it can be noticed that the change in nanoparticles' concentration toward a sparser monolayer causes the displacement of the dispersion curves to the region of higher frequencies with the simultaneous narrowing of the occupied bandwidth. Calculations show that the decrease in the concentration of surface particles causes the dispersion curves grouping near the surface wave dispersion curve, which is typical for the free surface (Fig. 2.8). It should also be noted that the dispersion curves disappeared after some critical particle concentration point is reached, since there were no solutions to the dispersion equation in this case.

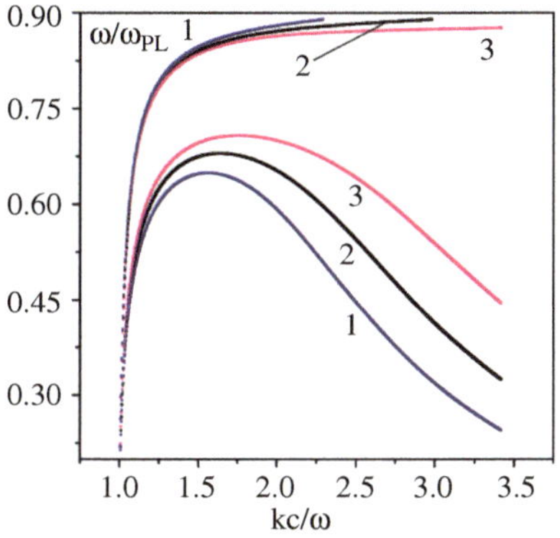

Figure 2.9 Influence of prolate ellipsoidal nanoparticles on *p*-polarized wave dispersion (concentration of particles in the monolayer is $n = N_S$). Semi-axis of the nanoparticles are (1) $h_x = 0.71$ nm, $h_z = 4$ nm; (2) $h_x = 1$ nm, $h_z = 2$ nm; (3) $h_x = 1.26$ nm, $h_z = 1.26$ nm. Reprinted from Ref. [21], Copyright 2008, with permission from Elsevier.

Let us analyze the influence of the prolate shape of nanoparticles on the lower branches of the *p*-polarized surface wave dispersion curves (Fig. 2.9). One can see that the high-frequency branch, common for prolate nanoparticles, is higher than the corresponding branch for spherical nanoparticles, and when the shape of the nanoparticle approaches the spherical one, it converges from higher frequency values, whereas the low-frequency branch behaves vice versa, approaching the corresponding curve from lower frequency values. In the case of oblate nanoparticles, the dispersion curves shift in other direction than the spherical nanoparticles. In Figs. 2.7 and 2.9, an interesting effect of the deviation of the dispersion curves is observed with the growth of the wave vector toward the decrease in frequencies. This shows that such waves are characterized by a negative dispersion, which physically signifies the opposite directions of the phase and the surface wave energy propagation. It is known that waves of this type can exist in a solid [38].

The obtained results show that the laws of SPP dispersion, in the case of nanostructured coating on the metal surface, essentially depend on the nanoparticles' shape, type, and surface concentration. In particular, the dispersion curve shifts toward the lower frequencies when the shape of the nanoparticles changes from the prolate ellipsoids ($h_{\perp}/h_{||} = 2$) to the spherical nanoparticles at constant concentration (Fig. 2.9, upper curves 2 and 3).

2.2.2.3 Peculiarities of SPR study to account the 3D polarization factor of the molecules

As was mentioned before, molecular surface concentration N_s and components of molecular susceptibility $\chi_{jl}(\omega)$ can be determined from the SPR experiments instead of effective layer thickness and refractive index, which are usually estimated. For calculating the molecular layer reflection coefficient, it is needed to know the initial molecules' susceptibility value $\chi_{ij}(\omega)$, which describes linear response on the local field by a single molecule located on the surface. Since the SPR simulation will be taken out for immunoglobulin biomolecules, one can assume that the molecule on the surface has a shape close to the ellipsoidal one. The linear response of the ellipsoidal particle on the surface was calculated earlier [39]. If h_x, h_y, h_z are the semi-axes of the molecule and V_p is the volume of the molecule, then the plane component of the molecular susceptibility is equal to

$$\tilde{\chi}_{ii} = \varepsilon_a V_p \frac{(\varepsilon_p - \varepsilon_a)}{\varepsilon_a + (\varepsilon_p - \varepsilon_a)m_i} L_{||},\ i = x, y, \tag{2.28}$$

where the plane local field factor is

$$L_{||} = \left[1 + \frac{(\varepsilon_a - \varepsilon_s)(\varepsilon_p - \varepsilon_a)}{3(\varepsilon_a + \varepsilon_s)(\varepsilon_a + (\varepsilon_p - \varepsilon_a)m_i)} \vartheta \right]^{-1}, \tag{2.29}$$

where ε_p is the dielectric constant of particle; ε_s and ε_a are dielectric constants of the substrate and the environment, respectively; m_i is the depolarization factor; $\vartheta = h_x h_y h_z/(2z_p)^3$ is the local field factor; and z_p is the z-coordinate of the center of the molecule.

The normal part of the molecular susceptibility is

$$\tilde{\chi}_{zz} = \varepsilon_a V_p \frac{(\varepsilon_p - \varepsilon_a)}{\varepsilon_a + (\varepsilon_p - \varepsilon_a)m_i} L_\perp, \tag{2.30}$$

where

$$L_\perp = \left[1 + \frac{(\varepsilon_a - \varepsilon_s)(\varepsilon_p - \varepsilon_a)}{3(\varepsilon_a + \varepsilon_s)(\varepsilon_a + (\varepsilon_p - \varepsilon_a)m_i)} 2\vartheta\right]^{-1} \tag{2.31}$$

Depolarization factors for a molecule with prolate shape, where $h_z > h_x = h_y$ (see Fig. 2.10a), are as follows:

$$m_z = \frac{1-\eta^2}{\eta^3}\left(\frac{1}{2}\ln\frac{1+\eta}{1-\eta} - \eta\right), \; m_x = m_y = \frac{1}{2}(1 - m_z), \tag{2.32}$$

where $\eta = (1 - \zeta^2)^{1/2}$, $\zeta = h_x/h_z$.

Depolarization factors for a molecule with oblate shape, where $h_x = h_y > h_z$ (see Fig. 2.10b), are as follows:

$$m_z = \frac{1+\nu^2}{\nu^3}(\nu - \arctan \nu), \; m_x = m_y = \frac{1}{2}(1 - m_z), \tag{2.33}$$

where $\nu = (\zeta^2 - 1)^{1/2}$. It should be noted that the interaction between the molecule and the surface can lead to the phenomenon of the local field amplification, which causes a significant increase in molecular polarizability [40].

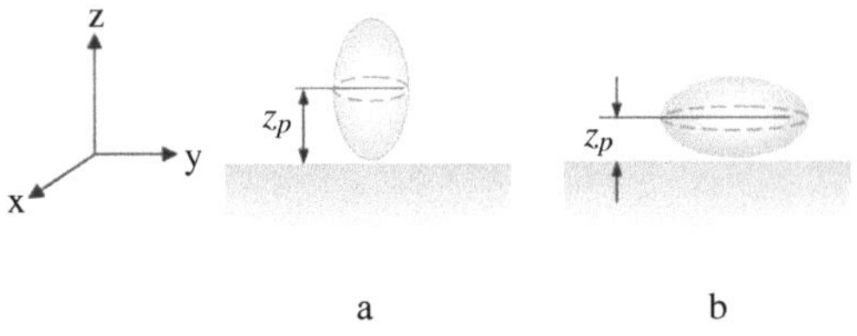

Figure 2.10 Protein molecule on the surface: (a) as a prolate ellipsoid; (b) as an oblate ellipsoid. Reprinted from Ref. [36], Copyright 2008, with permission from Elsevier.

To simulate the SPR experiment schematically shown in Fig. 2.6b, it is necessary to calculate the reflection coefficient for the system consisting of an ATR glass prism, a thin gold film, and a liquid with an adsorbed molecular layer. If one uses the described approach, then it is obvious that the molecular layer is located on the surface of a metallic film, which is located on the ATR glass prism. Using Eq. (2.25), one can calculate the SPR curves related to different molecule shapes. From Eqs. (2.22) and (2.24), it is seen that the reflection coefficient, which defines the SPR curve, depends on the concentration and shape of the molecules. The reflection coefficient was calculated using a specially designed software, according to Eq. (2.25) for different values of the parameter $\zeta = h_{||}/h_{\perp}$ (where $h_{||}$ and $h_{\perp}$ are the semi-axes of ellipsoid, parallel (||) and perpendicular to ($\perp$) to the substrate surface plane), which defines the shape of the molecule of similar mass. It turned out that SPR curves corresponding to molecules that have the shape of a prolate ellipsoid are characterized by a rather strong shift with the change in the parameter ζ (see Fig. 2.11a). Molecules characterized by the shape of an oblate ellipsoid show very small shifts for different values of the parameter ζ. For example, a change in the value of ζ from 1.1 to 10 results in a change in the angle of the minimum of the SPR curve from $\theta_{min} = 64.121°$ to $\theta_{min} = 64.262°$.

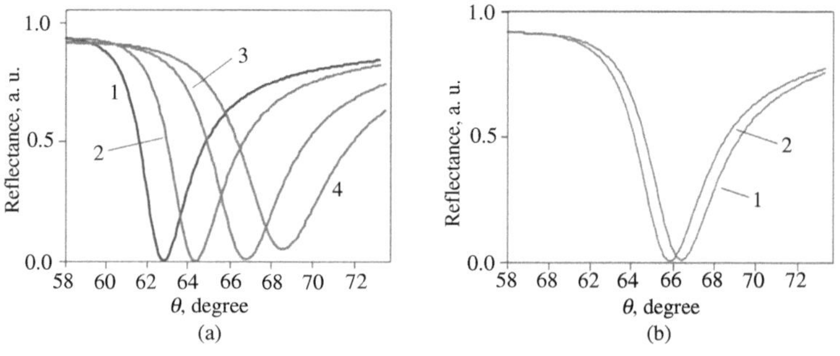

Figure 2.11 (a) Calculated SPR curves dependent on the shape of molecules. (1) Free surface, $\theta_{min} = 62.747°$; (2) oblate molecules $\zeta = 2.0$, $\theta_{min} = 64.262°$; (3) prolate molecules $\zeta = 0.12$, $\theta_{min} = 66.585°$; (4) prolate molecules $\zeta = 0.11$, $\theta_{min} = 68.302°$. (b) Calculated dependences of SPR curves on the composition of the molecular film consisting of prolate ($\zeta = 0.12$) and oblate ($\zeta = 2.0$) molecules. Part of prolate molecules: $f = 1$ (curve 1, $\theta_{min} = 66.282°$) and $f = 0.5$ (curve 2, $\theta_{min} = 65.777°$). Reprinted from Ref. [36], Copyright 2008, with permission from Elsevier.

Calculations show that the shift in the minimum of the SPR curve with increasing concentration of molecules is obvious. In particular, an increase in molecular concentration (for molecules that are characterized by $\zeta = 0.15$) by 5% leads to an angular shift of $\Delta\theta = +0.1°$. This means that the proposed approach may be useful for expanding the informativity of experimental results obtained with SPR measurements for the evaluation of molecular coatings or layers, because this approach may determine the surface concentration of molecules having their own molecular characteristics, such as the polarizability of one molecule or its shape.

It should be noted that this approach allows the possibility of considering the two-component coating on the surface of the SPR sensor. In particular, it provides an option to calculate the SPR curves for molecular layers consisting of molecules with prolate and oblate ellipsoid shapes. The dependences of the SPR curves on the molecular film composition consisting of prolate ($\zeta = 0.12$) and oblate ($\zeta = 2.0$) molecules are shown in Fig. 2.11b. It is seen that the change in the percentage of the prolate molecules from $f = 1$ (corresponding to the molecular layer, which consists only of prolate molecules) to $f = 0.5$ leads to the minimum shift in SPR curve of about 0.505°. This result shows strong dependence of the molecular layer dispersion properties on the shape of molecules.

2.3 Localized SPR Phenomenon and Theoretical Background for Its Application in Sensing

2.3.1 Theoretical Background of Localized SPR Method

2.3.1.1 Sensitive element model for LSPR sensor

Localized SPR is one of the peculiar optical properties of noble metal nanoparticles that occurs when the light falling on them is resonant with the collective oscillations of free electrons in the nanoparticle (Fig. 2.12). This electronic response causes unusual optical properties of nanosized metal—a peak appears in the light extinction spectrum of noble metal nanoparticles (Fig. 2.13), which is not observed in the corresponding spectrum for bulk material. The shape of the spectrum and the position of the LSPR peak depend on the shape of the nanoparticle [41, 42], its size [43, 44], the interparticle distance [45], and the dielectric properties of the material of the nanoparticles

[43, 46], as well as the dielectric properties of the environment [47–50] and the charge of the molecules contained therein [51], which is important for sensor applications.

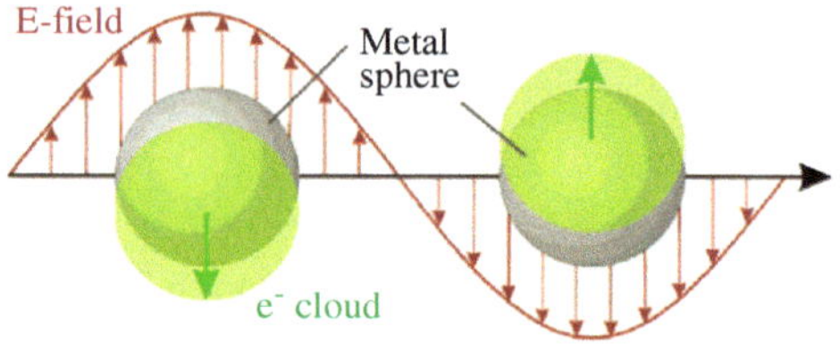

Figure 2.12 Excitation of a dipolar localized surface plasmon by an electric field of an incident light wave. Reprinted with permission from Ref. [52], Copyright 2003, American Chemical Society.

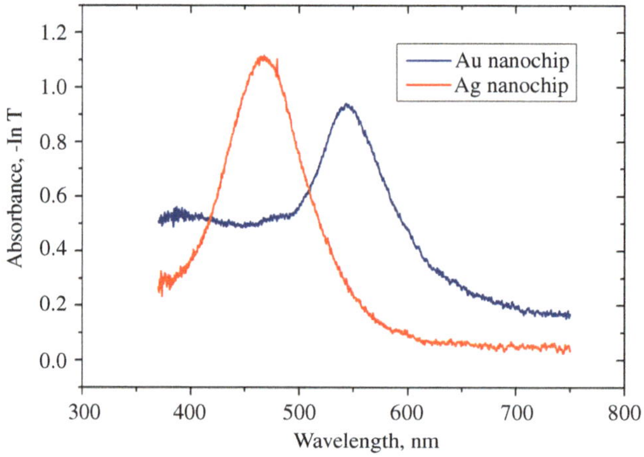

Figure 2.13 Light extinction spectra of random gold and silver nanoparticle arrays on glass substrates (nanochips).

Since the operating principle of LSPR sensors is based on the change in the optical properties of absorption and scattering of light by high-conductive metal nanoparticles when the adsorption of molecules or molecular process occurs on their surface, for the development of a theoretical background for describing the operation of the LSPR sensor, an important stage is the development of a model for the "nanoparticle–molecule" system, which forms the sensitive element of the sensor. To do this, one needs to specify the optical and geometric parameters of the nanoparticle, the molecular component, and the ambient environment.

2.3.1.2 Optical constants of gold nanoparticles

It is known that the optical properties of nanostructured and bulk materials differ, because the so-called size effects exist [53] due to the dependence of the dielectric permittivity $\varepsilon(\omega, R)$ on the size of the nanoparticle. Reducing the size of the metallic nanoparticle leads to an increase in the influence of the classical (a reduction in the mean free path of electrons [53], a decrease in the concentration of free electrons in a nanoparticle due to the spill-out effect [53]) and quantum mechanical (Landau damping [54], the interaction of plasmon oscillations with individual discrete electronic states [55]) size effects. The typical size of a nanoparticle, which the size effects begin to affect the dielectric constant at, varies: for the reduction in the mean free path of electrons, this size is smaller or comparable to the free path of electrons in a bulk material; for the spill-out effect, it is less than 10 nm, and quantum-mechanical size effects are important when considering only very small metal clusters [53]. Since the typical nanoparticulate materials for use in LSPR sensors are gold and silver, for which the electron mean free path is 42 and 52 nm [53], respectively, the most significant effect on the optical properties of nanoparticles produces an effect of reduction in the electron mean free path. Thus, this model, which is most commonly used to specify the optical constants of nanoparticles, is described below.

Optical constants of bulk gold were taken from Table II (pp. 290–295), Part II in Ref. [56] and approximated using 7–9 order polynomials with wavelength steps in the range of 0.1 to 1 nm (depending on the system under consideration). The dielectric function of gold has been modified in accordance with the size of the nanoparticle according to the model of reducing the mean free path of electrons. This modification was carried out by introducing an effective electron relaxation time

$$\tau_{\text{eff}}(R) = \left(\tau_{\text{bulk}}^{-1} + A\frac{V_{\text{F}}}{R} \right)^{-1}, \tag{2.34}$$

where $\tau_{\text{bulk}} = 9.3 \times 10^{-15}$ s [57] is the electron relaxation time for bulk gold, $V_{\text{F}} = 1.4 \times 10^6$ m/s [58] is the Fermi velocity, R is the spherical nanoparticle radius, and A is a constant, which can be set to 1 for spherical nanoparticles studied and isotropic surface electron scattering [53, 59]. The size-dependent electron relaxation time was further used to modify the values of the dielectric function in the Drude–Lorentz model [60, 61]:

$$\varepsilon_1(\omega,R)=\varepsilon_{1\text{bulk}}(\omega)+\frac{\omega_p^2}{\omega^2+\frac{1}{\tau_{\text{bulk}}^2}}-\frac{\omega_p^2}{\omega^2+\frac{1}{\tau_{\text{eff}}^2(R)}},$$

$$\varepsilon_2(\omega,R)=\varepsilon_{2\text{bulk}}(\omega)+\frac{\omega_p^2}{\omega}\left(\frac{\tau_{\text{eff}}(R)}{\omega^2\tau_{\text{eff}}^2(R)+1}-\frac{\tau_{\text{bulk}}}{\omega^2\tau_{\text{bulk}}^2+1}\right), \quad (2.35)$$

where ε_1 and ε_2 stand for the real and imaginary parts of the dielectric permittivity, ω is the angular frequency of light, and $\omega_p = 1.37 \times 10^{16}$ rad/s [61] is the plasma frequency for bulk gold. Optical constants adjusted for the size of the gold nanoparticles were calculated according to the following equations [62]:

$$n_1(\omega,R)=\sqrt{\frac{1}{2}\left(\varepsilon_1(\omega,R)+\sqrt{\varepsilon_1^2(\omega,R)+\varepsilon_2^2(\omega,R)}\right)},$$

$$k_1(\omega,R)=\sqrt{\frac{1}{2}\left(-\varepsilon_1(\omega,R)+\sqrt{\varepsilon_1^2(\omega,R)+\varepsilon_2^2(\omega,R)}\right)}. \quad (2.36)$$

For non-spherical gold nanoparticles, which were also considered a sensitive element of LSPR sensor, the optical constants of bulk gold were used without modification. This approach was used due to the fact that the geometric size of the studied non-spherical nanoparticles was predominantly greater than the mean free path of electrons in gold, so that the influence of dimensional effects on the optical constants of nanoparticles can be neglected.

2.3.1.3 Optical constants of the molecular component and environment

Introduction to the model of the sensitive element of the LSPR sensor of the molecular component was carried out by adding one or two dielectric layers of a certain thickness, which formed a coating on the surface of the gold nanoparticle. Two approaches were used to specify the optical properties of molecular layers. The first approach supposes the specification of the optical constants of the layer irrespective of the geometric dimensions of the molecules and the nanoparticle, i.e., the molecular layer was considered a uniform dielectric layer with a certain refractive index. In the framework of this approximation, the refractive index of the layer was considered both real and complex, as well as both dispersive and non-dispersive. The refractive indexes of the environment and the substrate (in the case of an array of nanostructures on the solid surface) were chosen as real and independent of the light wavelength.

In the second approach, which was realized only for spherical gold nanoparticles, the molecular layer was interpreted as a saturated monolayer of globular molecules, which are approximated with solid spheres (Fig. 2.14). To describe the shell consisting of a densely packed monolayer of globular molecules, a symmetric Bruggeman effective medium theory (EMT) [59] was used, which specified the effective value of the refractive index of the shell n_2 as a solution to the equation

$$f\frac{n_{\mathrm{m}}^2 - n_2^2}{n_{\mathrm{m}}^2 + 2n_2^2} + (1-f)\frac{n_0^2 - n_2^2}{n_0^2 + 2n_2^2} = 0\,, \tag{2.37}$$

where f is the shell fill factor by molecules, n_{m} is the molecules' refractive index, which was chosen to be equal to 1.46 as for biomolecular species [64, 65] regardless of the wavelength of light, n_0 is the refractive index of the environment (water), which was calculated by the formula (λ in nanometers) [66]

$$n_0 = 1.32334 + \frac{3479}{\lambda^2} - \frac{5.111\times 10^7}{\lambda^4}. \tag{2.38}$$

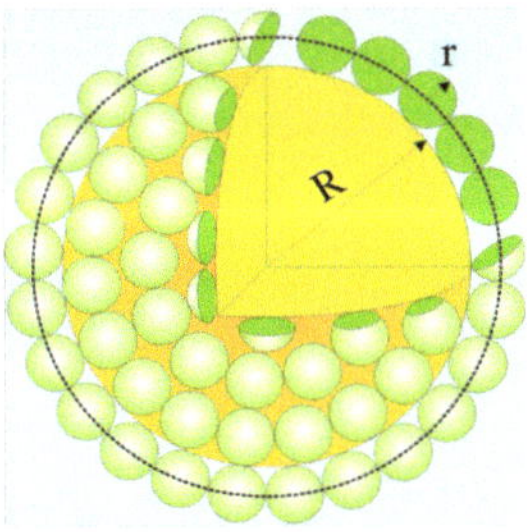

Figure 2.14 Schematic representation of a sensitive element model for LSPR sensor with a molecular layer in the form of a dense monolayer of globular molecules. Reprinted with permission from Ref. [63], Copyright 2011, IEEE.

Obviously, the fill factor for a saturated monolayer of spherical molecules depends on the ratio between the diameter of the nanoparticle and the shell thickness. This dependence for a spherical nanoparticle was obtained with the following considerations. Let us imagine the surface of the sphere of the radius $R + r$ (where r is the radius of the molecule), where the centers of the molecules are located on, and calculate the number of cross sections of the molecules with this surface. If one approximates the arrangement of these cross sections with a dense square grid of circles of radius r on a plane, one can obtain an expression for the number of molecules on the surface of a nanoparticle:

$$N_{\mathrm{m}} \approx \frac{S_{\mathrm{plane}}}{S_{\mathrm{square}}} = \frac{4\pi(R+r)^2}{4r^2} = \frac{\pi(R+r)^2}{r^2}. \tag{2.39}$$

The fill factor is calculated as a ratio between the volume occupied by molecules and the total volume of the shell:

$$f(R,r) = \frac{V_{\mathrm{m}}}{V_{\mathrm{shell}}} = \frac{N_{\mathrm{m}}\frac{4}{3}\pi r^3}{\frac{4}{3}\pi\left[(R+2r)^3 - R^3\right]} = \frac{\pi r(R+r)^2}{(R+2r)^3 - R^3} \tag{2.40}$$

2.3.1.4 Method of optical properties calculation based on the Mie scattering theory for LSPR sensor sensitive element

To simulate the optical properties of light absorption, scattering, and extinction for individual gold nanoparticles, the Mie theory of light scattering on a spherical nonmagnetic particle with a shell [59] was used. Within this approach, cross sections of extinction, scattering, and absorption of light (the rates of energy totally lost, scattered, and absorbed, respectively, divided by the incident light intensity) are expressed in the following form:

$$\sigma_{\mathrm{ext}} = \frac{2\pi}{|\vec{k}|^2}\sum_{l=1}^{L}(2l+1)\mathrm{Re}(a_l + b_l), \tag{2.41}$$

$$\sigma_{\mathrm{sca}} = \frac{2\pi}{|\vec{k}|^2}\sum_{l=1}^{L}(2l+1)(|a_l|^2 + |b_l|^2), \tag{2.42}$$

$$\sigma_{\mathrm{abs}} = \sigma_{\mathrm{ext}} - \sigma_{\mathrm{sca}}, \tag{2.43}$$

$$a_l = \frac{\psi_l(y)[\psi_l'(m_2y) - A_l\chi_l'(m_2y)] - m_2\psi_l'(y)[\psi_l(m_2y) - A_l\chi_l(m_2y)]}{\xi_l(y)[\psi_l'(m_2y) - A_l\chi_l'(m_2y)] - m_2\xi_l'(y)[\psi_l(m_2y) - A_l\chi_l(m_2y)]}, \tag{2.44}$$

$$b_l = \frac{m_2\psi_l(y)[\psi_l'(m_2y) - B_l\chi_l'(m_2y)] - \psi_l'(y)[\psi_l(m_2y) - B_l\chi_l(m_2y)]}{m_2\xi_l(y)[\psi_l'(m_2y) - B_l\chi_l'(m_2y)] - \xi_l'(y)[\psi_l(m_2y) - B_l\chi_l(m_2y)]}, \tag{2.45}$$

$$A_l = \frac{m_2\psi_l(m_2x)\psi_l'(m_1x) - m_1\psi_l'(m_2x)\psi_l(m_1x)}{m_2\chi_l(m_2x)\psi_l'(m_1x) - m_1\chi_l'(m_2x)\psi_l(m_1x)}, \tag{2.46}$$

$$B_l = \frac{m_2\psi_l(m_1x)\psi_l'(m_2x) - m_1\psi_l(m_2x)\psi_l'(m_1x)}{m_2\chi_l'(m_2x)\psi_l(m_1x) - m_1\psi_l'(m_1x)\chi_l(m_2x)}, \tag{2.47}$$

where $\vec{k}$ is the wave vector of light in the environment; L is the number of considered multipole modes; $m_1 = (n_1 + ik_1)/n_0$ and $m_2 = n_2/n_0$ are complex refractive indices of core and shell relative to the environment; $x = |\vec{k}|\,a$, $y = |\vec{k}|\,b$, $a = R$, and $b = R + 2r$ are the core and shell radii; $\psi_l(z)$, $\xi_l(z)$, and $\chi_l(z)$ are the Riccati–Bessel functions. The parameter L value was calculated according to the relationship indicated in Ref. [67]:

$$L = \left[y + 4y^{\frac{1}{3}} + 2 \right], \tag{2.48}$$

where the square brackets [ ] mean rounding to the nearest integer. To precisely determine the wavelength positions of maxima of the simulated cross sections spectra, parabolic approximation in the vicinity of absolute maximum position of the spectrum was used.

2.3.2 Influence of "Nanoparticle–Molecular Layer" System Parameters on the Optical Response of LSPR Sensor

2.3.2.1 Comparison of LSPR and SPR sensors response

The main method used in LSPR-based sensing in noble metal nanoparticles is the measurement of light extinction. Therefore, the response (specifically, the shift in the wavelength of the LSPR peak position in the extinction spectrum) was calculated for a single spherical gold nanoparticle immersed in water upon a thickening molecular monolayer using the equations given in Subsection 2.3.1. Figure 2.15 shows a simulated LSPR response of gold nanoparticles

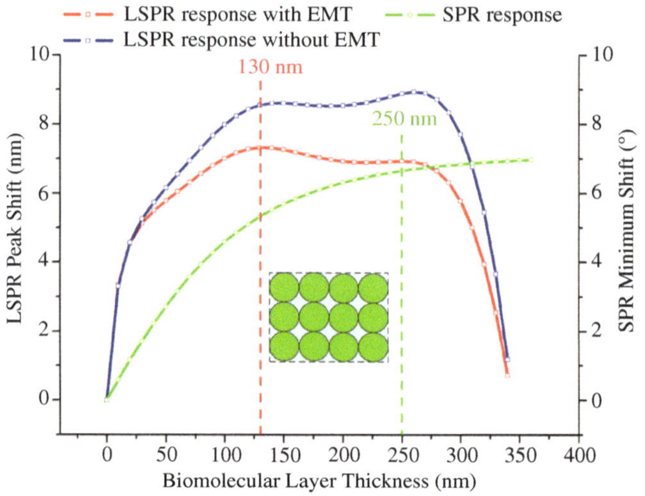

Figure 2.15 Response of LSPR sensor on the basis of a gold nanoparticle with a radius of 25 nm and SPR sensor based on a gold film with a thickness of 50 nm, depending on the thickness of the molecular layer. Calculations of the response of SPR sensor were carried out according to the theoretical model developed in Subsection 2.2.1 for the wavelength of 650 nm. The refractive index of a homogeneous layer for SPR and LSPR simulations without EMT was 1.398, corresponding to a flat square grid (inset) of molecules (refractive index 1.46) with a fill factor of π/6, and the refractive index of the environment (water) was 1.331. Reprinted with permission from Ref. [63], Copyright 2011, IEEE.

with a radius of 25 nm with increasing thickness of the coating according to the models, which treated a shell as homogeneous as well as composed of globular molecules (according to Eqs. (2.37), (2.38), (2.40)), compared with the response of a usual SPR sensor based on a 50 nm thick gold film. Usually, a small range of shell thicknesses (up to 30 nm [68, 69]) is studied, in which the shift in the LSPR peak rapidly increases with an increase in the layer thickness on the nanoparticle surface. A further increase in the thickness of the shell results in a slower change in LSPR response and is usually not studied. Indeed, an LSPR sensor is theoretically able to sense much thicker molecular coatings; in particular, for a nanoparticle with a radius of 25 nm, the LSPR response is saturated when the thickness of the molecular layer reaches 130 nm (as can be seen from Fig. 2.15). Further increase in the thickness of the shell does not show significant growth of response. This behavior of the LSPR response with saturation at thick molecular layers is similar to the response in the slowly decaying electric field of the nano-pyramidal particle, which was considered in Ref. [41]. This is expected, since the spherical nanoparticle with excited surface plasmons does not have sharp edges typical for pyramids, in which the electric field is the most intense, but also rapidly decaying (so-called "hot spots"). This value of the sensitivity "upper limit," which is equal to 130 nm, is comparable to the value for the SPR sensor (250 nm, see Fig. 2.15). Figure 2.15 also shows that the application of the effective medium theory reduces the response of the sensor due to a decrease in the effective refractive index of the shell with increasing diameter of the molecule.

2.3.2.2 Influence of the sensor element size on the response of the LSPR sensor

For further studying the LSPR sensitivity of spherical gold nanoparticles with molecular shell, extinction simulations were performed for nanoparticles of different sizes and molecular layers of different thicknesses. First, the above-described theoretical model was applied to the system mentioned in Ref. [69], specifically, a spherical gold nanoparticle with a radius of 5, 10, 15, and 20 nm coated with a homogeneous layer of polymethylmethacrylate (PMMA) with thickness of 0–30 nm located in air. Results of LSPR response simulation are presented in Fig. 2.16. In the thickness range of 0–15 nm, results similar to those presented in Ref. [69] were obtained. However, the behavior of the LSPR response at shell thicknesses larger than 15 nm differs from the expected: it is obvious that the shift in the LSPR peak in the extinction spectrum does not indicate a plateau. In addition, for a nanoparticle with a radius of 5 nm, the response begins to decrease when the thickness of the shell exceeds 20 nm.

To clarify this special behavior of the dependence of the LSPR peak shift on the thickness of the molecular layer, the LSPR response of a spherical gold nanoparticle with a wide range of radii, namely 1, 5, 20, and 50 nm, was simulated. The molecular coating on the surface of the nanoparticle was considered as a densely packed monolayer of globular molecules, as described in Subsection 2.3.1. The size of the molecules, and the shell thickness, respectively, varied in the range from 0 to 20 nm, which includes the typical sizes of most globular molecules studied by LSPR. Such a molecular layer was treated as a homogeneous shell with an index of refraction n_2, obtained according to the effective medium theory provided by Eqs. (2.37), (2.38), and (2.40). First it should be noted that the properties of LSPR of small gold nanoparticles (for example, with a radius of 1 nm) are not described in the literature as widely as the properties of nanoparticles of larger size. This may be due to experimental difficulties, as well as the complexity of the theoretical description of such system due to the significant influence of quantum and other size effects. Simulated shifts in the extinction peak are shown in Fig. 2.17. There are several typical features that should be noticed.

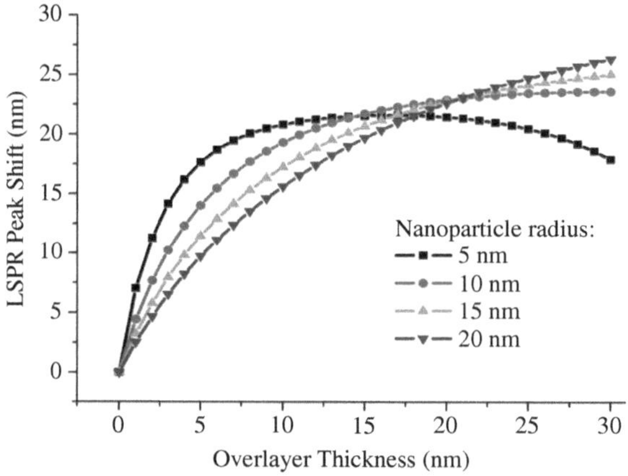

Figure 2.16 Response of LSPR sensor based on a gold nanoparticle with a radius of 5, 10, 15, and 20 nm depending on the thickness of the homogeneous PMMA shell (n = 1.496). Reprinted with permission from Ref. [63], Copyright 2011, IEEE.

First, the growth rate of the LSPR shift for the small shell thickness depends on the size of the nanoparticle: the smaller nanoparticle provides faster growth of the LSPR response, which is consistent with the results in Ref. [69]. However, the response for a nanoparticle with a radius of 50 nm grows faster than a response for a nanoparticle with a radius of 20 nm, which does not coincide with the general trend. Second, nanoparticles with a radius of 5, 20, and 50 nm show an ordinary response, which tends to saturation for $d > R$, where $d = 2r$ is the thickness of the molecular layer. As for the LSPR response of a gold nanoparticle with a radius of 1 nm, two significant features are observed: very fast growth at a thickness of 0 to 2 nm compared with nanoparticles of other sizes and a significant decrease in the large thicknesses of the coating, even to the negative shifts in the wavelength position of the LSPR peak.

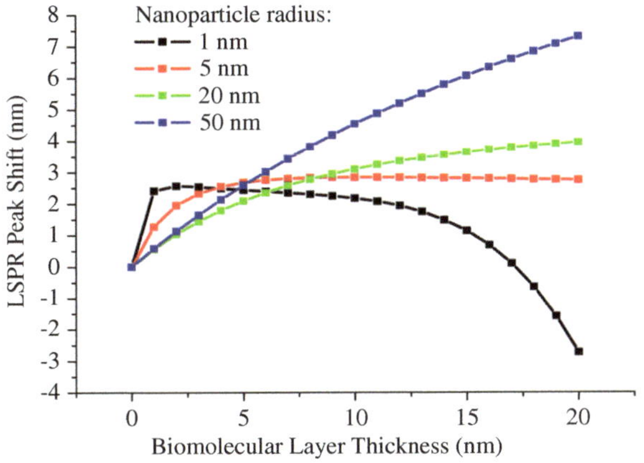

Figure 2.17 Response of the LSPR sensor on the basis of a gold nanoparticle with a radius of 1, 5, 20, and 50 nm depending on the thickness of the molecular layer. Reprinted with permission from Ref. [63], Copyright 2011, IEEE.

2.3.2.3 Features of the response of LSPR sensor based on small-size gold nanoparticles

Due to the importance of detecting small-size molecules such as toxins or pesticides, the dependence of the LSPR shift on the size of a molecule for a nanoparticle with a 1 nm radius is most interesting; therefore, the features shown in Fig. 2.17 were studied more precisely. For the complete study of the first feature, the LSPR response of small gold nanoparticles with a radius of 0.5–1 nm and a thickness of the molecular layer of 0–5 nm (Fig. 2.18a) was calculated.

Figure 2.18a shows that when the size of the nanoparticle is reduced, the initial LSPR response at $d < 1$ nm increases significantly, reaching a maximum value of about 5 nm for a nanoparticle with a radius of 0.5 nm, which is almost twice larger than for a nanoparticle with a radius of 1 nm. The reason for such increase in response was found after the same simulation with values of optical constants of gold nanoparticles same as of bulk gold. This approach results in the dependence of the LSPR peak shift shown in Fig. 2.18b. Obviously, the aforementioned growth in response in this figure is absent; so the growth of the LSPR response for small gold nanoparticles with a thin molecular coating is a consequence of the modification of the gold optical constants by taking into account the size-dependent electron relaxation time. The effect of this significant shift in the LSPR wavelength can be used for the better detection of small molecules with the size of about 1 nm using LSPR sensor based on nanoparticles with a radius of up to 1 nm.

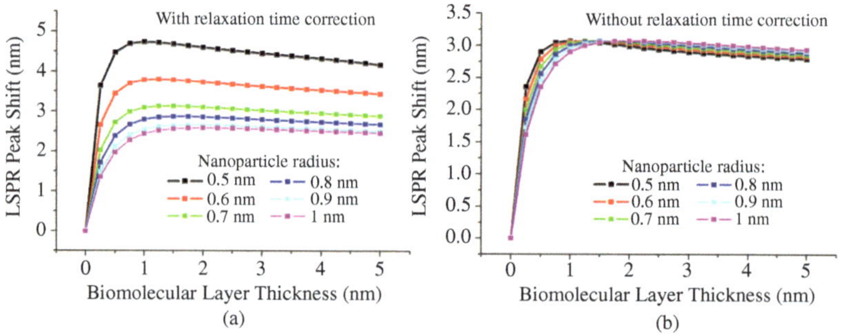

Figure 2.18 Response of the LSPR sensor on the basis of a gold nanoparticle with a radius of 0.5–1 nm depending on the thickness of the molecular layer: The simulation is carried out (a) with and (b) without correction of the electron relaxation time (using the optical constants of bulk gold). Reprinted with permission from Ref. [63], Copyright 2011, IEEE.

As for the decrease of LSPR response for a nanoparticle with a radius of 1 nm with molecular layer thickness increasing (see Fig. 2.17), it was discovered that the reason for this phenomenon is the ratio between the components that contribute to extinction. It is known that the total extinction of light is the sum of the scattering and absorption of light by a nanoparticle [59, 70]. For the simulation of the scattering and absorption cross-sectional spectra of a gold nanoparticle with a radius of 1 nm (Fig. 2.19a), Eqs. (2.42) and (2.43) of the Mie theory were used. Peak shifts in absorption and extinction cross-sectional spectra were also plotted depending on the thickness of the molecular layer (Fig. 2.19b).

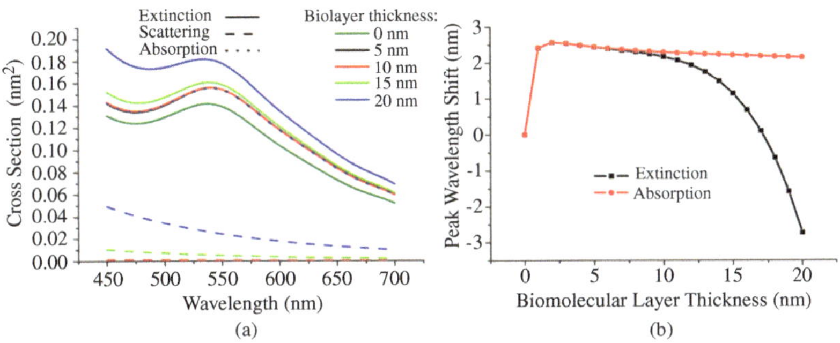

Figure 2.19 Spectra of light extinction, scattering, and absorption cross sections by a gold nanoparticle with a radius of 1 nm with a molecular coating thickness of 0, 5, 10, 15, 20 nm; (b) wavelength shifts of extinction and absorption peaks for a gold nanoparticle with a radius of 1 nm depending on the thickness of the molecular layer. Reprinted with permission from Ref. [63], Copyright 2011, IEEE.

Figure 2.19a shows that the contribution of scattering to the total extinction increases with the molecular shell thickness. This also can be observed in Fig. 2.19b where the change in the absorption peak shift matches with the change in the shift of the extinction peak only for thin molecular layers. With increasing thickness of molecular layers, the contribution of scattering for short wavelengths is greater than for long ones (Fig. 2.19a). Since the absorption contribution in this case is almost independent of the shell thickness, this results in an extinction peak shift toward shorter wavelengths. With the molecule size larger than 17 nm, the extinction peak is located at a wavelength even shorter than the initial position of the extinction peak of the uncoated nanoparticle.

2.3.2.4 Dependence of LSPR sensor response on the ratio of extinction components (scattering and absorption)

The distribution of contributions from scattering and absorption to extinction for larger gold nanoparticles with a radius of 5, 20, and 50 nm (Fig. 2.20a–f) was also investigated. For these nanoparticles, the scattering cross-sectional spectrum reveals a peak whose position also depends on the thickness of the molecular coating on the nanoparticle. Figure 2.20 shows the dependences between the cross-sectional spectra (and corresponding peak shifts) and the thickness of the molecular layer. Obviously, the contribution of

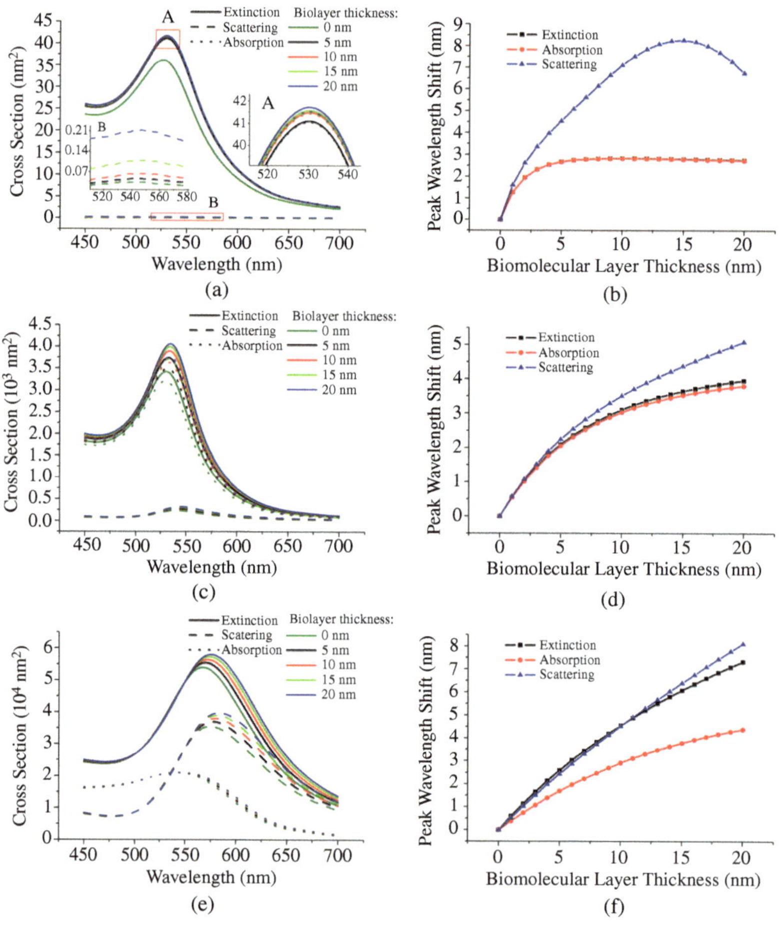

Figure 2.20 Spectra of light extinction, scattering, and absorption cross sections by a gold nanoparticle with a radius of (a) 5, (c) 20, and (e) 50 nm with a molecular coating thickness of 0, 5, 10, 15, 20 nm, and corresponding extinction, scattering, and absorption peak shifts for a gold nanoparticle with a radius of (b) 5, (d) 20, and (f) 50 nm depending on the thickness of the molecular layer. Reprinted with permission from Ref. [63], Copyright 2011, IEEE.

scattering to total extinction depends on the size of the nanoparticle: the larger the nanoparticle, the stronger the influence of scattering. The contribution of scattering to extinction for a gold nanoparticle with a radius of 5 nm is not significant (Fig. 2.20a). However, the scattering spectrum has a peak with position that is much more sensitive to the change in the thickness of the molecular layer than the absorption and extinction peaks: the peak of scattering shifts by almost 8 nm in comparison with saturated displacements of absorption peaks and extinction with a value near 2.5 nm (Fig. 2.20b). For a nanoparticle with a radius of 20 nm, the contribution of scattering to full extinction does not exceed 7–8% (based on a comparison of peak height) (Fig. 2.20c). It has a small effect on the position of the peak of extinction; the extinction peak shift slightly prevails over the absorption peak shift. The scattering peak shift for this nanoparticle exceeds the extinction peak shift by a value of 28% (for a thickness of the molecular layer at 20 nm) (Fig. 2.20d). For a nanoparticle with a radius of 50 nm, scattering becomes the main component (Fig. 2.20e). This leads to the fact that the shifts of extinction and scattering peak position show similar behavior and differ from each other by no more than 11% in the investigated thickness range (Fig. 2.20f). It is clearly seen from the graphs presented that the scattering peak shifts faster with the growth of the thickness of the molecular layer than the absorption and extinction peaks for the core–shell configurations investigated. This fact indicates that high-sensitivity sensors based on the scattering of light by nanostructures can be created, although they may require more complex experimental equipment than sensors based on extinction measurements of light due to the weak light intensities scattered by small nanoparticles.

2.3.2.5 LSPR sensor response description using the number of molecules and surface concentration parameters

Using the effective medium theory allows the introduction of such parameters of the molecular layer as the number of molecules in the layer, which can be calculated according to Eq. (2.39). The dependences of the LSPR extinction peak shift on the thickness of the molecular shell and the number of molecules for gold nanoparticles of different radii are shown in Fig. 2.21.

It is also possible to determine the surface concentration of molecules on the surface of a nanoparticle, which for a saturated monolayer is equal to

$$N_{\mathrm{S}} = \frac{N_{\mathrm{m}}}{4\pi R^2} = \frac{(R+r)^2}{4r^2R^2}. \qquad (2.49)$$

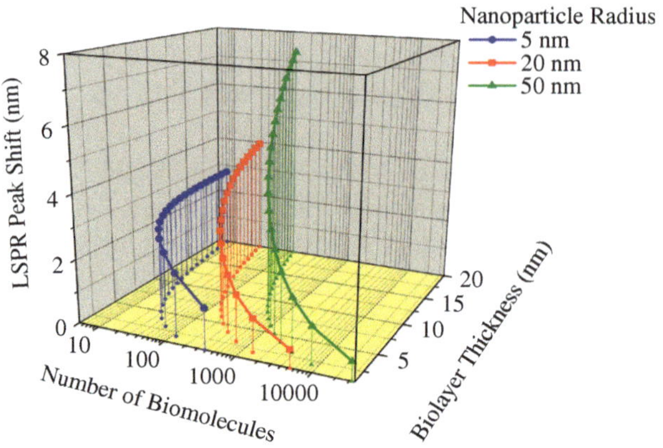

Figure 2.21 LSPR extinction peak shift dependence on the thickness of the molecular shell and the number of molecules for a gold nanoparticle with a radius of 5, 20, and 50 nm. Reprinted with permission from Ref. [63], Copyright 2011, IEEE.

These parameters can be used to describe submonolayer molecular coatings.

2.3.3 Comparative Analysis of LSPR Sensor Optical Response Measurement Modes

During the development of LSPR sensors, there is a problem of choosing the mode of registering the LSPR response that will provide an optimal performance. The main question is the selection of a suitable method for processing the LSPR spectrum to obtain the maximum response value for the same process being investigated. Usually two approaches are used to measure the LSPR response: determining the wavelength shift of the LSPR extinction peak position [71–75] and determining the change in the magnitude of extinction at the selected wavelength located within the peak [72, 76–78]. Both methods have advantages and disadvantages that limit their range of applications. The first approach provides the possibility of reliable detection of the molecular process, because the characteristic shift of light extinction peak position during the formation of a non-absorbing molecular coating on a metal nanoparticle can be caused only by the LSPR phenomenon, and it is not sensitive to the errors of the spectrophotometer calibration. On the other hand, if the molecular layer is quite thin, then recording the shift in the peak position may be complicated. The second method provides a more sensitive response to molecular events in certain situations and can be easily implemented for continuous monitoring [79], but it lacks insensitivity to the calibration errors inherent in the first approach. In addition, to maximize the measured response for this method, it is necessary to search for the optimal wavelength position at which the measurement will be performed. Therefore, this subsection presents results of a comparative analysis of possible modes of measurement of LSPR response upon the formation of saturated molecular layers on the surface of a spherical gold nanoparticle used as a sensitive element of LSPR sensor. The analysis was performed based on the theoretically calculated dependencies of LSPR response on the size of the nanoparticle and the thickness of the molecular coating using the equations given in Subsection 2.3.1.

The LSPR response measurement modes discussed in this subsection are shown in Fig. 2.22. In addition to the two aforementioned response measurement modes, new methods for measuring the LSPR response were introduced and reviewed, which describe the difference between LSPR spectra of molecule-covered and uncovered nanoparticles.

First, special points were selected on the LSPR extinction spectrum for uncoated gold nanoparticle, which would serve as a basis for measuring the response. As the criterion for choosing these special points, the values of the derivative of the extinction efficiency $\frac{dQ_{\text{ext}}}{d\lambda}$ were used (the extinction efficiency $Q_{\text{ext}} = \frac{\sigma_{\text{ext}}}{\pi R^2}$ is a dimensionless value, which is used due to the necessity of comparing the extinction spectra for nanoparticles with sizes that vary in a wide range). According to this criterion, the points for measuring the LSPR response were set at the wavelengths corresponding to $\frac{dQ_{\text{ext}}}{d\lambda} = 0$ and where dependence $\frac{dQ_{\text{ext}}}{d\lambda}$ has extrema within the LSPR peak; these wavelengths correspond to the peak maximum and two points on the left and right slopes of the LSPR extinction peak. Second, LSPR responses were marked for each of the special points as the distance between the point on the extinction spectrum corresponding to the uncoated gold nanoparticle and the point on the same spectrum corresponding to the coated nanoparticle in the "horizontal" (i.e., along the wavelength axis, indicated as H_{left}, H^-_{top}, H^+_{top}, and H_{right}) and "vertical" (i.e., along the extinction efficiency axis, marked as V_{left}, V_{top}, and V_{right}) directions (see Fig. 2.22). The commonly used extinction peak shifts $\Delta\lambda_{max}$ and maximum extinction shift across the wavelength range V_{max} were also calculated for their comparison with the previously described LSPR responses.

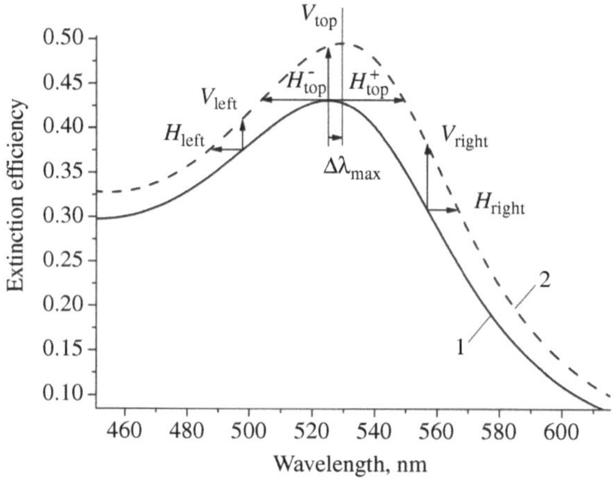

Figure 2.22 LSPR sensor response measurement modes upon the appearance of a molecular coating on the surface of a gold nanoparticle: 1 is the extinction efficiency spectrum for an uncoated gold nanoparticle with a radius of 5 nm located in water, 2 is the extinction efficiency spectrum for a gold nanoparticle with a radius of 5 nm coated with a 20 nm thick molecular layer and located in water. Reprinted from Ref. [80] under a Creative Commons Attribution-NoDerivatives 4.0 International License. Figure caption was adapted.

Also, the transformation of the LSPR extinction spectrum with the increase in gold nanoparticle size was investigated. It was found that when the nanoparticle radius reached 80 nm, a new peak appeared in the extinction spectrum, which corresponds to the quadrupole LSPR excitation. This peak increases with increasing the size of the nanoparticle, and when the radius of the nanoparticle increases to 125 nm, it exceeds the dipole LSPR peak by intensity (see Fig. 2.23). According to the aforementioned results, it was verified if the quadrupole peak could provide a higher LSPR response for large gold nanoparticles used as sensitive elements of the LSPR sensor compared to the dipole peak.

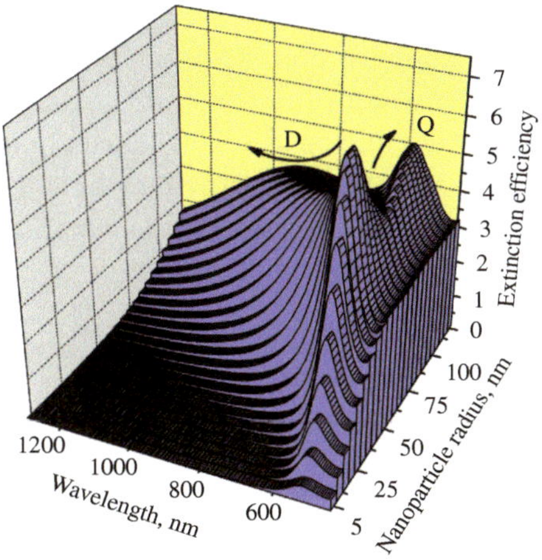

Figure 2.23 Evolution of the extinction efficiency spectrum upon an increase in the size of uncoated gold nanoparticle located in water. The arrows indicate the transformation of the extinction peaks: D is the dipole peak, and Q is the quadrupole peak. Reprinted from Ref. [80] under a Creative Commons Attribution-NoDerivatives 4.0 International License. Labels were added to the figure and figure caption was adapted.

During the analysis of the aforementioned modes of LSPR response for a gold nanoparticle with a radius ranging from 5 to 125 nm coated with a saturated molecular layer with a molecule size of 2, 5, 10, 15, and 20 nm, the following features were revealed. As for the "horizontal" modes of response H_{left}, H_{top}^{-}, H_{top}^{+}, H_{right}, and $\Delta\lambda_{max}$ for the dipole and quadrupole LSPR peaks, it was found that the quadrupole peak gave lower response values compared with the dipole peak for the same shell configurations (for example, in Fig. 2.24, the values of the response $\Delta\lambda_{max}$ for the dipole and quadrupole peaks are compared). For the dipole peak, the maximum response values for all studied sizes of the molecules were H_{top}^{-} and H_{top}^{+} (Fig. 2.25), taking into account their opposite signs, but the maximum absolute response value was provided by the response measurement mode H_{top}^{+}. This result means that there is an alternative method for determining the "horizontal" LSPR response (H_{top}^{+}) that can give a significantly greater response than the usual measurement method $\Delta\lambda_{max}$, which will improve the sensitivity of the LSPR sensor. For example, gold nanoparticle with a radius of 125 nm produces $\Delta\lambda_{max} = 2.1$ nm and $H_{top}^{+} = 35$ nm when coated with a molecular layer with a thickness of 2 nm, which corresponds to near 17 times response amplification. It should be noted that the maximum LSPR response in the studied range of nanoparticle sizes for all the coating thicknesses considered in the measurement modes H^-_{top}, H^+_{top} and $\Delta\lambda_{max}$ is provided by a gold nanoparticle with a radius of 125 nm, which shows the advantage of large-sized nanoparticles as a basis of the LSPR sensor.

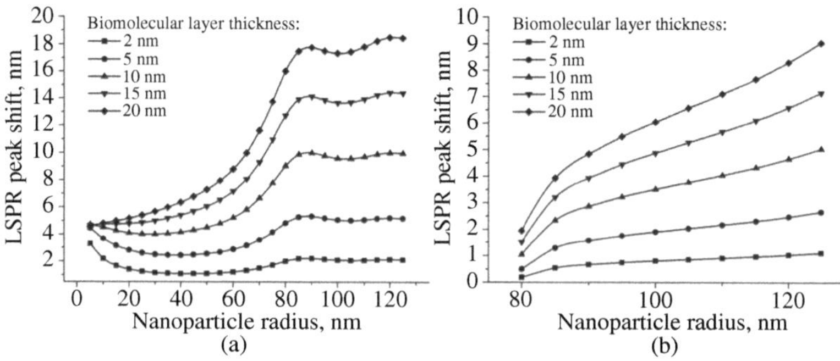

Figure 2.24 The LSPR extinction peak shift $\Delta\lambda_{max}$ depending on the radius of the nanoparticle for different thicknesses of the molecular coating: (a) dipole LSPR peak, (b) quadrupole LSPR peak. Reprinted from Ref. [80] under a Creative Commons Attribution-NoDerivatives 4.0 International License. Figure caption was adapted.

During the study of "vertical" LSPR responses, the problem of choosing the measurement unit appeared. The theoretical modeling gives the dimensionless extinction efficiency, while experimental measurements produce a signal in optical density units, which depends on the sample structure. This fact led to introducing the new relative measurement unit for the LSPR response, the relative extinction difference unit (REDU), which would serve as the basis for measuring the LSPR responses V_{left}, V_{top}, V_{right}, and V_{max}. A response V_{right} for a gold nanoparticle with a radius of 10 nm and a 1 nm thick molecular layer, equal to 0.04631 in units of extinction efficiency, was set as one that equals to 1 REDU. As a result of the comparison of "vertical" LSPR responses, it was found that the maximum response is provided by the V_{right} measurement mode for both dipole and quadrupole peaks for all investigated sizes of molecules.

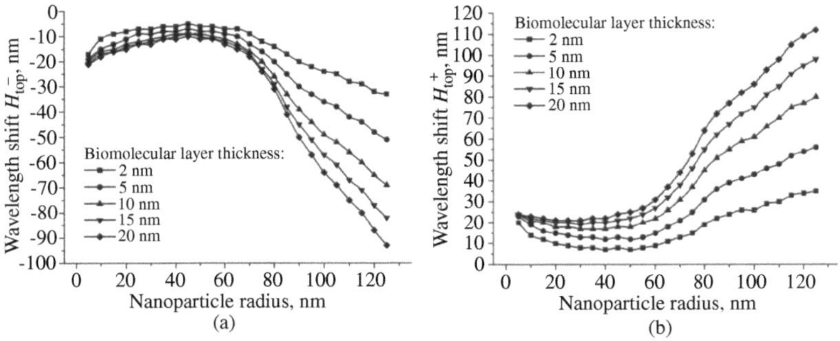

Figure 2.25 LSPR responses (a) H^-_{top} and (b) H^+_{top}, measured on the dipole LSPR peak, depending on the radius of the nanoparticle for different thicknesses of the molecular coating. Reprinted from Ref. [80] under a Creative Commons Attribution-NoDerivatives 4.0 International License. Figure caption was adapted.

Dependences of V_{right} on the nanoparticle size for different molecular coating thicknesses are shown in Fig. 2.26. As it can be seen, in the case of a "vertical" spectrum shift, an absolute maximum response for all the discussed molecule sizes is provided by the V_{right} response measured at the dipole LSPR peak. However, with a radius of nanoparticles larger than 90 nm, the V_{right} response on the quadrupole peak exceeds the response to the dipole peak. It is worth noting that the response curves for both the dipole and quadrupole LSPR peaks have maxima (see Fig. 2.26) that correspond to the optimal size of the nanoparticle for detecting a molecular layer that produces a maximum LSPR response. The corresponding radii of nanoparticles are about 40 nm for the dipole LSPR peak and near 100–105 nm for the quadrupole one.

An interesting fact is the successful selection of the wavelength point on the right slope of the LSPR peak to measure the V_{right} response. This point was selected close to the point at which the maximum extinction shift V_{max} is produced. This result indicates a simple method for choosing the optimal point within the LSPR peak in the mode of measuring the "vertical" response; with acceptable accuracy, it can be selected as the extremum point of the spectrum derivative on the right side of the peak.

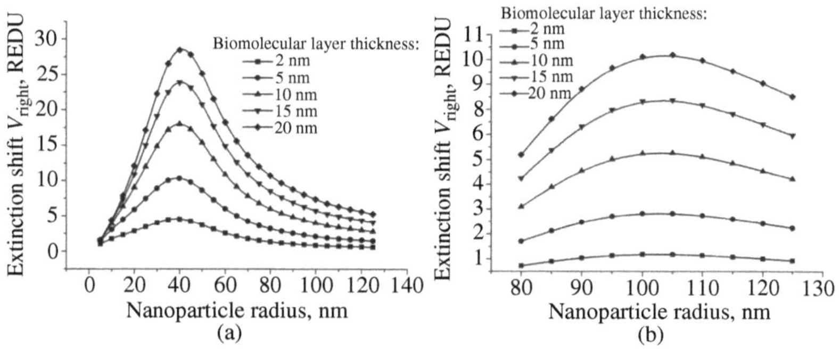

Figure 2.26 LSPR response V_{right} depending on the nanoparticle radius for different molecular coating thicknesses: (a) dipole LSPR peak, (b) quadrupole LSPR peak. Reprinted from Ref. [80] under a Creative Commons Attribution-NoDerivatives 4.0 International License. Figure caption was adapted.

2.3.4 Optical Response of LSPR Sensor to Formation of Absorbing Dielectric Layers

Most of the research on the optical response of LSPR sensors considers the optically transparent analytes. However, there is a class of molecules and chemical compounds, important from the point of view of sensing, that absorb light in the visible and near-infrared regions of the spectrum (for example, cytochrome proteins, organic dyes, organometallic compounds, etc.). Their molecular resonances can overlap with the LSPR bands of nanoparticles, which leads to an unusual response of the LSPR sensor, including the enhancement of its response [81–83]. Therefore, it is necessary to investigate the dependence of the response of the LSPR sensor to the optical properties of this type of analytes for sensitivity optimization of the sensor.

Several papers reported the modeling of LSPR response upon the interaction of nanoparticles with molecules that resonantly absorb light [84–87]. The scaled refractive index values were used to calculate the LSPR spectra of nanoparticles, which were obtained from an analysis of the experimental electronic absorption spectra of the corresponding molecules by using the Kramers–Kronig relations with variable geometric parameters of metal nanoparticles. Such an approach looks weakly applicable for the analysis of the response of LSPR sensor when the optical properties of molecules that absorb light (for example, spectral position, intensity, and width of molecular resonance) change. In addition, the scaling of the refractive index, which was used to better match the results of the simulation with the experimental LSPR response, makes the simulation results unclear.

Therefore, this subsection presents the results of a theoretical study of the LSPR sensor performance during the formation of a light-absorbing coating on the surface of a spherical gold nanoparticle. Coating optical constants were modeled in two versions: as dispersionless and as having dispersion in the approximation of a Lorentz dipole oscillator with one resonant frequency. The dependences of the LSPR response on the values of the real and imaginary parts of the refractive index of the coating (for dispersionless case) and the parameters of the resonant absorption of the coating (for dispersion case) were analyzed.

As the system under investigation, a spherical gold nanoparticle with a radius of 20 nm immersed in water was chosen, and the LSPR response $\Delta\lambda_{ext}$ (wavelength shift of the peak position in the extinction spectrum) was determined upon the formation on the surface of a nanoparticle of a uniform light-absorbing coating of 5 nm thickness (Fig. 2.27). The light extinction spectra of a gold nanoparticle coated with a homogeneous coating were calculated using the relationships given in Subsection 2.3.1.

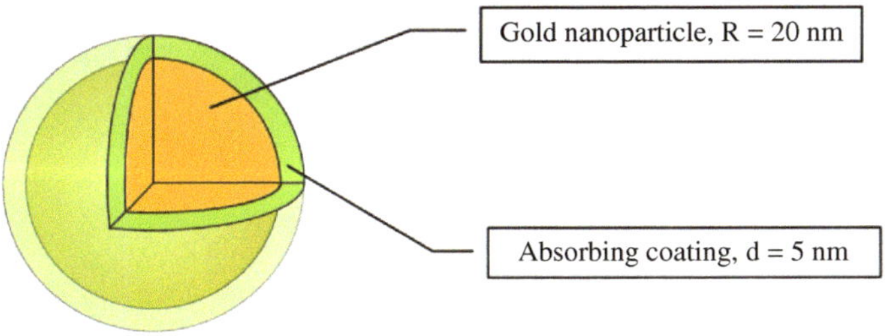

Figure 2.27 Configuration of the system under study.

First, the LSPR sensor response was studied when a dispersionless coating formed on the nanoparticle surface, the optical constants of which did not depend on the light wavelength. For this purpose, the light extinction spectra of uncoated and coated nanoparticles were modeled, with coating optical constants varying within n_{shell} = 1.35 –

1.5 and $k_{shell} = 0 - 0.1$. Figure 2.28 shows the calculated dependence of the LSPR response on the values of the optical constants of the shell.

The results shown in Fig. 2.28 indicate that the LSPR extinction peak undergoes a shift with changes both in the real and imaginary parts of the refractive index of the coating. Dependences of the LSPR response on the real ($\Delta\lambda_{ext}(n_{shell})$) and imaginary ($\Delta\lambda_{ext}(k_{shell})$) parts of the refractive index of the coating are linear in the range of values studied. It was found that the increase in the real part of the refractive index of the coating causes approximately a twofold shift in the position of the extinction peak than the same increase in the imaginary part of the refractive index of the coating. In particular, the slope of the dependence $\Delta\lambda_{ext}(n_{shell})$ is from 33 nm/unit (at ($k_{shell} = 0$) to 30.8 nm/unit (at $k_{shell} = 0.1$). In the case of dependence $\Delta\lambda_{ext}(k_{shell})$, this value ranges from 17.9 nm/unit (at $n_{shell} = 1.35$) to 14.6 nm/unit k_{shell} (at $n_{shell} = 1.5$). Thus, the change in the value of the real part of the refractive index of the coating has a major role in the formation of the LSPR response when the absorbing layer forms on the surface of the nanoparticle (with the proportional magnitude of the changes in n_{shell} and k_{shell}).

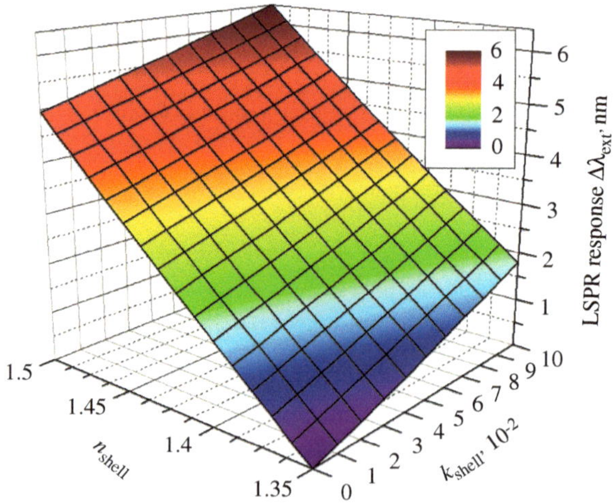

Figure 2.28 Simulated LSPR sensor response upon formation of a dispersionless coating with a thickness of 5 nm on a gold nanoparticle with a radius of 20 nm depending on the values of coating optical constants.

The LSPR response of the sensor upon the formation of a dispersive absorbing layer on the nanoparticle surface was modeled as follows. The optical properties of the absorbing layer were simulated within the framework of the Lorentz dipole oscillator model with a single resonance frequency ω_0 [61]:

$$\varepsilon'_{\text{shell}}(\omega) = \varepsilon^{\infty}_{\text{shell}} + \frac{\omega'^2_{\text{p}}(\omega_0^2 - \omega^2)}{(\omega_0^2 - \omega^2)^2 + (\gamma\omega)^2},$$
$$\varepsilon''_{\text{shell}}(\omega) = \frac{\omega'^2_{\text{p}}\gamma\omega}{(\omega_0^2 - \omega^2)^2 + (\gamma\omega)^2}, \qquad (2.50)$$

where $\varepsilon'_{\text{shell}}$ and $\varepsilon''_{\text{shell}}$ are the real and imaginary parts of the dielectric function of the absorbing coating, $\varepsilon^{\infty}_{\text{shell}}$ is the high-frequency dielectric permittivity, ω'_{p} is the plasma frequency for the absorbing coating material, and γ is the damping parameter for the absorbing coating material. Using the relation (2.50) with the input values of $\lambda_0 = 2\pi c/\omega_0 = 500$ nm (resonance wavelength), $\lambda'_{\text{p}} = 2\pi c / \omega'_{\text{p}} = 6000$ nm (the plasma wavelength for the absorbing coating material), $\varepsilon^{\infty}_{\text{shell}} = 1.9$ and $\Gamma = 2\pi c/\gamma = 25$ nm (the half-width of the absorption peak) and using Eq. (2.36), the spectra of the optical coating constants n_{shell} and k_{shell} were simulated (Fig. 2.29). These optical constants were used to model the extinction spectra of a coated nanoparticle as follows. The spectra of coating optical constants were shifted along the wavelength axis in such a way that the position of the absorption peak acquired values from 450 to 700 nm, without recalculating the values of the constants using Eq. (2.50). For each value of the position of the absorption peak, the extinction spectrum and the specific position of its LSPR peak were obtained. Such a restrictive approach was used to isolate, at the given stage of the research, the influence on the LSPR response only of the positions of the absorption peak of the coating, and not the amplitude of the change in optical constants in the absorption peak region, since this amplitude is not constant when the values of the optical constants are recalculated using Eqs. (2.50) and (2.36) with new ω_0 values.

The dependence of the LSPR sensor response on the position of the absorption peak of the coating is shown in Fig. 2.30. Obviously, the dependence has an oscillating character, as well as the spectrum of the real part of the refractive index of the absorbing coating. At the same time, the asymmetric nature of the oscillations is inherent to this dependence, in contrast to the spectrum of the real part of the refractive index. The obtained result agrees with the experimental data presented in Refs. [84–86].

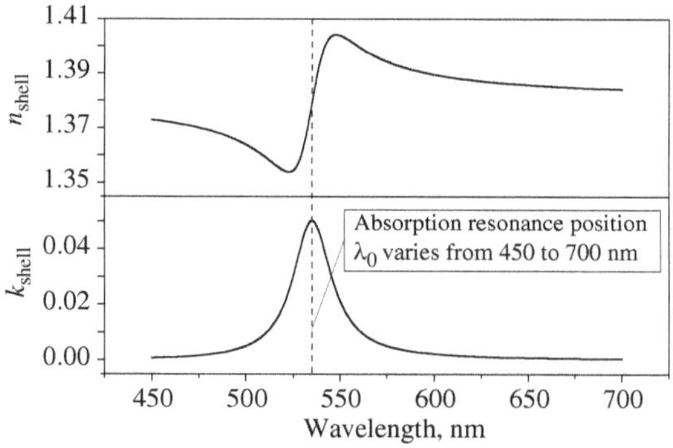

Figure 2.29 Modeled optical constants spectra of a dispersive absorbing coating.

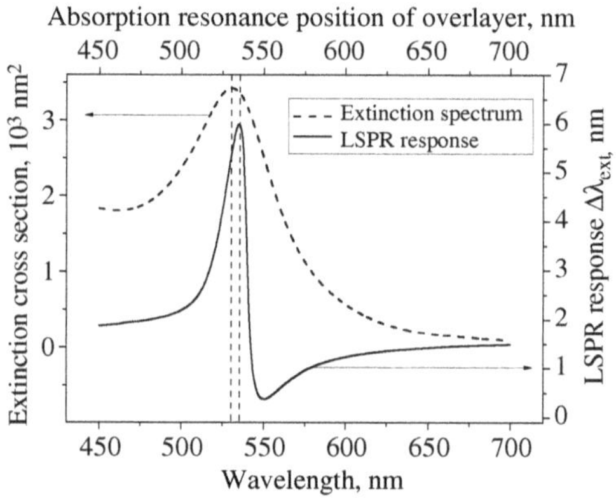

Figure 2.30 Dependence of the LSPR response of the sensor on the position of the absorption peak of the coating in comparison with the extinction spectrum of the uncoated gold nanoparticle.

An important feature of the obtained dependence is a significant increase in the LSPR response when the positions of the LSPR peak in the extinction spectrum of the uncoated nanoparticle and the absorption peak of the coating approach (see Fig. 2.30, where the peak positions are indicated by vertical dashed lines). The obtained result indicates that the largest LSPR response occurs when the position of the absorption peak of the coating is shifted with respect to the LSPR peak in the extinction spectrum of the nanoparticle over a distance of several nanometers. This feature can also be explained by the combined effect on the value of the LSPR response of the real and imaginary parts of the refractive index of the dispersive absorbing shell. Specifically, Fig. 2.29 shows the spectra of the real and imaginary parts of the refractive index of the absorbing coating with the position of the absorption peak, which causes the largest LSPR response in the considered nanoparticle-coating configuration. The configuration of peak positions for this case is as follows: the LSPR peak of the uncoated nanoparticle is at 530.5 nm; the LSPR peak of coated nanoparticles is at 536.5 nm; the peak absorption of the coating is at 535 nm; the peak in the spectrum of the real part of the refractive index of the coating is at 548 nm (which corresponds to $(535 + \Gamma/2)$ nm). Thus, the position of the LSPR peak of the coated nanoparticle is in the region between the peaks in the spectra of the real and imaginary parts of the refractive index of the coating; and its position and, consequently, the resultant LSPR response are due to a combination of two differently directed processes in this area: increase in n_{shell} and decrease in k_{shell}. Since the nature of the dependences n_{shell} and k_{shell} on the wavelength depends on the parameters entered into the relations (2.50), the magnitude of the maximum LSPR response and the magnitude of the above-mentioned shift between the absorption peak of the coating and the LSPR peak in the extinction spectrum of the nanoparticle needed to reach the maximum LSPR response will depend on these parameters of the absorbing coating. Namely, these include the half-width of the absorption peak of the coating, the value of the real part of the refractive index of the coating outside the absorption region and the plasma frequency of the coating material.

In particular, an easy-to-determine experimental parameter of absorbing molecules is the half-width of the absorption peak. Therefore, the LSPR response was modeled depending on the position of the absorption peak of the coating using Eq. (2.50) with the same input parameters as in the previous case, but with a variable half-width of the absorption peak Γ in the range from 5 to 50 nm. Figure 2.31 shows the dependence of the LSPR response enhancement arising when the absorption peak of the coating and the LSPR peak converge, calculated according to the ratio $RE = \Delta\lambda_{\text{ext}}^{\text{max}} / \Delta\lambda_{\text{ext}}^{700\,\text{nm}}$, on the value of the parameter Γ. Here, the index "max" denotes the maximum value in the investigated range of positions of the absorption peak of the coating of 450–700 nm, and the expression in the upper index "700 nm" denotes the value at the position of the absorption peak of the coating equal to 700 nm, where the LSPR response practically corresponds to the situation of the non-absorbing coating. It should be noted that the value of the half-width of the absorption peak directly affects the amplitude of the oscillations in the spectral dependences n_{shell} and k_{shell}; consequently, it influences the obtained value of the LSPR response amplification. Thus, the significant LSPR response shown in Fig. 2.31 for small values of Γ is associated with an increase in the amplitude of the oscillations in spectra n_{shell} and k_{shell}, which, in particular, indicates the prospects of using narrow peaks in the absorption spectra of molecules to improve the sensitivity of the LSPR sensor. In addition, as can be seen from Fig. 2.31, the obtained dependence is well fitted to the exponential-type function $RE = a \cdot e^{-\Gamma/b} + c$ (the adjusted coefficient of determination is 0.99).

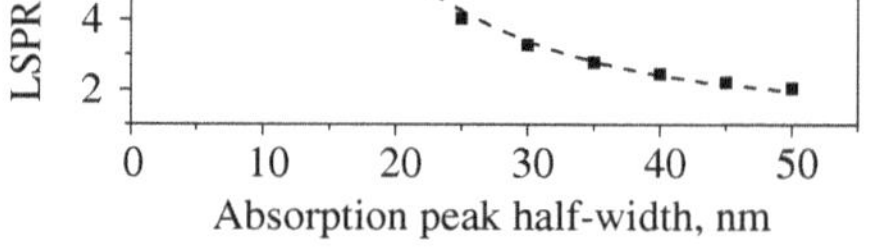

Figure 2.31 Dependence of the amplification of the LSPR response on the half-width of the coating absorption peak.

Figure 2.32 shows the dependence of the shift between the LSPR peak for the uncovered nanoparticle and the position of the absorption peak of the coating, which provides the maximum LSPR response (i.e., the shift between resonances optimal from the point of view of the sensitivity of the LSPR sensor), on the half-width of the absorption peak of the coating. From this dependence it is seen that in order to achieve the maximum response of the LSPR sensor, it is necessary to use as a sensitive element nanoparticles that have an LSPR peak in the light extinction spectrum that lies within ±5 nm (for molecules with wide absorption peaks (Γ = 25 – 50 nm)) or is shifted toward smaller wavelengths up to a distance of 20 nm (for molecules with narrow absorption peaks (Γ = 5 – 20 nm)) with respect to the absorption peak of molecules. In addition, as can be seen from Fig. 2.32, the obtained dependence is well fitted with an exponential function of the type $\Delta\lambda = a \cdot e^{-\Gamma/b} + c$ (the adjusted coefficient of determination is 0.98).

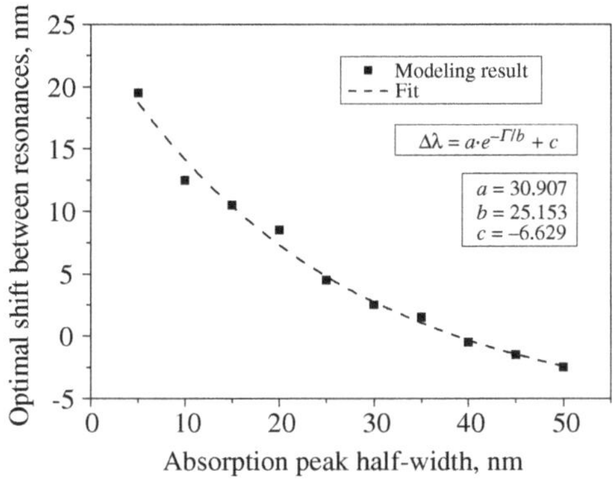

Figure 2.32 Dependence of the shift between the plasmon and molecular resonances, optimal for the LSPR sensor performance, on the half-width of the absorption peak of the coating.

2.4 Conclusion

Theoretical background and simulation results described in this chapter demonstrate the vast potential of SPR and LSPR techniques as a basis for highly sensitive detection and characterization of monolayer and submonolayer molecular coatings. Particularly, methods based on SPR enable the high-sensitive measurements of adsorption and other molecular interactions, quantification of molecular surface density, elucidation of molecular shape and orientation distributions, as well as characterization of optical properties of both transparent and light-absorbing analytes.

Common theoretical approaches for the description of SPR phenomenon in thin metal films and LSPR phenomenon in metal nanoparticles were presented. The concept of Green's function for the characterization of SPR sensor response was considered. The Mie scattering theory and its applications for characterization of LSPR sensor response were exemplified. Configurations of generic SPR and LSPR sensor elements have been discussed by demonstrating the use of computer simulations to study the peculiarities of sensor response formation.

One of the studies proposed the theoretical background for 3D quantification of the surface concentration of molecules from SPR data. The novel approach presented allows an estimation of the influence of the spatial form of a molecule on the SPR angle position, which may be useful to determine or estimate the orientation of molecules relative to the surface of SPR sensors for a more comprehensive understanding and quantitative estimation of molecular layers.

Another study proposed an approach for calculating the dispersion of surface electromagnetic waves in the framework of Green's function using the effective susceptibility concept. For the metal surface covered by ellipsoidal particles, it was shown that the shape of the covering particles influences the dispersion of surface plasmons. Namely, the interference of the surface plasmon and configurational resonances of the particle, arising in response to the external field, leads to the occurrence of new surface waves. These waves can be excited by s- and p-polarized light. Consequently, using the SPR sensor method, one can measure dispersion curves and determine the shape of the molecules and, in that way, their type.

Several theoretical studies based on the Mie theory related to sensing applications of LSPR phenomenon in gold nanoparticles have been considered. One of these studies shows that an LSPR sensor based on spherical gold nanoparticle is capable of detecting a thick molecular coating and has a comparable (twice less) sensitivity limit to that of an SPR sensor based on a thin gold film at comparable size parameters (specifically, the nanoparticle diameter and thickness of the metal film, which are equal to 50 nm). It was found out that the gradient of the shift of LSPR peak position decreases when the coating thickness increases. Additionally, a significant LSPR peak position shift was demonstrated for nanoparticles with a radius of less than 1 nm coated with a molecular shell with about 1 nm thickness, which shows the possibility of developing a high-sensitive sensor for the detection of small molecules. It was found that for small gold nanoparticles with a radius of up to 5 nm, the LSPR response can be both positive and negative due to the scattering contribution to the total light extinction. It was also found that the contribution of scattering to full extinction depends on the ratio of the nanoparticle size and the coating thickness, and the gradient of shift of scattering spectrum peak with molecular layer thickness increasing is more significant than for the absorption and extinction peaks, which makes it perspective to create high-sensitive sensors based on the measurement of light scattering by nanostructures. The equations for the globular molecules number and their surface concentration in the case of a saturated molecular monolayer on the nanoparticle surface were developed for molecular layer description according to the effective medium theory. These parameters can be used, in particular, when studying submonolayer molecular coatings.

In another study, a new model for calculating the extinction spectrum wavelength shift was proposed for obtaining a maximum LSPR response and it was shown that it is more effective compared to the traditionally used extinction peak shift $\Delta\lambda_{max}$. It has been found that approaches based on the measurement of a wavelength shift give a larger LSPR response when large-scale nanoparticles are used as a sensitive sensor element, and response measurements are performed on the dipole LSPR peak. It has been shown that among the various approaches to measuring the extinction shift, the most productive is "vertical" response V_{right} and the optimal nanoparticles radii were found for this mode, which were about 40 nm for dipole and about 100–105 nm for quadrupole LSPR peaks, respectively.

The last considered study demonstrated that the LSPR extinction peak is shifted with changes both in the real and imaginary parts of the coating refractive index. It was also shown that at proximity of the LSPR peak in the nanoparticle extinction spectrum and the peak in the coating absorption spectrum, it is possible to amplify the maximum LSPR peak position shift in the sensor extinction spectrum from 2 to 15 times and decrease it compared to the case when these resonance peaks are spaced apart on the wavelengths axis. It was found that in order to achieve maximum amplification of the LSPR sensor response, one needs to use a sensitive element, which is based on the nanoparticles with LSPR peak in the light extinction spectrum that lies near or shifted toward smaller wavelengths with respect to the molecular absorption peak.

References

1. Edel, J. B., Kornyshev, A. A., Kucernak, A. R., and Urbakh, M. (2016). Fundamentals and applications of self-assembled plasmonic nanoparticles at interfaces, *Chem. Soc. Rev.*, **45**, pp. 1581–1596.
2. Boardman, A. D. (1982). *Electromagnetic Surface Modes* (Wiley, New York).
3. Bozhevolnyi, S. I., Martin-Moreno, L., and Garcia-Vidal, F. (Eds.) (2017). *Quantum Plasmonics* (Springer, Switzerland).
4. Schasfoort, R. B. (Ed.) (2017). *Handbook of Surface Plasmon Resonance*, 2nd Ed. (The Royal Society of Chemistry, UK).
5. Pendry, J. (1999). Playing tricks with light, *Science*, **285**, pp. 1687–1688.
6. Ren, F., Li, M., Gao, Q., Cowell, W., Luo, J., Jen, A. K. and Wang, A. X. (2015). Surface-normal plasmonic modulator using sub-wavelength metal grating on electro-optic polymer thin film, *Opt. Commun.*, **352**, pp. 116–120.
7. Ozbay, E. (2006). Plasmonics: merging photonics and electronics at nanoscale dimensions, *Science*, **311**, pp. 189–193.
8. Zhang, Y., Manjavacas, A., Hogan, N. J., Zhou, L., Ayala-Orozco, C., Dong, L. et al. (2016). Toward surface plasmon-enhanced optical parametric amplification (SPOPA) with engineered nanoparticles: A nanoscale tunable infrared source, *Nano Lett.*, **16**, pp. 3373–3378.
9. Smolyaninov, I. I. (2005). A far-field optical microscope with nanometre-scale resolution based on in-plane surface plasmon imaging, *J. Opt. A: Pure Appl. Opt.*, **7**, S165.
10. Zeng, S., Hu, S., Xia, J., Anderson, T., Dinh, X. Q., Meng, X. M. et al. (2015). Graphene–MoS_2 hybrid nanostructures enhanced surface plasmon resonance biosensors, *Sens. Actuators B Chem.*, **207**, pp. 801–810.
11. Rifat, A. A., Mahdiraji, G. A., Sua, Y. M., Shee, Y. G., Ahmed, R., Chow, D. M. et al. (2015). Surface plasmon resonance photonic crystal fiber biosensor: A practical sensing approach, *IEEE Photon. Technol. Lett.*, **27**, pp. 1628–1631.
12. Getsko, O. M., Demidenko, Y. V., and Lozovski, V. Z. (2001). Mixed surface plasmon-phonon modes on an adsorption of molecules on the

surface of semiconductor. 2. Absorption of electromagnetic field by a molecular film, *Ukr. J. Phys.*, **46**, pp. 355–359.

13. Bernardi, M., Mustafa, J., Neaton, J. B., and Louie, S. G. (2015). Theory and computation of hot carriers generated by surface plasmon polaritons in noble metals, *Nat. Commun.*, **6**, 7044.
14. Jiang, S., Zhang, Y., Zhang, R., Hu, C., Liao, M., Luo, Y. et al. (2015). Distinguishing adjacent molecules on a surface using plasmon-enhanced Raman scattering, *Nat. Nanotechnol.*, **10**, pp. 865–869.
15. Amendola, V., Pilot, R., Frasconi, M., Marago, O. M., and Iati, M. A. (2017). Surface plasmon resonance in gold nanoparticles: A review, *J. Phys. Condens. Matter*, **29**, 203002.
16. Homola, J., Yee, S. S., and Gauglitz, G. (1999). Surface plasmon resonance sensors, *Sens. Actuators B Chem.*, **54**, pp. 3–15.
17. Caucheteur, C., Guo, T., and Albert, J. (2015). Review of plasmonic fiber optic biochemical sensors: Improving the limit of detection, *Anal. Bioanal. Chem.*, **407**, pp. 3883–3897.
18. Mauriz, E., García-Fernández, M. C., and Lechuga, L. M. (2016). Towards the design of universal immunosurfaces for SPR-based assays: A review, *Trends Anal. Chem.*, **79**, pp. 191–198.
19. Nguyen, H. H., Park, J., Kang, S., and Kim, M. (2015). Surface plasmon resonance: A versatile technique for biosensor applications, *Sensors*, **15**, pp. 10481–10510.
20. Dmitruk, N. L., Litovchenko, V. G., and Strizhevskii, V. L. (1989). *Surface Polaritons in Semiconductors and Dielectrics* (Naukova Dumka, Kiev) (in Russian).
21. Chegel, V., Demidenko, Y., Lozovski, V., and Tsykhonya, A. (2008). Influence of the shape of the particles covering the metal surface on the dispersion relations of surface plasmons, *Surf. Sci.*, **602**, pp. 1540–1546.
22. Kretschmann, E. (1971). The determination of the optical constants of metals by excitation of surface plasmons, *Z. Phys.*, **241**, pp. 313–324 (in German).
23. Rzhanov, A. V., Svitashev, K. K., Semenenko, A. I., Semenenko, L. V., and Sokolov, V. K. (1979) *Principles of Ellipsometry* (Nauka, Novosibirsk) (in Russian).
24. Born, M. and Wolf, E. (1980). *Principles of Optics: Electromagnetic Theory of Propagation, Interference and Diffraction of Light*, 6th Ed. (Pergamon Press, Oxford).

25. Azzam, R. M. A. and Bashara, N. M. (1999). *Ellipsometry and Polarized Light* (North-Holland, Amsterdam).
26. Löfås, S., Malmqvist, M., Rönnberg, I., Stenberg, E., Liedberg, B., and Lundström, I. (1991). Bioanalysis with surface plasmon resonance, *Sens. Actuators B Chem.*, **5**, pp. 79–84.
27. Matsubara, K., Kawata, S., and Minami, S. (1988). Optical chemical sensor based on surface plasmon measurement, *Appl. Opt.*, **27**, pp. 1160–1163.
28. Advincula, R., Aust, E., Meyer, W., and Knoll, W. (1996). In situ investigations of polymer self-assembly solution adsorption by surface plasmon spectroscopy, *Langmuir*, **12**, pp. 3536–3540.
29. Johansen, K., Lundström, I., and Liedberg, B. (2000). Sensitivity deviation: Instrumental linearity errors that influence concentration analyses and kinetic evaluation of biomolecular interactions, *Biosens. Bioelectron.*, **15**, pp. 503–509.
30. Borschagovski, E. G., Gecko, O. M., Lozovski, V. Z., and Hudik, B. I. (1989). Ellipsometry of ultrathin films, *Opt. Spectrosc.*, **66**, pp. 1345–1348.
31. Bobbert, P. A. and Vlieger, J. (1986). Light scattering by a sphere on a substrate, *Physica A*, **137**, pp. 209–242.
32. Shirshov, Y. M., Chegel, V. I., Subbota, Y. V., Matsas, E. P., Rachcov, A. E., and Sergeeva, T. A. (1995). Determination of dielectric constant and thickness of thin biological layers using surface plasmon resonance, *Proc. SPIE*, **2648**, pp. 118–123.
33. Chegel, V. I. (2002). An influence of dielectric characteristics of ambient and external factors on parameters of physical and biological sensors based on surface plasmon resonance, PhD Thesis, Institute of Semiconductors Physics of National Academy of Science of Ukraine, Kyiv, Ukraine (in Ukrainian).
34. Lozovski, V. (2001). Susceptibilities of nano-particles at the surface of a solid, *Physica E*, **9**, pp. 642–651.
35. Keller, O. (1996). Local fields in the electrodynamics of mesoscopic media, *Phys. Rep.*, **268**, pp. 85–262.
36. Chegel, V., Chegel, Y., Guiver, M. D., Lopatynskyi, A., Lopatynska, O., and Lozovski, V. (2008). 3D-quantification of biomolecular covers using surface plasmon-polariton resonance experiment, *Sens. Actuators B Chem.*, **134**, pp. 66–71.
37. Baryakhtar, I. V., Demidenko, Y. V., Kriuchenko, S. V., and Lozovskii, V. Z. (1995). Electromagnetic waves in the molecular layers adsorbed on the surface of a solid, *Surf. Sci.*, **323**, pp. 142–150.

38. Agranovich, V. M. and Mills, D. L. (1982). *Surface Polaritons. Electromagnetic Waves at Surfaces and Interfaces* (North-Holland, Amsterdam).

39. Evlyukhin, A. B. and Bozhevolnyi, S. I. (2005). Surface plasmon polariton scattering by small ellipsoid particles, *Surf. Sci.*, **590**, pp. 173–180.

40. Román-Velázquez, C. E., Noguez, C., and Barrera, R. G. (2000). Substrate effects on the optical properties of spheroidal nanoparticles, *Phys. Rev. B*, **61**, pp. 10427–10436.

41. Haes, A. J., Zou, S., Schatz, G. C., and Van Duyne, R. P. (2004). A nanoscale optical biosensor: The long range distance dependence of the localized surface plasmon resonance of noble metal nanoparticles, *J. Phys. Chem. B*, **108**, pp. 109–116.

42. Liu, Y., Yang, L., and Shen, Y. (2018). Hydrothermal synthesis of gold nanoplates and their structure-dependent LSPR properties, *J. Mater. Res.*, **33**, pp. 2671–2679.

43. Khlebtsov, N. G. (2008). Determination of size and concentration of gold nanoparticles from extinction spectra, *Anal. Chem.*, **80**, pp. 6620–6625.

44. Derkachova, A., Kolwas, K., and Demchenko, I. (2016). Dielectric function for gold in plasmonics applications: Size dependence of plasmon resonance frequencies and damping rates for nanospheres, *Plasmonics*, **11**, pp. 941–951.

45. Kim, I. H., Kim, J. H., Choi, J. Y., Shin, C. H., Kim, J. H., Bae, G. T. et al. (2018). Tuning the interparticle distances in self-assembled gold nanoparticle films with their plasmonic responses, *Chem. Phys. Lett.*, **715**, pp. 91–99.

46. Lee, K. S. and El-Sayed, M. A. (2006). Gold and silver nanoparticles in sensing and imaging: Sensitivity of plasmon response to size, shape, and metal composition, *J. Phys. Chem. B*, **110**, pp. 19220–19225.

47. Yu, J., Zhang, J., Zhang, L., Jia, J., Xu, W., Wang, J. et al. (2016). Size and dielectric-environment dependence of transversal resonance modes of localized surface plasmons in silver nanorods, *Appl. Opt.*, **55**, pp. 4871–4876.

48. Murray, W. A., Auguié, B., and Barnes, W. L. (2009). Sensitivity of localized surface plasmon resonances to bulk and local changes in the optical environment, *J. Phys. Chem. C*, **113**, pp. 5120–5125.

49. Forcherio, G. T., Blake, P., Seeram, M., DeJarnette, D., and Roper, D. K. (2015). Coupled dipole plasmonics of nanoantennas in discontinuous,

complex dielectric environments, *J. Quant. Spectrosc. Radiat. Transf.*, **166**, pp. 93–101.

50. Miller, M. M. and Lazarides, A. A. (2005). Sensitivity of metal nanoparticle surface plasmon resonance to the dielectric environment, *J. Phys. Chem. B*, **109**, pp. 21556–21565.
51. Chegel, V., Rachkov, O., Lopatynskyi, A., Ishihara, S., Yanchuk, I., Nemoto, Y. et al. (2012). Gold nanoparticles aggregation: Drastic effect of cooperative functionalities in a single molecular conjugate, *J. Phys. Chem. C*, **116**, pp. 2683–2690.
52. Kelly, K. L., Coronado, E., Zhao, L. L., and Schatz, G. C. (2003). The optical properties of metal nanoparticles: The influence of size, shape, and dielectric environment, *J. Phys. Chem. B*, **107**, pp. 668–677.
53. Kreibig, U. and Vollmer, M. (1995) *Optical Properties of Metal Clusters* (Springer-Verlag, Berlin).
54. Weick, G., Ingold, G. L., Jalabert, R. A., and Weinmann, D. (2006). Surface plasmon in metallic nanoparticles: Renormalization effects due to electron-hole excitations, *Phys. Rev. B*, **74**, 165421.
55. Kawabata, A. and Kubo, R. (1966). Electronic properties of fine metallic particles. II. Plasma resonance absorption, *J. Phys. Soc. Jpn.*, **21**, pp. 1765–1772.
56. Palik, E. D. (Ed.) (1998). *Handbook of Optical Constants of Solids*, Volume 1 (Academic Press, San Diego, CA).
57. Johnson, P. B. and Christy, R. W. (1972). Optical constants of the noble metals, *Phys. Rev. B*, **6**, pp. 4370–4379.
58. Lide, R. D. (Ed.) (2004). *Handbook of Chemistry and Physics*, 84th Ed. (CRC Press, Boca Raton, FL).
59. Dmitruk, N. L., Goncharenko, A. V., and Venger, E. F. (2009). *Optics of Small Particles and Composite Media* (Naukova Dumka, Kyiv).
60. Westcott, S. L., Jackson, J. B., Radloff, C., and Halas, N. J. (2002). Relative contributions to the plasmon line shape of metal nanoshells, *Phys. Rev. B*, **66**, 155431.
61. Fox, M. (2001). *Optical Properties of Solids* (Oxford University Press, New York).
62. Bass, M. (Ed.) (1995). *Handbook of Optics* (McGraw-Hill, New York).
63. Lopatynskyi, A. M., Lopatynska, O. G., Guo, L. J., and Chegel, V. I. (2011). Localized surface plasmon resonance biosensor—Part I: Theoretical study of sensitivity—Extended Mie approach, *IEEE Sens. J.*, **11**, pp. 361–369.

64. Tokuhisa, H., Zhao, M., Baker, L. A., Phan, V. T., Dermody, D. L., Garcia, M. E. et al. (1998). Preparation and characterization of dendrimer monolayers and dendrimer–alkanethiol mixed monolayers adsorbed to gold, *J. Am. Chem. Soc.*, **120**, pp. 4492–4501.

65. Weisser, M., Tovar, G., Mittler-Neher, S., Knoll, W., Brosinger, F., Freimuth, H. et al. (1999). Specific bio-recognition reactions observed with an integrated Mach–Zehnder interferometer, *Biosens. Bioelectron.*, **14**, pp. 405–411.

66. Khlebtsov, N. G., Bogatyrev, V. A., Khlebtsov, B. N., Dykman, L. A., and Englebienne, P. (2003). A multilayer model for gold nanoparticle bioconjugates: Application to study of gelatin and human IgG adsorption using extinction and light scattering spectra and the dynamic light scattering method, *Colloid J.*, **65**, pp. 622–635.

67. Bohren, C. F. and Huffman, D. R. (1983) *Absorption and Scattering of Light by Small Particles* (Wiley-Interscience, New York).

68. Yan, S., Wang, Y., Wen, T., and Zhu, J. (2006). A study on the optical absorption properties of dielectric-mediated gold nanoshells, *Physica E*, **33**, pp. 139–143.

69. Xu, H. and Käll, M. (2002). Modeling the optical response of nanoparticle-based surface plasmon resonance sensors, *Sens. Actuators B Chem.*, **87**, pp. 244–249.

70. Kreibig, U., Schmitz, B., and Breuer, H. D. (1987). Separation of plasmon-polariton modes of small metal particles, *Phys. Rev. B*, **36**, pp. 5027–5030.

71. Li, X., Jiang, L., Zhan, Q., Qian, J., and He, S. (2009). Localized surface plasmon resonance (LSPR) of polyelectrolyte-functionalized gold-nanoparticles for bio-sensing, *Colloids Surf. A Physicochem. Eng. Asp.*, **332**, pp. 172–179.

72. Chen, Y. Q. and Lu, C. J. (2009). Surface modification on silver nanoparticles for enhancing vapor selectivity of localized surface plasmon resonance sensors, *Sens. Actuators B Chem.*, **135**, pp. 492–498.

73. Hall, W. P., Anker, J. N., Lin, Y., Modica, J., Mrksich, M., and Van Duyne, R. P. (2008). A calcium-modulated plasmonic switch, *J. Am. Chem. Soc.*, **130**, pp. 5836–5837.

74. Wang, Y., Deng, J., Di, J., and Tu, Y. (2009). Electrodeposition of large size gold nanoparticles on indium tin oxide glass and application as refractive index sensor, *Electrochem. Comm.*, **11**, pp. 1034–1037.

75. Park, J. H., Byun, J. Y., Shim, W. B., Kim, S. U., and Kim, M. G. (2015). High-sensitivity detection of ATP using a localized surface plasmon resonance (LSPR) sensor and split aptamers, *Biosens. Bioelectron.*, **73**, pp. 26–31.

76. Cheng, C. S., Chen, Y. Q., and Lu, C. J. (2007). Organic vapour sensing using localized surface plasmon resonance spectrum of metallic nanoparticles self-assemble monolayer, *Talanta*, **73**, pp. 358–365.

77. Shang, L., Liu, C., Watanabe, M., Chen, B., and Hayashi, K. (2017). LSPR sensor array based on molecularly imprinted sol-gels for pattern recognition of volatile organic acids, *Sens. Actuators B Chem.*, **249**, pp. 14–21.

78. Chegel, V., Lucas, B., Guo, J., Lopatynskyi, A., Lopatynska, O., and Poperenko, L. (2009). Detection of biomolecules using optoelectronic biosensor based on localized surface plasmon resonance. Nanoimprint lithography approach, *Semicond. Phys. Quantum Electron. Optoelectron.*, **12**, pp. 91–97.

79. Nath, N. and Chilkoti, A. (2002). A colorimetric gold nanoparticle sensor to interrogate biomolecular interactions in real time on a surface, *Anal. Chem.*, **74**, pp. 504–509.

80. Lopatynskyi, A., Lopatynska, O., and Chegel, V. (2011). Comparative analysis of response modes for gold nanoparticle biosensor based on localized surface plasmon resonance, *Semicond. Phys. Quantum Electron. Optoelectron.*, **14**, pp. 114–121.

81. Guo, L., Jackman, J. A., Yang, H. H., Chen, P., Cho, N. J., and Kim, D. H. (2015). Strategies for enhancing the sensitivity of plasmonic nanosensors, *Nano Today*, **10**, pp. 213–239.

82. Zhao, J., Das, A., Schatz, G. C., Sligar, S. G., and Van Duyne, R. P. (2008). Resonance localized surface plasmon spectroscopy: Sensing substrate and inhibitor binding to cytochrome P450, *J. Phys. Chem. C*, **112**, pp. 13084–13088.

83. Das, A., Zhao, J., Schatz, G. C., Sligar, S. G., and Van Duyne, R. P. (2009). Screening of type I and II drug binding to human cytochrome P450-3A4 in nanodiscs by localized surface plasmon resonance spectroscopy, *Anal. Chem.*, **81**, pp. 3754–3759.

84. Haes, A. J., Zou, S., Zhao, J., Schatz, G. C., and Van Duyne, R. P. (2006). Localized surface plasmon resonance spectroscopy near molecular resonances, *J. Am. Chem. Soc.*, **128**, pp. 10905–10914.

85. Zhao, J., Sherry, L. J., Schatz, G. C., and Van Duyne, R. P. (2008). Molecular plasmonics: Chromophore–plasmon coupling and single-particle nanosensors, *IEEE J. Sel. Top. Quant. Electron.*, **14**, pp. 1418–1429.

86. Zhao, J., Jensen, L., Sung, J., Zou, S., Schatz, G. C., and Van Duyne, R. P. (2007). Interaction of plasmon and molecular resonances for rhodamine 6G adsorbed on silver nanoparticles, *J. Am. Chem. Soc.*, **129**, pp. 7647–7656.

87. Chen, L., Wu, B., Guo, L., Tey, R., Huang, Y., and Kim, D. H. (2015). A single-nanoparticle NO_2 gas sensor constructed using active molecular plasmonics, *ChemComm*, **51**, pp. 1326–1329.

Chapter 3

Plasmonic Nanochips Development and Applications

3.1 Introduction

The phenomenon of plasmonics is widely used in optical devices [1], imaging microscopy [2], biosensing [3–5], and medical diagnostics [6–8]. Improvement in sensitivity, even down to single-molecule detection limits, is needed in many applications, and this problem demands a solution at the present moment. One of the possible ways to obtain general sensitivity enhancement for multiple applications is to fabricate nanopatterned plasmonic substrates (nanochips) capable of generating strong local electromagnetic fields or, in other words, offering significant plasmonic enhancement (PE), due to the occurrence of localized surface plasmon resonance (LSPR) phenomenon in highly conductive metal nanoparticles. It was shown both theoretically and experimentally that enhanced local field provides signal amplification for LSPR [9–11], surface-enhanced Raman scattering (SERS) [10, 12], surface-enhanced fluorescence (SEF) [13–15], and surface-enhanced infrared absorption (SEIRA) [16, 17] techniques. The peculiarity of PE accompanying LSPR is that the enhanced field is concentrated in confined space with nanometer dimensions ("hot spots") [18]—a phenomenon that depends on nanostructure size, shape, and material properties [19, 20].

In this chapter, different approaches for the fabrication of plasmonic nanochips based on noble metal thin films and

Molecular Plasmonics: Theory and Applications
Volodymyr I. Chegel and Andrii M. Lopatynskyi

ISBN 978-981-4800-65-5 (Hardcover), 978-0-429-29511-9 (eBook)
www.jennystanford.com

nanostructure arrays are discussed, including a method based on gold island film deposition with subsequent thermal annealing and nanoimprint lithography (NIL) technique. The application of latter techniques for the preparation of plasmonic nanochips is comparatively described in more detail by analyzing their structural, spectral, sensing, and plasmonic enhancement properties. Additionally, this chapter presents the results of theoretical and experimental studies on application of plasmonic nanochips for SEIRA- and SEF-based research.

3.2 Fabrication of Plasmonic Nanochips Based on Noble Metal Thin Films and Nanostructure Arrays

3.2.1 Fabrication of Thin Films with Surface Roughness

As it was mentioned before, continuous thin films of high-conductive metals can be used both as a surface exhibiting plasmonic enhancement phenomenon and a sensitive element of surface plasmon resonance (SPR) sensor. However, the optimal parameters of metal films in terms of film thickness and surface roughness are quite different for these applications.

For plasmonic enhancement applications, thin and rough metal films are optimal. For example, gold films with a thickness of 20–40 nm have been exploited in SEIRA experiments [17, 21]. To fabricate such substrates, the protocol based on vacuum deposition of 99.999 pure Au upon glass supports (TF-1 glass, 20×20 mm) via an intermediate adhesive Cr layer was used. Before Au deposition, glass surface was cleaned by $NH_4OH:H_2O_2:H_2O$ and $HCl:H_2O_2:H_2O$ solution, subsequently, both 1:2:2 by volume concentration during 5 min at boiling temperature. Then it was rinsed in bidistilled water and dried in a flow of pure nitrogen. The gold was evaporated from molybdenum heater and deposited at a rate of 1.0–1.5 nm/s on room temperature substrate. The Cr interlayer thickness did not exceed 1–1.5 nm. The gold surface just after deposition looked like hydrophobic surface with wetting angle close to 80° and random roughness about 5 nm (Fig. 3.1a).

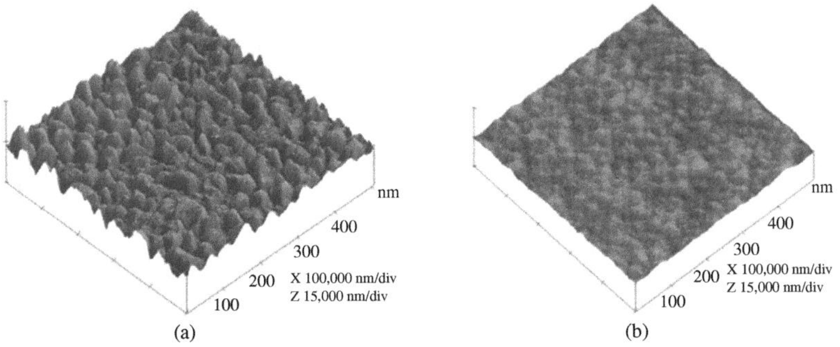

Figure 3.1 Atomic force microscopy (AFM) images of gold film surface (a) before and (b) after chemical polishing by piranha solution. Adapted by permission from Ref. [21], Copyright 2004, Springer Nature.

For SPR sensor application, larger thickness and lower surface roughness of metal film are preferable. Gold films, 45–50 nm thick, produced by thermal vacuum evaporation method are usually exploited as plasmon-polariton oscillation bearers in SPR sensors [22–24]. To enhance the adhesion, a thin (up to 5 nm) chromium layer was evaporated on the glass slide before the evaporation of gold. Thermal evaporation was performed in vacuum (10^{-6} mmHg) on the substrates at the room temperature. Annealing of films for 30 min at 120°C took place after the evaporation to decrease the surface roughness [22]. To further improve the surface evenness, chemical etching by piranha solution was applied (Fig. 3.1b). This is a crucial stage in order to obtain a surface suitable for biomodification relying on an oriented immobilization of biomolecules, for example, immunoglonulins (see Section 4.2 in Chapter 4).

3.2.2 Fabrication of Random Nanostructure Arrays

3.2.2.1 Surface nanopartening of random-fashion nanostructures using colloidal nanoparticles

Among several approaches, colloidal Au and Ag nanoparticles are used for the preparation of chip-based structures in array format by immobilization of nanoparticles on the solid support. Several articles from different groups have presented chip-based format optical biosensors, in which gold or silver nanostructures are immobilized in a random fashion on an optically transparent substrate [25–28]. This approach opens a way for high-throughput biosensor control and multiplexed analysis of biomolecular interactions with minimal consumption of reagents. There are several different approaches to fabrication of chip-based structures based on colloidal plasmonic nanoparticles. Liu et al. [27] demonstrated simple fabrication of large-area gold nanostructures using thiol-stabilized gold nanoparticles without complicated lithography and vacuum evaporation techniques involved in the fabrication process. Gold nanoparticles having a mean size of about 5 nm with a distribution range from 2 to 8 nm were obtained using a modernized method of Murray et al. [29]. In this work, hexanethiol was used as the stabilizing functional ligand to cover the gold nanoparticles to reach good dispersion in organic solvents such as xylene and toluene. The Au nanoparticle colloidal solution was spin-coated onto the 10 mm × 10 mm indium–tin-oxide glass substrate at a speed of 2000 rpm for 30 s and then annealed at temperatures in the range of 200–550°C on air. As a result of the melting of Au nanoparticles, nanostructures with different morphologies were formed (Fig. 3.2). The high limit of temperature (550°C) was found due to the observation of undesired structural changes in the transparent glass support. The regulation of temperature during the annealing process and changing the concentration of the Au nanoparticle colloids allow the tunability of the optical response. In addition, a concentration ranging from 40 to 120 mg/ml was recommended by authors for the large-area fabrication of size-optimized gold nano-island structures. It should be noted that a disadvantage of this method is a relatively wide LSPR absorbance peak of prepared structures, which is explained by large size dispersity of nanostructures.

Malynych and Chumanov [28] proposed the method of forming stable random-fashion monolayers of colloidal Ag nanoparticles on the glass substrate with a layer of polymer maintaining only the lower part of the nanoparticles. This enables LSPR oscillations at the top part of particles, which are free from the polymer. Importantly, arrays of Ag nanoparticles prepared in such a way exhibit extra narrow peaks in the extinction spectra, which can be used to develop high-sensitive biosensors. To produce such a nanoparticle array, a self-assembly of 100 nm silver particles on glass or silicon

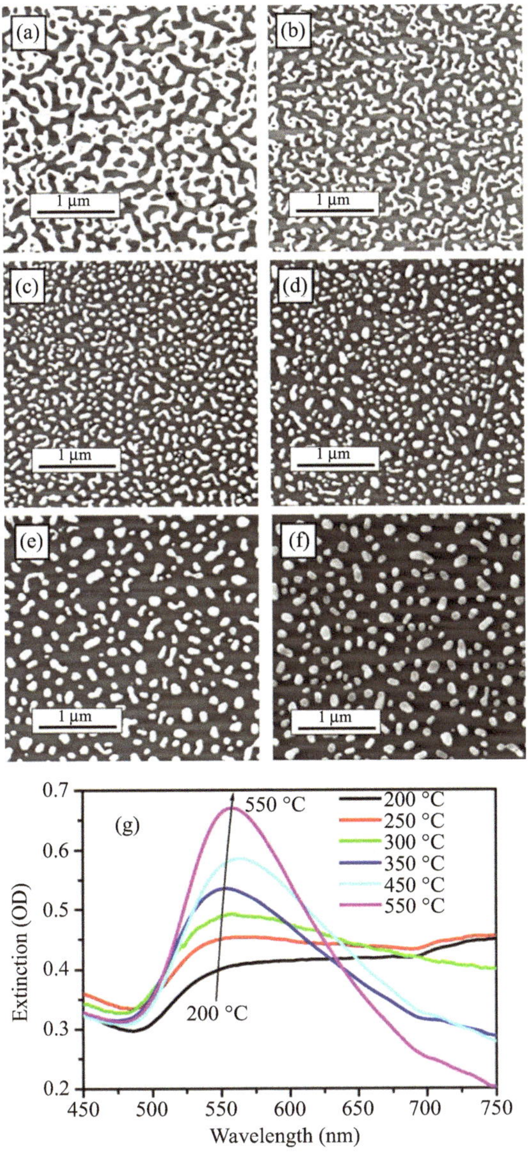

Figure 3.2 (a)–(f) Scanning electron microscopy (SEM) images of the samples of gold nanostructures that have been fabricated using an annealing temperature of 200, 250, 300, 350, 450, and 550°C, respectively. (g) The corresponding optical extinction spectra. Adapted from Ref. [27], Copyright 2010, with permission from Elsevier.

substrates coated with poly(vinylpyridine) (PVP) was created. PVP acts as an efficient surface modifier for the immobilization of metal nanoparticles due to its capability of simultaneous attachment to different substrates via hydrogen bonding and interaction with metal particles due to metal–ligand interactions of the nitrogen atom of the pyridyl group [30]. The PVP-treated substrates were immersed into a colloidal solution of silver nanoparticles in deionized water at low ionic strength. Low ionic strength is needed to support long-range electrostatic repulsion between silver nanoparticles to yield two-dimensional arrays of non-touching particles. By adjusting the exposure time, arrays with various surface densities can be produced. If the average interparticle distance for such arrays becomes of the same order as the nanoparticle diameter, the light extinction spectrum changes drastically with the appearance of a new sharp peak located at 436 nm (Fig. 3.3, black curve).

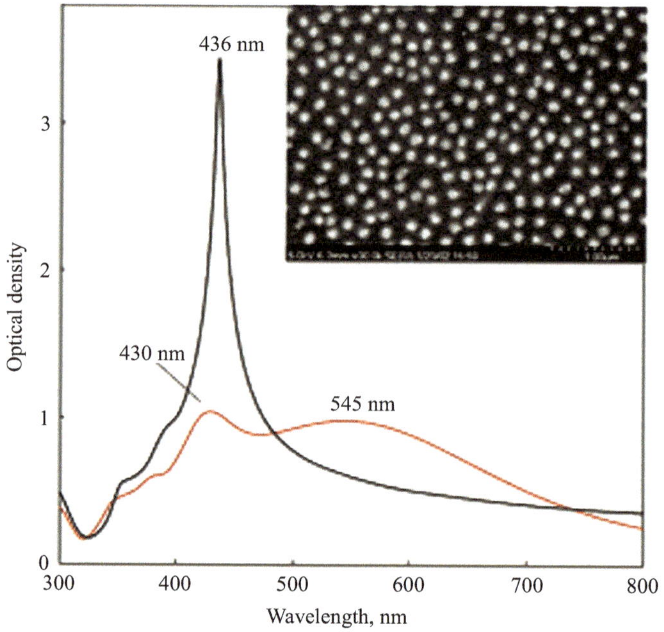

Figure 3.3 Extinction spectra of 100 nm Ag particles in water (red curve) and the same particles assembled into a closely spaced 2D array imbedded in poly(dimethylsiloxane) (PDMS) (black curve). Inset: electron microscopy image of this 2D array. Reprinted with permission from Ref. [28]. Copyright (2003) American Chemical Society.

In several papers, techniques of utilizing colloidal nanoparticles were extended to obtain nanorod chip-based structures. Two chip-based methods (slow and rapid) have been proposed by Geddes' group [31] for the deposition of silver nanorods onto glass substrates using colloidal Ag nanoparticles. Before nanorod deposition, the surface of glass slides was modified in several steps. At first, glass slides were treated with "piranha solution" (3:7 30% hydrogen peroxide/concentrated sulfuric acid) for at least 2 h. After that, the substrates were rinsed copiously with deionized water and dried under a stream of dry N_2 gas. The pretreated slides were silanized by dipping into a 2% ethanolic solution of 3-(aminopropyl) triethoxysilane (APS) for 2 h. Then, the APS-coated glass substrates were rinsed in ethanol and water with further sonication in ethanol for 30 s. Subsequently, the glass slides were rinsed with water and dried under a stream of dry N_2 gas. Furthermore, the technology developed by Murphy et al. [32] for the preparation of silver nanorods in solution was modernized for chip-based variant. Briefly, Murphy proposed to use a prepared beforehand silver seed to stimulate growth in silver nanorods by chemical reduction of a silver salt. To fabricate nanorods and nanowires of different aspect ratio, $AgNO_3$ salt was reduced by ascorbic acid in the presence of silver seed, cetyltrimethylammoniun bromide (CTAB) surfactant, and NaOH. The rod-like surfactant micelles in solution promoted silver nanorod growth. As a result, it became possible to reproducibly fabricate silver nanorods having varying aspect ratio of 2.5–15 (with 10–15 nm short axes) and nanowires 1–4 micrometer long with 12–18 nm short axes. Subsequently, in the slow method by Geddes, silver nanorods were precipitated onto APS-coated glass slides by ordinary immersion into the silver nanorod solution. The adsorption of silver nanorods on the surface of glass slides from the solution continued for a few days, and the light absorption at 550 nm reached only 20% that of the silver nanorods solution. In the rapid method, spherical silver seeds chemically bound to the glass surface were grown into silver nanorods due to a cationic surfactant and silver ions present in the solution. The length of nanorods was determined by the number and duration of immersions of silver-seed-coated glass slides into a growth solution and ranged from tens of nanometers to several micrometers. The formation of silver nanorods on the glass substrates was evident after 10 min of immersion due to the color change (clear to green) on the glass slide and in the solution. To increase the concentration of silver nanorods on the surface, the silver nanorod-coated glass slides were immersed in similar fresh growth solution containing CTAB, $AgNO_3$, ascorbic acid, and NaOH again. This procedure can be repeated till the needed loading of the silver nanorods onto the glass substrate is reached (Fig. 3.4). Interestingly, that minimal change in the process of preparation (immersing the silver-seed-coated glass slides in 40 mL of 0.80 M CTAB solution for 1–3 min [33] instead of 5 min) results in the growth of triangular structures. These properties of surfactant-based technology are of importance, because the shape of nanostructures sometimes plays a crucial role in their further applications in biosensors. For instance, silver triangles are more suitable for SERS applications, whereas silver nanorods are preferable for biomolecular sensing [34]. The major limitations of the above-mentioned technologies are related to nanoparticle shape, disordering, monodispersity, and reproducibility, which negatively influence their extensive application.

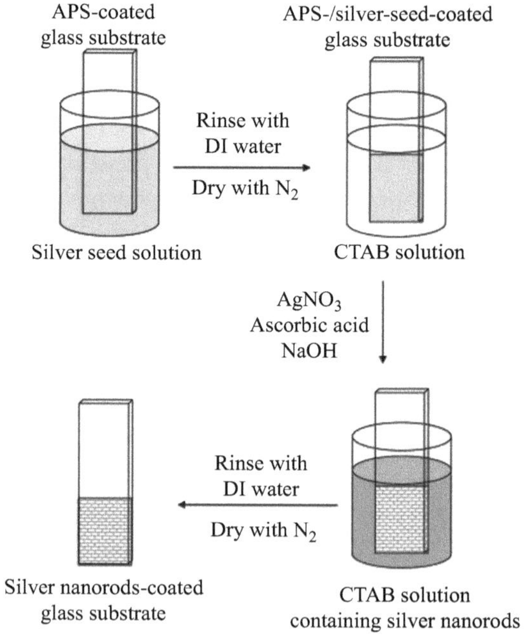

Figure 3.4 Rapid deposition of silver nanorods on a glass substrate. Reprinted with permission from Ref. [31]. Copyright (2005) American Chemical Society.

3.2.2.2 Oblique angle deposition method

There are many difficulties in easily preparing robust, metal-coated substrates with the desired surface morphology, which provide repeatability and high sensitivity for biosensor applications. Substrates patterned by silver nanorod arrays for biosensing and, particularly, for SERS and SEF can be prepared by the recently developed method of oblique angle deposition (OAD) by Dluhy et al. [35, 36]. This nanofabrication technique offers a flexible, easy, and inexpensive method for the fabrication of integrated nanostructured substrates for high-sensitive biological spectral applications. The substrates with nanorods produced by OAD have the advantage of large area and allow preparation of nanopatterned structures for plasmonic biosensors rapidly, accurately, and cost-effectively to detect, for instance, extremely low levels of viruses [37]. In this work, SERS substrates were prepared by electron beam/sputtering evaporation. In the OAD technique, the angle between metal vapor and the normal of the substrate surface is fixed at 86° and about 500 nm Ag thin film is deposited. During deposition, randomly located but uniformly aligned nanorod arrays raise on the substrate. The length of the nanorods is proportional to the deposition time, and the nanorods are inclined with respect to the normal of the substrate surface. SEM micrographs (Fig. 3.5) show the average rod length and diameter of the nanorod arrays to be (868 ± 95) nm and (99 ± 29) nm, respectively. The surface density of the nanorods was estimated to be (13.3 ± 0.5) rods μm^{-2} with an average tilt angle of (71.3 ± 4.0)°.

3.2.2.3 Colloidal lithographies

Colloidal lithography, an economical alternative to the common scanning beam lithography techniques, utilizes self-assembly of colloidal particles (usually polystyrene) on the solid substrate surface to form a mask for subsequent evaporation and/or etching processes in order to fabricate metallic nanostructure arrays. The main advantage of this group of techniques is the possibility to produce both ordered and random nanostructured patterns over large areas, not achievable by conventional nanofabrication methods, resulting in relatively low fabrication costs and high throughput. In addition, narrow nanostructure size distribution (less than 5%) can be achieved if polystyrene colloids with high monodispersity are used. Depending on the structure and composition of the colloidal mask formed, colloidal lithography methods are subdivided into sparse colloidal lithography (SCL), hole-mask colloidal lithography (HCL), and nanosphere lithography (NSL).

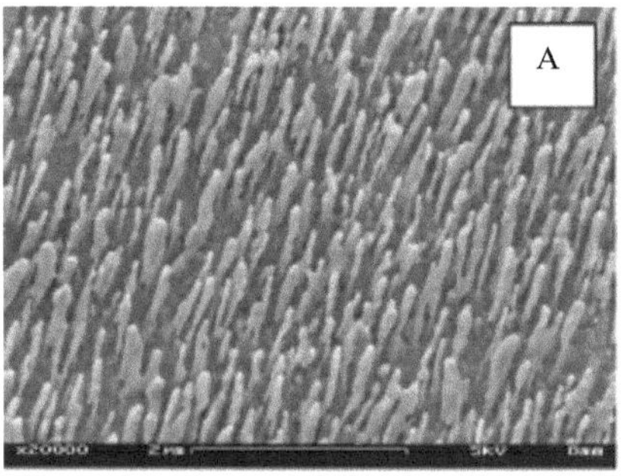

(a) $h = 868$ nm

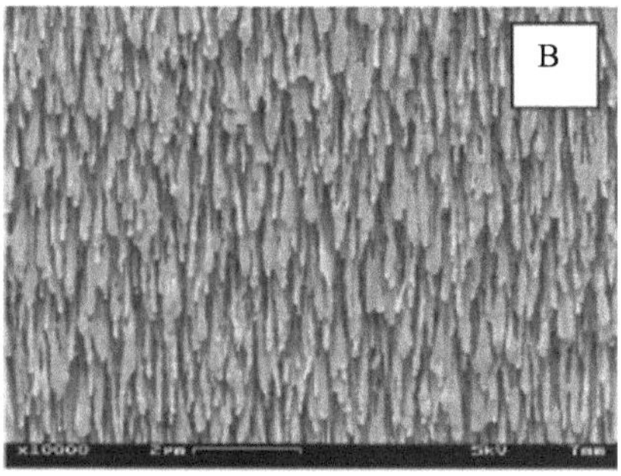

(b) $h = 2080$ nm

Figure 3.5 Representative scanning electron micrographs of the Ag nanorod arrays deposited with different lengths, (a) h = 868 nm and (b) h = 2080 nm. The typical SERS substrate used for virus detection is represented in (a), i.e., 870 nm. Reprinted with permission from Ref. [37]. Copyright (2006) American Chemical Society.

The SCL technique involves adsorbing colloidal polystyrene spheres onto an oppositely charged substrate via electrostatic self-organization into a sparse random pattern with interparticle distance defined by the particle–particle repulsion, which can be adjusted by the concentration and ionic strength of colloidal solution [38]. Adsorbed polystyrene particles are further used as a mask to produce different types of nanostructured metal surface with feature sizes down to 20 nm [38], which are defined by the size of polystyrene spheres used. Single nanoholes and random nanohole arrays in thin metal films are produced by metal evaporation on the colloidal mask and subsequent lift-off of the polystyrene particles [39–43]. Biosensing potential of such nanostructured thin gold and silver films has been demonstrated for selective sensing of antigens such as cancer antigen 19-9 (of less than 1 pg on a 0.1 mm^2 probing area, after surface functionalization with respective antibodies) [40], detection of NeutrAvidin with a detection limit of <0.1 ng/cm^2 [41], sensing of lipid-membrane-mediated biorecognition reactions in the zeptomole regime [42], and for probing the formation of macroscopic and laterally mobile supported lipid bilayers and their binding with proteins [43]. Fabrication of metallic nanorings with similar technique involves an additional technological step; Ar^+ ion-beam etching is used to deposit a metal shell on the polystyrene spheres due to secondary sputtering process. After the removal of polystyrene spheres, a random array of nanorings is formed [44, 45]. Gold nanorings were shown to exhibit a redshifted LSPR that can be tuned over a wide wavelength range in the NIR region by varying the ratio of the ring thickness to its radius [44] and the "bulk" refractive index sensitivities of up to 880 nm per refractive index unit (RIU) (observed for 150 nm diameter rings formed with a 20 nm sacrificial gold layer), which is substantially (>5 times) larger than those of nanodiscs with similar diameters [45], suggesting nanorings as potential ultrasensitive refractive index biosensor platforms. The SCL technique can be also used to produce nanodiscs in an approach when colloidal mask is deposited on top of a metal film and acts as a protecting layer during the subsequent etching of the metal. After the removal of colloidal mask, a random array of metallic nanodiscs is formed [46]. More oblate disc shapes were demonstrated to have higher refractive index sensitivity, making them of interest as substrates for optimizing optical biosensing methods at the nanometer scale [46]. Common drawbacks of the SCL approach are limitations in producing nanostructures composed of materials with low etching selectivity, necessity of the reactive oxygen treatment for the polystyrene mask removal and the restriction of nanostructure dimensions to the polystyrene sphere size [47]. In biosensor application, one may face an issue with insufficient sensor chip reproducibility of SCL-fabricated nanopatterned metallic films due to the random nature of these nanostructure arrays.

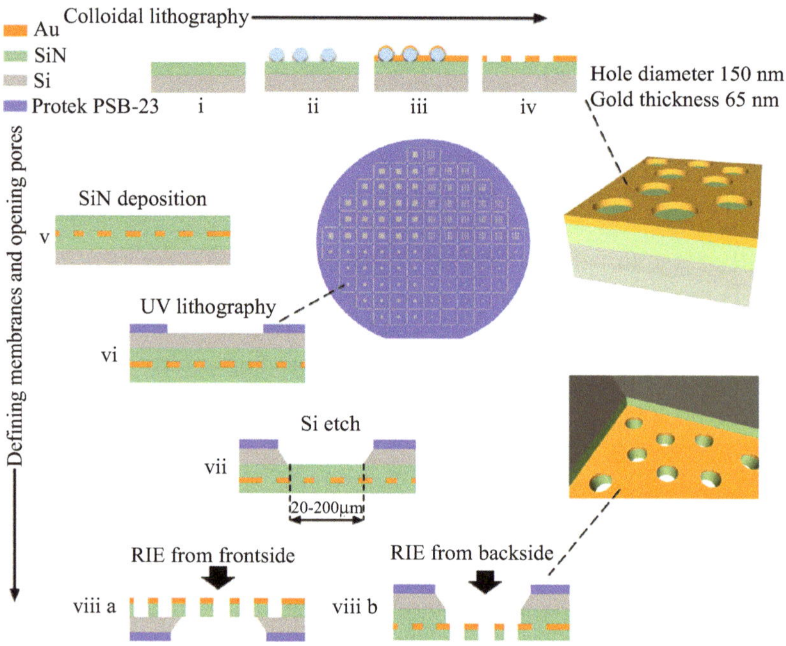

Figure 3.6 Schematic image illustrating the fabrication of sensor chips with membranes penetrated by short-range ordered nanoplasmonic pores. Note that the schematic illustration is not to scale. Adapted with permission from Ref. [48]. Copyright (2010) American Chemical Society.

A related technique based on SCL combined with UV lithography and reactive-ion etching (RIE) steps has been developed to fabricate short-range ordered nanoplasmonic pores penetrating through a thin (around 250 nm) multilayer membrane composed of gold and silicon nitride (SiN), which is supported on an Si wafer (Fig. 3.6) [48]. At first, a thin metal film with nanoholes is fabricated by colloidal lithography on an Si wafer coated with SiN (Fig. 3.6, i–iv). In the subsequent step, the whole nanostructure is covered with a second SiN layer (Fig. 3.6, v) for the purpose of protection during wet etching of Si wafer. UV lithography is applied to prepare areas where Si removal will take place (Fig. 3.6, vi), and Si wafer is subsequently etched in tetramethyl-ammonium-hydroxide (TMAH) (Fig. 3.6, vii). During the final step, the holes in metal film are converted into pores by etching the SiN with RIE. Here, the gold film acts as an etch mask. It is possible to apply RIE from the front side (Fig. 3.6, viii a) or the backside (Fig. 3.6, viii b) of the sample. In the latter case, the gold will be accessible only in the regions of the membranes. Flow-through nanoplasmonic sensing of specific biorecognition reactions has been demonstrated using this nanohole membrane with a signal-to-noise ratio of around 50 at a temporal resolution below 190 ms and molecular uptake at least 1 order of magnitude faster than under stagnant conditions. Additionally, a high-throughput fabrication scheme that enables parallel production of multiple (more than 50) separate sensor chips or more than 1000 separate nanoplasmonic membranes on a single wafer has been presented.

The HCL method extends the approach of the SCL technique by introducing a sacrificial resist layer combined with a thin film mask with nanoholes (a so-called "hole-mask") [47]. Figure 3.7 shows the common fabrication steps of the HCL. A sacrificial film (usually poly(methyl methacrylate)) is deposited onto a substrate. Similar to SCL, a colloidal solution of polystyrene beads is deposited onto the charged-substance pretreated sacrificial surface, and a short-ordered polystyrene nanoparticle array is formed due to electrostatic interaction between a colloid and a surface. After quick stimulated drying of the surface, a thin film, which is resistant to sacrificial film etchant, is deposited, and polystyrene particles are removed, thus forming a hole-mask in the etching-resistant mask. Subsequently, sacrificial resist layer is selectively etched through the hole-mask. After the above-mentioned preparation stages, the hole-mask can then either be used as a deposition or etch mask, or both, to produce a variety of metallic nanostructures on the substrate surface. The HCL technique has been used to fabricate Au and Ag nanodiscs [47, 49–51], embedded nanodiscs [47], metal-dielectric nanodiscs [52–54], and nanocones [47]. Au nanodiscs have been employed for plasmon-enhanced colorimetric enzyme-linked immunosorbent assay with single-molecule sensitivity against horseradish peroxidase [49] and to achieve extremely low limit of detection for bacterial and cancer diagnostics (down to several pg/cm^2) [50] based on LSPR optical label-free biodetection. An interesting nanoplasmonic biosensor chip consisting of gold nanodiscs on a silicon solar cell with integrated electrical detection was successfully used to monitor a specific biorecognition reaction in real time [51]. An enhanced refractive index sensitivity of metal nanodiscs when placed on dielectric nanopillars has been demonstrated [54]. Ag nanocones have been shown as suitable substrates for SERS applications with even single nanocones yielding appropriate signal intensity and showing excellent reproducibility of the SERS features [47]. Though the HCL technique removes certain restrictions of the SCL method, it is still prone to sensor chip reproducibility issues due to the unordered fashion of the produced metallic nanoparticle arrays.

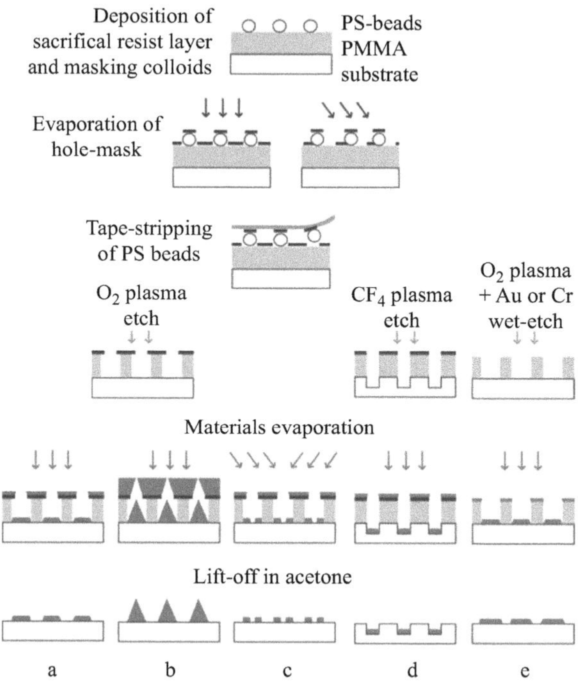

Figure 3.7 Diagram illustrating the basic process steps and resulting structures produced with HCL nanofabrication. Resulting structures are (a) arrays of nanodiscs and oriented elliptical nanostructures, (b) nanocone arrays, (c) (binary) arrays of nanodisc pairs, (d) embedded nanodiscs, and (e) discs with fine-tunable diameters. Reprinted with permission from Ref. [47], Copyright 2007, John Wiley and Sons.

3.2.3 Fabrication of Ordered Nanostructure Arrays

3.2.3.1 Surface nanopatterning of periodic ordered nanostructures using colloidal nanoparticles

Two-dimensional patterned substrates with nanosize features have attracted great interest due to the necessity for high-resolution devices in biosensor research. The manufacturing of periodically ordered Au and Ag nanostructures has obvious advantages in comparison with disordered nanoarrays due to higher possibility for cooperative oscillation of free electrons in the ordered nanoparticle array, so-called "cooperative plasmon resonance." Different methods have been demonstrated to fabricate periodic plasmonic nanostructures. Electron beam lithography with subsequent evaporation and lift-off is commonly used to produce one- or two-dimensional ordered nanoarrays. The obvious disadvantages of this method are small working range (<200 μm), low speed, and high costs. Interference lithography is a high-speed method for fabrication of large-area nanostructures with possibilities to change the periodicity of array. A drawback is the necessity to perform the complicated nanomanufacturing lift-off procedure.

Zhang et al. [55] proposed an interesting and one of the simplest methods of manufacturing similar structures based on high-speed combination of colloidal gold synthesis and interference lithography. In this interesting work, the authors proposed an alternative method for fabricating periodic ordered nanostructures, demonstrating advantages of simplicity, high throughput, and low cost. In this technique, a one-dimensional photoresist (PR) grating on top of an indium tin oxide (ITO) glass substrate is produced initially by interference lithography. Subsequently, a gold nanoparticle colloidal suspension is spin-coated onto the PR mask. The average size of the nanoparticles was about 1.5 nm in diameter, and the ITO layer has a thickness of about 210 nm. Bearing in mind the low melting temperature of the gold nanoparticles, the sample was heated to above 200°C for about 5 min and subsequently cooled to room temperature. With temperature higher than the melting point, the gold colloidal solution aggregated into the grating grooves of the ITO substrate due to the strong wetting on the ITO surface and the constraint by the groove structures. The processes of annealing of gold nanoparticles and their accumulation in the grooves only results in the formation of nanowires, which are distributed periodically over an area of up to 20 mm^2. The mask-determined parameters of the gold nanowire structure, such as a period below 500 nm and a width from 100 to 300 nm, were obtained (Fig. 3.8). The height of the nanowires was dependent on the concentration of the gold colloid and the depth of the mask grating. The main advantage of this promising method is the ordering of nanostructures and the absence of lift-off process; however, the rough surface of the prepared nanowires due to the nanocrystal size distribution broadens the plasmon resonance spectrum.

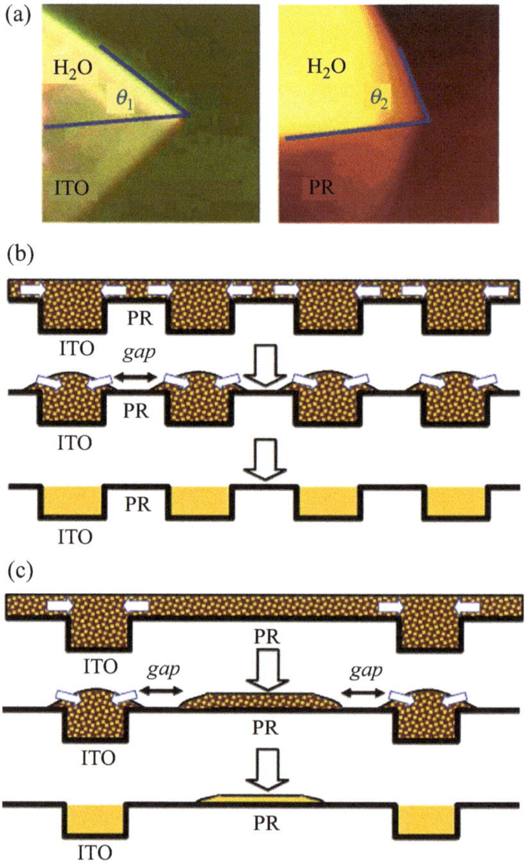

Figure 3.8 (a) Measurements of contact angles of water on the ITO and on the PR surfaces: θ_1 = 44–45°, θ_2 = 73–75°. (b) Mechanisms for the confinement of gold nanoparticles into the grating grooves when the PR channel is small enough. (c) Two gaps will form for structures with large PR channels. Reprinted with permission from Ref. [55]. Copyright (2006) American Chemical Society.

3.2.3.2 Anodic porous alumina membranes

The usage of well-known anodic porous alumina (Al_2O_3) nanomembranes was established five decades ago as a convenient and powerful tool for nanomanufacturing, starting from the pioneer works of Possin with the preparation of nanowires within mica template [56]. The subsequent works of Martin [57] and Moscowits [58] using anodized aluminum oxide (AAO) as a template for metal nanowire fabrication initiated wide utilization of this method for different applications. Among the advantages of utilization of such AAO template are simplicity, possibility to use cheap equipment, tunability in preparation of nanostructures with different size and interparticle distance, and possibility to work with different materials [59, 60]. The AAO membrane method allows better control of nanowire parameters than other protocols for the synthesis of similar nanostructures. Because AAO is also highly suitable for manufacturing nanostructure arrays that provide high surface-to-volume ratio and good access for biomolecules due to vertical alignment [61–64], we will consider this method in more detail.

The main approaches for the preparation and utilization of AAO were presented in an excellent review by Lei et al. [65]. The nano-template from anodic porous alumina is a highly ordered nanopore array with controllable geometry. Using different acid solutions (sulfuric for small (10–30 nm), oxalic for medium (30–80 nm), and phosphoric for larger than 80 nm pore sizes) in the electrochemical treatment of thin Al foil (or evaporated film), careful control of pore diameter is possible [65, 66]. The size of pores can be regulated also by anodization voltage in the range of 10–200 nm [65]. It should be noted that good tuning of nanopore diameter is observed for small and medium pore size, whereas difficulties occur for large pores. Usually, the quality of nanopore array ordering is better on the bottom side of the membrane, so Masuda and Satoh [67] proposed that the anodization process be performed in two steps, in which the top alumina layer is removed from the Al foil, leaving the highly ordered bottom layer. The next stage of the anodization process results in a pore array with high regularity for both sides of the AAO membrane. Subsequent annealing and polishing improves pore regularity and membrane surface smoothness. The area of pores free of defects typically does not exceed several square micrometers, so the "pretexturing" procedure was proposed by Masuda et al. [67], in which a single silicon carbide (SiC) mold was used as a stamp before anodization procedure, which allows to extend the defect-free ordered areas up to millimeter size. Disadvantages of thick (more than 1 μm) AAO membranes are difficulties in manufacturing zero-dimensional nanostructures and preparing nanostructures directly on the substrate. To overcome this problem, a novel nanopatterning technology for the preparation of nanostructures using ultra-thin alumina mask (UTAM) was developed [68, 69]. Nanoparticle or nanohole arrays using AAO can be prepared on almost any smooth substrate (Fig. 3.9). Two different types of membranes—attached (which is placed on the surface) and connected (fabricated on the surface)—can be used. An advantage of attached membranes is higher regularity of pore arrays; however, if the attachment with the substrate is not in good condition, detachments are possible. A thin (5 nm) Au layer covering the backside of the AAO usually serves as a working electrode and as an initial substrate for the deposition of materials within pore array. Due to the direct contact of electrolyte solutions with gold, deposition takes place mainly in the pores [70]. The geometry of AAO template defines the diameter of nanowires and the distance between them, whereas the charge passed during the deposition procedure is proportional to the length of the prepared nanowires. The process of preparation of nanowires is finished by dissolving the AAO template in 1 M acidic or basic solution for several hours [70]. The total area of such a mask can be up to a few square centimeters. An advantage of the AAO method, especially in biosensing, is the possibility to prepare nanostructures with regulated interparticle gap, which is of importance due to the role of the so-called "hot spots" (interspaces in the vicinity of plasmonic nanoparticles, where a strong electric field is generated). Decreasing the distance between nanoparticles increases the electric field and, mainly, results in higher response of a biosensor. Using AAO membranes, Mirkin et al. [71] developed the on-wire lithography (OWL) technique, which allows production of segmented gold nanowires with unique narrow regulated gaps (from 2.5 nm). By periodically changing the composition of electrolyte and the time of deposition, segmented nanowires were prepared. After membrane dissolution, an oxide (e.g., SiO_2) or metal backing was deposited on the wires by thermal evaporation. The etching procedure was performed to dissolve the segments between gold segments, whereas backing maintains spacing between them. Similar structures enhance, for instance, the SERS signal up to 10^8 times. Using the OWL technology, it is possible to produce nanostructures with distance between segments especially for biomolecules with known length. Besides the fabrication of one-dimensional nanostructures, AAO membranes can be used as original in the preparation of metal-replicated membranes [72, 73], and at present it is an excellent low-cost alternative method for manufacturing large-scale ordered arrays.

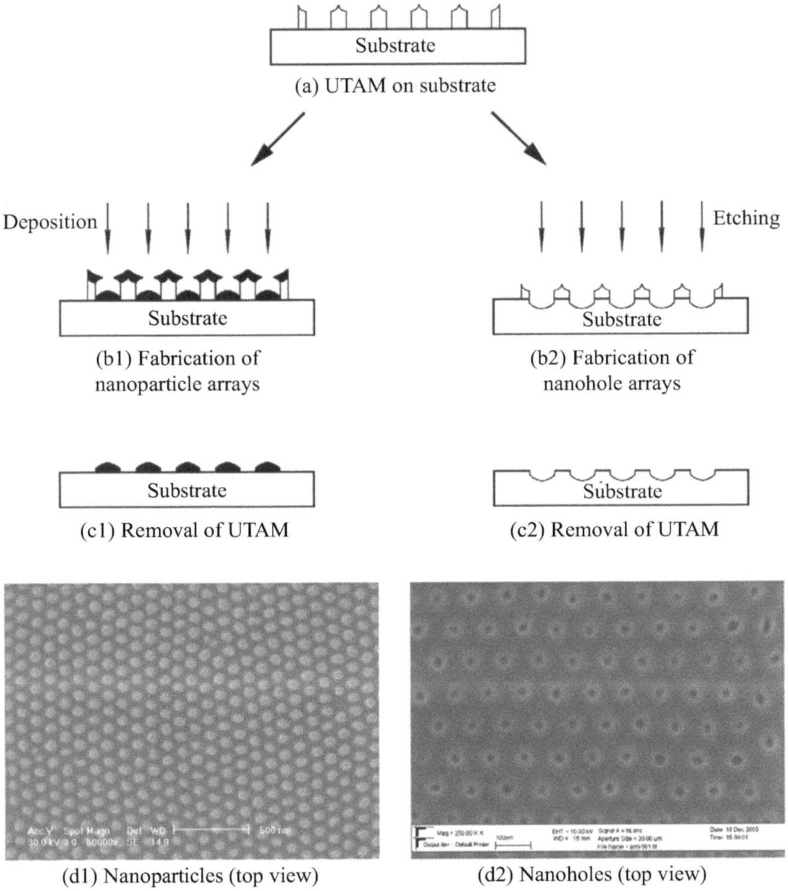

Figure 3.9 Schematic outline of the general fabrication processes of nanoparticles (a–b1–c1) and nanoholes (a–b2–c2). d1 and d2 are highly ordered nanoparticle (Pd) and nanohole arrays (etched using focused ion beam) on Si substrates, respectively. Reprinted from Ref. [65], Copyright 2007, with permission from Elsevier.

3.2.3.3 Scanning beam lithographies

Scanning beam techniques, such as electron beam lithography (EBL) and focused ion beam (FIB) lithography, are the most common techniques for producing metal nanostructures on the solid surface. The main advantage of these methods is the ability to fabricate nearly arbitrary one- and two-dimensional patterns with precise control of the size, shape, and interparticle distance of surface-bound plasmonic nanoparticles. This type of control is needed to meet the requirements of sensor chip reproducibility and optical properties tunability for biosensing, SERS, and SEF applications.

EBL uses a highly focused electron beam to scan across a thin layer of radiation-sensitive resist that is deposited on a substrate, which is subsequently used as a mask for etching and metal deposition procedures to yield the desired metallic nanopattern. EBL provides outstanding resolution down to sub-10 nm range, which is attainable through the exploitation of specialized resists such as hydrogen silsesquioxane [74, 75] or traditional poly(methyl methacrylate) resist with ultrasonically assisted development [76]. The EBL technique is often used when dense gold and silver nanoparticle arrays of a particular geometry and size are needed. This method has been used to fabricate highly uniform arrays of monomers [77–79], dimers [80], and oligomers [81] of discs (Fig. 3.10), elliptical discs [82], trigonal prisms [77, 83], spheroids [84], half spheroids [85], rods [86], split rings [87], L-shaped [88] and almost spherical particles [89] with varying size and interparticle distance. EBL has been also applied for the creation of gold nanogratings [90–92] and subwavelength apertures [93] in thin gold films. Such nanoparticle arrays have been exploited to achieve low detection limits for anti-biotin and streptavidin in LSPR biosensor (3 nM and 7 pM, respectively) [79], to create a multianalyte DNA sensor based on surface-enhanced resonance Raman spectroscopy [87], and to enhance the dye fluorescence both on periodic nanoantenna [82] and subwavelength aperture [93] arrays. Another important application of EBL is the preparation of reusable masks and master stamps for other lithographic techniques such as nanostencil lithography [94] and nanoimprint lithography [95–97]. However, limitations of fabrication area (<200×200 μm^2), low speed due to serial processing, and high costs entail the low throughput sample preparation with the EBL method.

Figure 3.10 SEM images of a typical gold heptamer sample fabricated by EBL. (Top) A normal view of the sample. The interparticle gap distance is 20 nm. The thickness of the gold nanoparticles is 80 nm. Reprinted with permission from Ref. [81]. Copyright (2010) American Chemical Society.

Another scanning beam lithography technique, FIB, allows writing both subtractive and additive patterns into a resist or directly onto the thin metal film by a highly focused beam of ions through the milling, ion-assisted etching, and ion-induced deposition processes [98, 99]. The attainable resolution of the FIB lithography method is <10 nm [100, 101], which is comparable to that of EBL. FIB techniques usually employ Ga^+ ions to form a scanning beam [102–104]; however, the use of other ions such as Ar^+ for ion-assisted lithography has also been reported [105]. In principle, FIB can be used to fabricate arbitrary two-dimensional metal nanostructures, but due to its relatively slow processing capabilities in common with EBL, it is mostly used for the production of holes in metal films rather than isolated metal nanoparticles. Various metallic (i.e., Au, Ag, Al) nanostructures such as single and arrayed circular [102–104, 106–109] and rectangular [108] nanoholes, circular double-holes [110], slits [90, 111, 112], circular slits [113], and V-grooves [114] with varying size and periodicity have been fabricated by means of FIB lithography for research in the field of plasmonics and, especially, for biosensor applications [102–104, 106, 107, 110]. Namely, biosensors based on nanohole arrays demonstrated detection of attomolar streptavidin concentrations [106], 6-fold improvement in response time for surface adsorption of mercaptoundecanoic acid when operated in flow-through format [103], and up to 6.5-fold enhancement of Rhodamine 6G single-molecule fluorescence detection as compared to the open solution [102].

3.2.3.4 Colloidal lithographies

NSL is a type of colloidal lithography that employs a densely packed two-dimensional colloidal crystal as a deposition/etching mask. As opposed to SCL and HCL techniques, NSL is capable of producing ordered nanoparticle arrays with well-controlled shape, size, and interparticle spacing [115]. Such a possibility is achieved by charging the substrate surface with the same charge sign that is carried by polystyrene nanospheres. When polystyrene colloidal solution is deposited on such a pretreated substrate and dried, capillary forces drag the nanospheres together and they crystallize in a hexagonally packed pattern (Fig. 3.11). This colloidal mask can be readily used for both additive and subtractive lithographic processes to produce metallic nanostructures with sizes down to 20 nm [116]. In one of the NSL approaches, a metal film is deposited through the mask to produce a metallic "film over nanosphere" (FON) structure. Enhanced electromagnetic fields generated by the surface roughness of the Ag FON structure have been employed to excite strong SERS signals for detection of *Bacillus subtilis* (a harmless analog of *Bacillus anthracis*, an anthrax causative agent) with limit of detection of 2.1×10^{-14} M [117], for in vivo glucose detection using an implantable SERS sensor [117] and for quantitative lactate detection in the clinically relevant range 0.5–22 mM [118]. Alternatively, if the nanosphere mask is removed after the metal deposition, typically by sonicating the sample in a solvent, a substrate can be obtained, which is covered with surface-confined metallic nanoparticles having triangular cross section, which corresponds to the shape of voids between polystyrene nanospheres. Such nanostructures, mostly made of Ag, have been widely exploited in LSPR and SERS biosensing [8, 119–121]. Ordered arrays of nanovoids in gold films have also been prepared by the electrochemical deposition of Au through the colloidal mask and subsequent removal of polystyrene nanospheres [122] and have been used as substrates for NIR-SERS [123]. Other types of nanostructures that have been prepared using NSL-based techniques include rhombic nanoparticles [124], nanopillars [125], anchored nanoparticles (Fig. 3.11) [126], "film over nanowell" structures [127], nanodiscs [128, 129], nanoholes [128], split rings [130], and micropatterned metallic films [131] for applications in LSPR biosensing and surface-enhanced spectroscopies. The main disadvantage of NSL is a variety of defects that arise in the process of crystallization as a result of nanosphere polydispersity, site randomness, point and line defects; typical defect-free domains have sizes in the 10–100 μm range. Dense nanosphere mask also imposes geometric constraints on the shape of the produced nanoparticles; specifically, only triangular or hexagonally shaped metal nanoparticles can be fabricated by metal deposition through the stacked layers of close-packed nanospheres. In order to tune the nanoparticle shape, additional processing is required, such as moving the sample or changing an angle during metal deposition, annealing, and electrochemical growth.

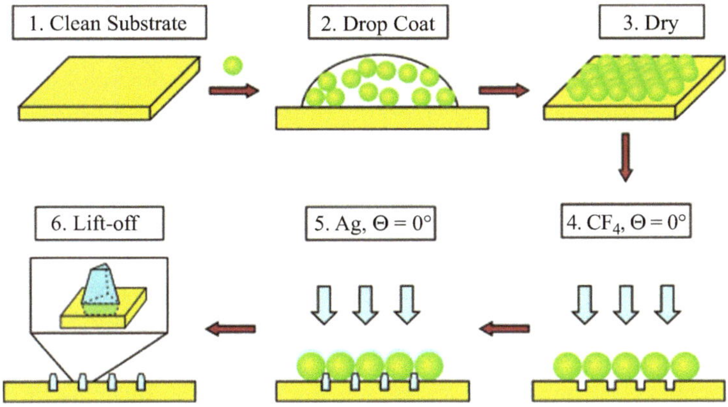

Figure 3.11 Schematic representation of the anchored Ag nanoparticle array fabrication using NSL. Reprinted with permission from Ref. [126]. Copyright (2007) American Chemical Society.

3.2.3.5 Nanoimprint lithography

Nanoimprint lithography is a nanopatterning technique that provides a tradeoff between fabrication throughput, precision, costs, and sample area, which allows eliminating drawbacks of scanning beam and colloidal lithographies for specific applications. This fast-developing technology provides means to replicate features on a hard or soft stamp in a thermoplastic or photocurable resist by embossing or molding. Subsequent deposition of a metal film on such a replica with or without further resist lift-off can be used to produce plasmonic structures resembling the stamp nanorelief for applications in chemical and biosensing. Among the advantages of NIL is the capability to create nanoreliefs over large sample areas in parallel mode with high pattern uniformity and low defect densities, which is crucial for mass production of reproducible sensor chips for biosensing instruments. Namely, NIL is capable of molding a variety of materials and pattern features with a sub-10 nm resolution on the cm^2 area scale [132–134]. Another advantage of NIL is a wide range of nanoparticle shapes that can be fabricated using specially prepared reusable NIL stamps (Fig. 3.12) [135], which is important in biosensing applications to optimize the performance of plasmonic nanoparticle arrays.

NIL has been employed to produce uniformly oriented and homogenous noble metal nanoparticle arrays using "nanoblock" molds fabricated from a collection of one-dimensional grating molds with different profiles (Fig. 3.13) [135]. Such anisotropic Ag and Au nanoparticle arrays have shown light polarization- and dimension-dependent plasmonic properties [135]. A similar approach has been applied to produce arrays of Au nanorectangles, which were shown to be a suitable basis for LSPR biosensor by detecting bovine serum albumin (BSA)–anti-BSA specific interaction [136]. High-density square arrays of free-standing cylindrical gold-coated polymer "nanofingers" fabricated using NIL have been demonstrated to be an efficient SERS substrate providing enhancement factors up to 2×10^{10} [137]. A two-dimensional array of cavities fabricated by NIL using a soft elastomeric mold in polyurethane, coated by 50 nm Au film, has demonstrated its sensing potential in light transmission experiments during the formation of a self-assembled monolayer of hexadecanethiol [138]. Similar quasi-3D plasmonic crystals have

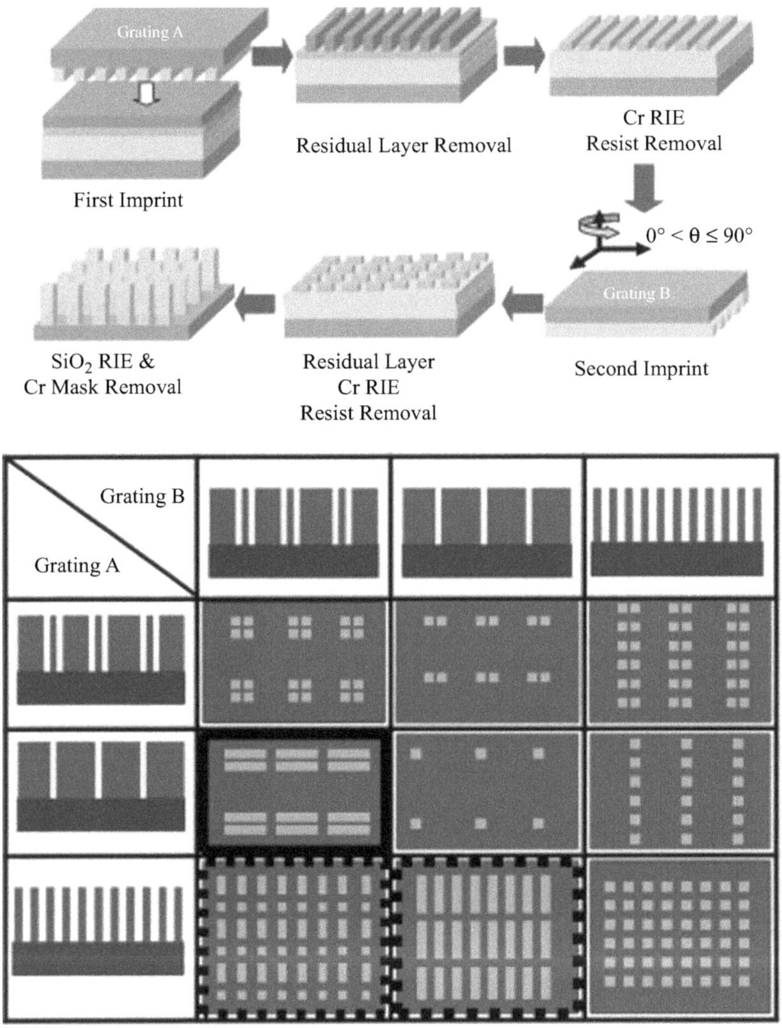

Figure 3.12 (Top) Illustration of the general process used to fabricate "nanoblock" NIL molds possessing different lattice and particle geometries. (Bottom) Matrix illustrating possible nanoparticle array configurations from nanoblock molds produced using a collection of one-dimensional grating molds with different profiles. The array bordered by a solid line is produced using the inverse profile of grating A. Similarly, the arrays bordered by broken lines are achieved using the inverse of grating B. Reprinted with permission from Ref. [135], Copyright 2008, John Wiley and Sons.

shown a refractive index sensitivity of 700–800 nm/RIU [139] and enabled full multi-wavelength spectroscopic and spatially resolved detection of biomolecular binding events with sensitivities that correspond to small fractions of a monolayer [139–142], for example, 400 pM limit of detection for anti-IgG/IgG binding [142]. Au dot and ring arrays produced by NIL were used to detect the binding of streptavidin to biotin, and it was shown that the sensitivity of the LSPR spectra to the binding of the biomolecules was enhanced as the ring width of Au rings was decreased [143]. Arrays of gold-coated nanodomes fabricated on glass substrates using a soft NIL revealed complex plasmonic resonances highly sensitive to the array dimensions, the thickness of the gold layer, and the refractive index of the surrounding medium [144]. Other metallic nanostructures fabricated using NIL, which are promising for biosensing applications, include arrays of Pt nanowires [145], Ag island films deposited on grated and pillared Si substrates [146], elliptical Au nanodisc arrays [147], and Au nanodot and nanowire arrays [148, 149].

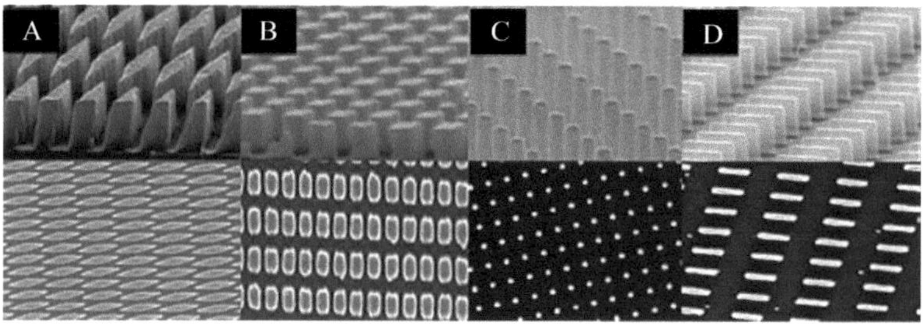

Figure 3.13 SEMs of nanoblock molds derived from one-dimensional gratings. Reprinted with permission from Ref. [135], Copyright 2008, John Wiley and Sons.

However, this attractive nanolithographic technique is still under development and has several challenges to overcome. One of the important problems to be solved is the useful lifetime of the mold; at present, nanoimprint stamps require replacement after about 50 imprints made [150] due to the wear produced by temperature variations and high pressures applied during the imprinting process. A possible way to partially solve this problem is to avoid high temperature by using special resist treatment, which allows imprinting at room temperature [151]. Thermally assisted imprinting also has limited throughput caused by time requirements of thermal cycling; however, employment of photocurable resist instead of thermoplastic or imprinting at room temperature and high pressures [152] can speed up the nanofabrication process. Other conditions that should be maintained in order to enable uniform pattern transfer over large areas include absence of air bubbles between the mold and resist, parallelism of rigid mold and resist surfaces, and applied pressure uniformity.

3.3 Properties of Plasmonic Nanochips Based on Nanostructure Arrays of Different Structure

The most evident advantage of the NIL technique is the exploitation of templates with relatively large linear dimensions and sub-10 nm resolution [133] for nanostructure preparation that makes NIL suitable for fabrication of uniformly oriented and homogeneous gold nanostructure arrays (NSA) with controlled nanoparticle size, shape, and spacing. However, due to the associated high costs for the NIL technique exploitation in terms of equipment needed for NIL template fabrication and the nanoimprinting process itself, the development of an alternative technique, which can be the basis for nanochip fabrication, is of considerable interest. In this section, a comparison of the two above-mentioned methods of Au NSA fabrication—gold island film deposition with subsequent thermal annealing for random NSA and nanoimprint lithography for ordered NSA—is presented from the point of view of LSPR spectral measurements, sensing, and PE modeling.

3.3.1 Morphological and Spectral Properties of Gold Nanostructure Arrays

The nanochip fabrication technique based on gold island film deposition with subsequent thermal annealing [153] was adapted, which is an affordable NSA preparation method yielding satisfactory results. Briefly, gold island films of varying thickness (5 to 15 nm) were obtained by thermal evaporation in vacuum on precleaned microscope glass substrates (n = 1.51). After island film deposition, samples were annealed at 550°C for 6 h in N_2 atmosphere. As a result of annealing, the gold film color changed from blue of different intensity to blue, violet, and pink (depending on the film thickness), which confirms the formation of separated gold nanostructures having different sizes depending on the initial island film thickness values (see AFM images in Fig. 3.14).

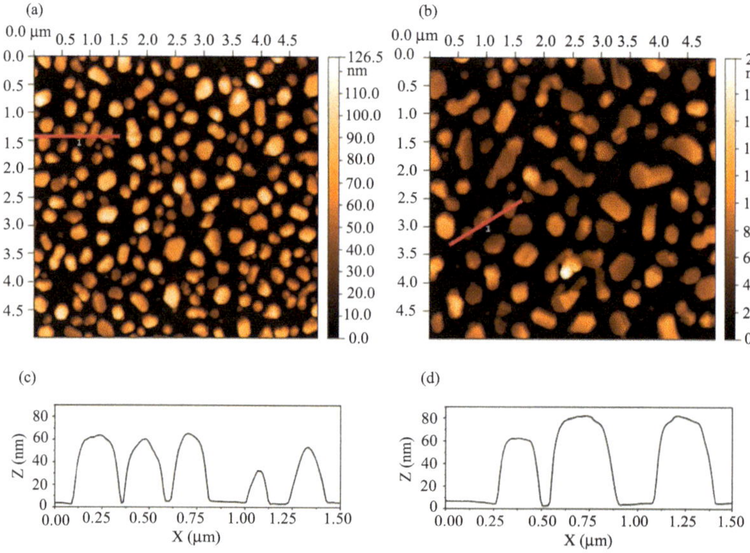

Figure 3.14 AFM characterization results for random gold NSA samples produced via thermal annealing of gold island films. (a, b) Top view AFM images and (c, d) corresponding AFM profiles for random gold NSA samples R6 and R7, respectively (see table in inset of Fig. 3.15). Reprinted from Ref. [154] under a Creative Commons Attribution 4.0 International License. Figure caption was adapted.

As a result of spectrophotometric measurements, the light extinction spectra of the samples in air atmosphere were obtained (Fig. 3.15). It was found that the peak position in the unpolarized light extinction spectrum, which corresponds to the occurrence of LSPR, shifts toward longer wavelengths with an increase in the initial gold island film thickness (see Fig. 3.15). Additionally, the direct relation between lateral dimensions and height of nanostructures produced after annealing and the initial gold island film thickness was revealed. Thus, it is possible to tune the LSPR spectral position in the wavelength range from 576 nm to 809 nm (see Fig. 3.15), and the associated PE spectral profile, by changing the initial gold island film thickness. Obviously, the considered nanochip fabrication technology based on gold island films with subsequent thermal annealing can be exploited while taking into account inherent technological limitations that hinder the preparation of geometrically ordered nanoparticle arrays.

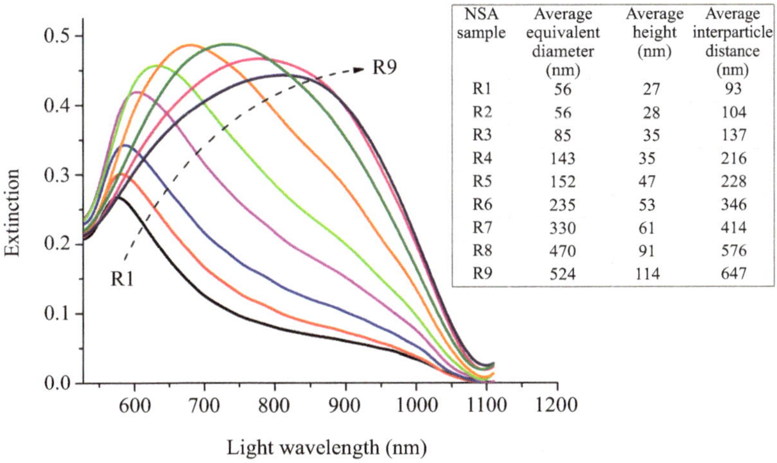

NSA sample	Average equivalent diameter (nm)	Average height (nm)	Average interparticle distance (nm)
R1	56	27	93
R2	56	28	104
R3	85	35	137
R4	143	35	216
R5	152	47	228
R6	235	53	346
R7	330	61	414
R8	470	91	576
R9	524	114	647

Figure 3.15 Measured extinction spectra of random NSA samples produced via thermal annealing of gold island films. Normal incidence unpolarized light extinction spectra for samples R1–R9 with different initial gold island film thickness exhibit peaks located in the wavelength range from 576 nm to 809 nm. Inset: the table with respective NSA geometrical parameters. Reprinted from Ref. [154] under a Creative Commons Attribution 4.0 International License. Figure caption was adapted.

Ordered gold nanoparticle arrays were produced by means of NIL technique according to the protocol described earlier [135, 136] (Fig. 3.12). Briefly, Pyrex glass substrates were cleaned in a 1:1 piranha solution (30% H_2O_2:29% NH_4OH), rinsed with a copious amount of DI:H_2O and dried using N_2. The resist was spin-coated to the appropriate thickness on the substrates, baked for solvent removal, and imprinted using the template. After nanoimprinting and sample separation, the residual polymer layer was removed using O_2 plasma reactive-ion etching. Au metallization was accomplished using an electron beam evaporator. Lift-off was performed by soaking the samples in acetone and using an ultrasonic bath. After the completion of lift-off, samples were rinsed with methanol and isopropanol and dried with N_2. This technique can be used to create a variety of structures by simply using different one-dimensional gratings (i.e., with various duty cycles or periods) and relative angular orientation of the gratings for successive imprints to create different NIL templates [135] (Fig. 3.13).

Structural characteristics of ordered NSA were investigated using the atomic force microscopy method. NIL-fabricated NSA samples of different geometry were studied. These samples were comprised of parallelepiped-shaped nanoparticles, located in an ordered array with square or rectangular lattice. AFM images of nanostructure arrays fabricated using the mentioned technique are shown in Fig. 3.16.

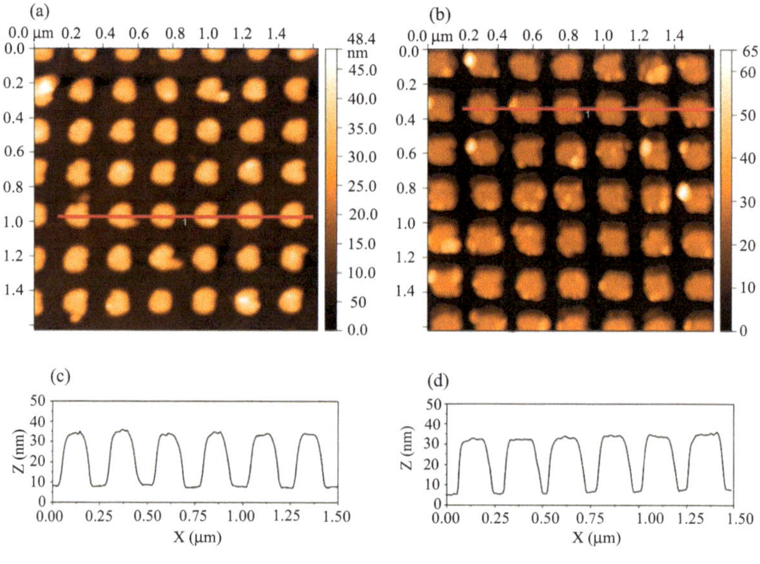

Figure 3.16 AFM characterization results for ordered gold NSA samples produced using NIL. (a, b) Top view AFM images and (c, d) corresponding AFM profiles for ordered gold NSA samples O2 and O4, respectively (see table in inset of Fig. 3.17). Reprinted from Ref. [154] under a Creative Commons Attribution 4.0 International License. Figure caption was adapted.

Spectral characteristics and structural parameters of ordered NSAs were obtained using the previous experimental conditions and yielded the following results (see Fig. 3.17). Unpolarized light extinction spectra for each of the samples exhibited bands having an expressed peak with different spectral positions (from 606 nm to 783 nm) and extinction intensities, which mainly depend on the geometrical parameters of the nanostructures. Significant spectral shift in the LSPR band also implies the possibility of wide-range wavelength tuning using NSA fabricated by the NIL technique. Strong variation in the extinction intensity should be noted as a disadvantage of such spectral tuning for both methods of NSA fabrication.

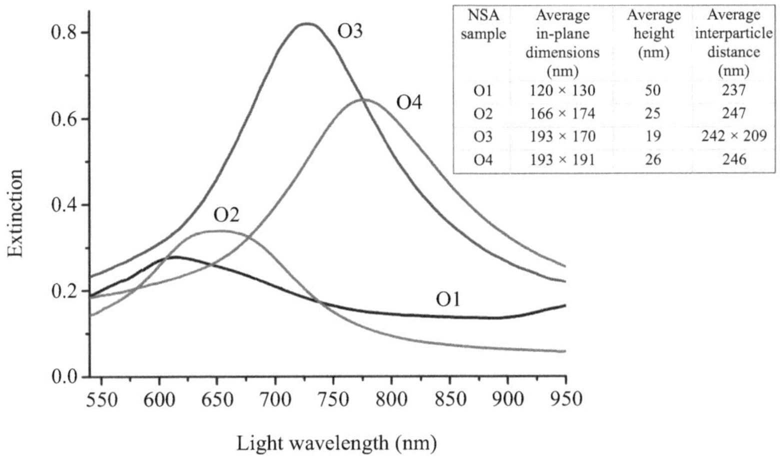

NSA sample	Average in-plane dimensions (nm)	Average height (nm)	Average interparticle distance (nm)
O1	120 × 130	50	237
O2	166 × 174	25	247
O3	193 × 170	19	242 × 209
O4	193 × 191	26	246

Figure 3.17 Measured extinction spectra of ordered NSA samples produced using NIL. Normal incidence unpolarized light extinction spectra for samples O1–O4 exhibit peaks located in the wavelength range from 606 nm to 783 nm. Inset: the table with respective NSA geometrical parameters. Reprinted from Ref. [154] under a Creative Commons Attribution 4.0 International License. Figure caption was adapted.

3.3.2 Theoretical Comparison of Sensing Properties of Gold Nanostructure Arrays

According to the AFM results presented in Subsection 3.3.1, the shape of the gold nanoparticles in NSA can be considered intermediate between a parallelepiped and a semi-ellipsoid. In order to determine the optimal nanoparticle shape in terms of biosensing sensitivity and to characterize the biomolecular layers formed during the study of biomolecular processes using the proposed NSA, the modeling of the optical response of the relevant LSPR biosensor was performed using the finite-difference time-domain (FDTD) method with the fitting to the experimental data. Namely, the non-polarized light extinction peak shift value (LSPR response $\Delta\lambda_{max}$) was determined upon the adsorption of BSA molecules onto the NSA (0.86 nm) and depending on the concentration of the anti-BSA in the PBS buffer solution (see Fig. 4.8 in Chapter 4). Two NSA models were constructed consisting of 4 × 4 nanostructures with dimensions of 120 nm × 130 nm × 50 nm (width × length × height) and a 240 nm period with the shape of parallelepiped and tri-axial semi-ellipsoids, respectively, located on the glass surface (n_S = 1.474) in a medium of buffer solution (n_0 = 1.334 [155]) (Fig. 3.18). Biomolecular layers were modeled as homogeneous dielectric layers on the surface of gold nanoparticles with a thickness corresponding to the diameter of the sphere equivalent to the volume of a biomolecule: 6 nm (for a monolayer of BSA molecules) [156] and 12 nm (for a monolayer of anti-BSA molecules) [157].

For both constructed models, the non-polarized light extinction spectra for uncovered NSA were simulated. After that, the non-polarized light extinction spectra for NSA coated with a layer of BSA were modeled, and the refractive index of the BSA layer varied from 1.34 to 1.39. From the obtained extinction spectra, the positions of their peaks and their shift relative to the peaks of the uncovered NSA extinction spectrum were determined, and a regression analysis of the dependence of this shift on the refractive index of the layer using linear and parabolic approximation was performed and the value of the adjusted determination coefficient reached more than 0.999. Using the results obtained, for the experimental LSPR response upon the adsorption of BSA molecules equal to 0.86 nm, the refractive indices of the BSA layer were defined as equal to 1.34701 (for semi-ellipsoid-based NSA) and 1.34772 (for parallelepiped-based NSA), which were used at the next stage of simulation.

The above-described approach was also used for the next layer of anti-BSA, which resulted in the sets of model spectra obtained (Fig. 3.19a, results for parallelepiped-based NSA only) and refractive indices of the anti-BSA molecule layer at different concentrations of the solution that were in accordance with the experimentally measured LSPR responses presented in Fig. 4.8. From the obtained values of the refractive index of the anti-BSA molecule layer, using Eq. (2.37) of the symmetric Bruggeman effective medium theory described in Chapter 2, the layer filling factors by anti-BSA molecules were determined, which are presented in Fig. 3.19b in terms of the density of the monolayer of globular biomolecules (with a maximum filling factor of $\pi/6$).

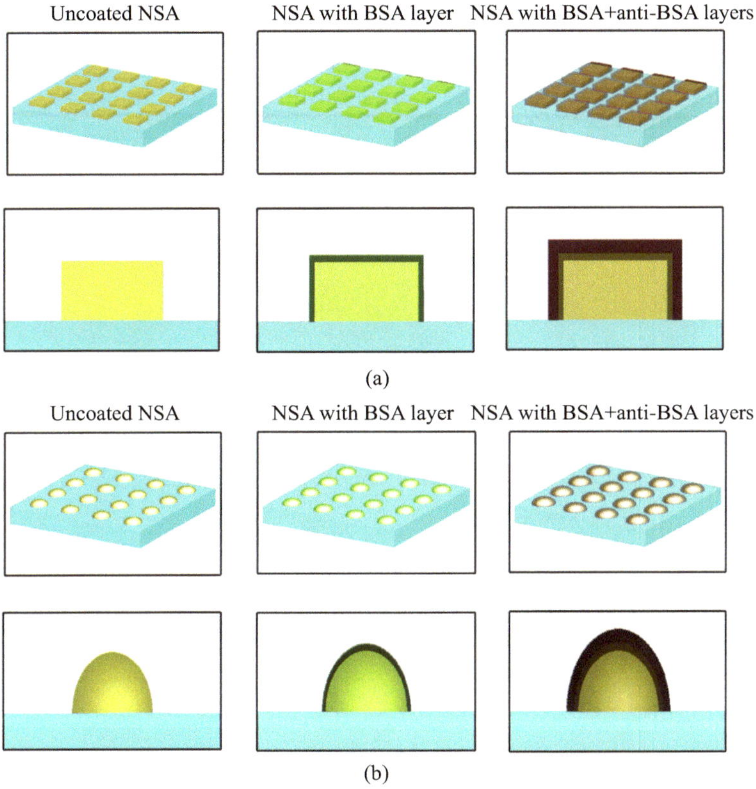

Figure 3.18 NSA models with biomolecular layers used for simulations: (a) parallelepiped-based NSA, (b) semi-ellipsoid-based NSA.

From Fig. 3.19b, it is evident that the developed NSA-based LSPR biosensor is capable of detecting submonolayer protein coatings, and the proposed model allows evaluating the possibility of registration of molecules with different molecular weights by specifying the necessary parameters of layers covering nanostructures. It is also obvious that the shape of nanoparticles in the NSA influences the sensitivity of the LSPR biosensor. Namely, the nanoparticles in the shape of parallelepipeds give the same LSPR response at densities of the biomolecular layer, which is 1.75–2 times lower than in the case of nanoparticles in the shape of semi-ellipsoids of the same size, suggesting a greater sensitivity of the sensor elements using parallelepiped-based NSA. In addition, the dynamic range of the biosensor in the case of parallelepiped-based NSA is wider, because at the concentration of anti-BSA solution of 250 μg/ml, the density of the anti-BSA monolayer for semi-ellipsoid-based NSA almost reaches 1, while for parallelepiped-based NSA, the monolayer is only half-saturated with molecules. Thus, using parallelepiped-based gold NSA as a sensor element is more appropriate and this should be considered during the production of NSA.

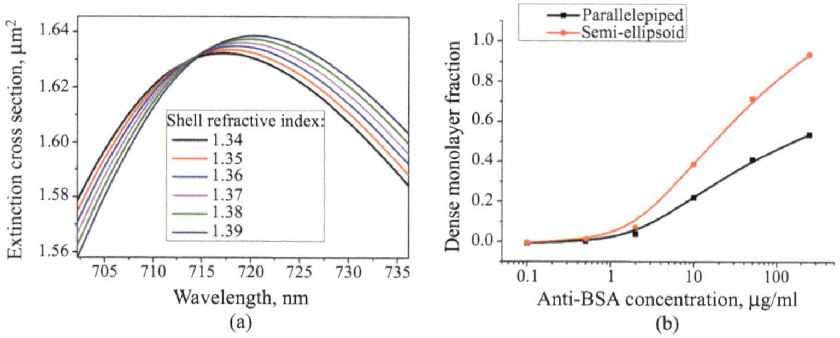

Figure 3.19 (a) Modeled non-polarized light extinction spectra of the NSA for different refractive indices of the anti-BSA layer in the "nanoparticle–BSA layer–anti-BSA layer" system for parallelepiped-based NSA. (b) Dependence of density of anti-BSA monolayer adsorbed onto NSA on the concentration of anti-BSA molecules in a buffer solution.

3.3.3 Theoretical Comparison of Plasmonic Enhancement Properties of Gold Nanostructure Arrays

Plasmonic nanochips enabling the PE effect can be fabricated using a multitude of methods [158]. A simple and commonly used approach involves highly conductive, continuous film possessing surface roughness as effective plasmonic amplifiers; however, this method does not yield the ability for spectral tuning and, consequently, the process of enhancement cannot be applied for matching with molecular resonances that is preferable for a number of spectroscopic techniques. The benefit of spectrally controlled nanostructured PE surfaces is obvious as only uniform surface-bound nanostructure arrays (PE nanochips) with known surface 3D geometry can provide a real possibility to perform preliminary estimation of final PE parameters when using this technique and ensure their stability and reproducibility.

To optimize the parameters required to achieve the highest PE response for the random and ordered NSAs, a theoretical study of intensity enhancement for the electric component of the electromagnetic field near the gold nanostructure surface depending on its shape and size was carried out. According to AFM results (see Fig. 3.14), nanoparticles of random NSA can be considered approximately as two-axial semi-ellipsoid in the case of smaller nanoparticles and more complex shapes in the case of larger ones. So a common nanoparticle model for random NSA in the shape of two-axial semi-ellipsoid with average dimensions presented in the inset of Fig. 3.15 was applied.

It is well known that the greatest electric field intensity enhancements associated with highly conductive metal nanoparticles are expected in "hot spots" on the nanoparticle surface, which are usually located on sharp corners, edges, or projections of nanoparticles [18–20]. As the semi-ellipsoid shape has no expressed regions for "hot spots," it was anticipated that the NIL method allows for the production of NSAs with the above-mentioned features resulting from nanoparticles of various shapes (including parallelepiped shape). To compare the influence of "hot spots" on the PE level, nanoparticles having the parallelepiped shape with a square base of the same area (i.e., having the same equivalent diameter) and same height (effective geometrical parameters (EGP)) as their semi-ellipsoid-shaped counterparts were included in the simulation. Due to the relatively sparse character of the investigated NSA samples (surface-to-surface distance between nanostructures lies within the 37 to 123 nm range as determined by AFM), the nanoparticles in the array were considered to be non-interacting (the strongest coupling between nanoparticles was observed at surface-to-surface distance close to 20 nm [159]) and the individual Au nanoparticles have been selected for theoretical calculations.

Finally, the Au nanoparticles of semi-ellipsoid and parallelepiped shape located on glass substrates in air environment have been considered a model system. Due to the lack of a general analytical model that fully describes the electromagnetic response of arbitrary-shaped nanoparticle, numeric methods should be used. Thus, to estimate the scattering and extinction of electromagnetic field by non-spherical Au nanoparticles, the FDTD method implemented in Lumerical's FDTD Solutions software was exploited. This allowed spectral distributions of local electromagnetic field intensity in the near field of the nanoparticle and light extinction spectrum that represent the excitation of LSPR in the nanoparticle to be obtained. The full 3D FDTD model was computed for linearly polarized incident light and wavelength ranging from 400 to 2000 nm with simulation mesh fixed at 2 × 2 × 2 nm. A linearly polarized light source was used, which is common for the case of symmetric nanoparticle shapes considered; therefore, results obtained can also be applied to unpolarized NSA illumination.

First, the light extinction properties of model nanoparticle systems have been simulated for the two considered shapes (i.e., semi-ellipsoid and parallelepiped). Figure 3.20 shows the typical light extinction spectra for both shapes with EGP corresponding to samples R1, R4, and R6 presented in the inset of Fig. 3.15. It is evident that the simulated extinction spectra follow the same trend of red-shifting and widening upon increase in nanoparticle dimensions as experimental extinction spectra, which can be related to the mutual influence of light scattering and secondary radiation of electrons (radiation damping effect) [160, 161]. Additionally, the extinction response of semi-ellipsoid and parallelepiped-shaped nanoparticles does not exhibit noticeable differences except for a permanent red-shift and marginally increased FWHM of parallelepiped-shaped nanoparticle spectrum (see Fig. 3.20), which can be attributed to the shape effect. The observed difference between the experimental and modeled extinction spectra, which is most evident for random NSA, can be explained by the growing mismatch between the model semi-ellipsoid and experimental nanostructure shapes with increasing nanostructure size. Additionally, the near-field interaction between nanostructures and their size distribution also contribute to the observed discrepancy between the experimental and calculated data. Nevertheless, the aforementioned trend of spectra red-shifting and widening is obvious.

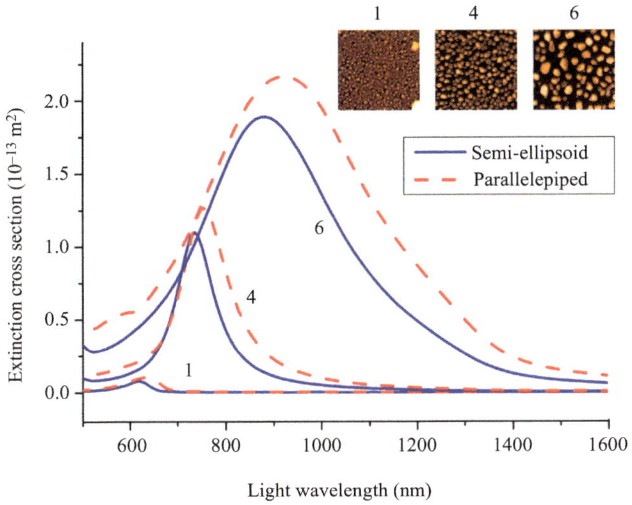

Figure 3.20 Simulated light extinction cross-sectional spectra for semi-ellipsoid- and parallelepiped-shaped Au nanoparticles on glass substrates. Insets show top view AFM images for random gold NSA samples. Numbers correspond to samples R1, R4, and R6, respectively. Effective geometrical parameters of Au nanoparticles used in FDTD modeling are the same for semi-ellipsoid and parallelepiped shapes and correspond to the data presented in the inset of Fig. 3.15 for samples R1, R4, and R6, respectively. Reprinted from Ref. [154] under a Creative Commons Attribution 4.0 International License. Figure caption was adapted.

The comparison of maximal local electric field distribution around the nanoparticles of different shapes shows that parallelepiped-shaped nanoparticles on glass substrates provide electric field intensity enhancements more than an order of magnitude greater than their semi-ellipsoidal counterparts (Fig. 3.21). For example, parallelepiped nanoparticles with equivalent diameters of 85 nm and height of 35 nm exhibit maximal incident field intensity enhancement up to approximately 1.2×10^4. In comparison, semi-ellipsoidal nanoparticles with the same EGP provide maximal electric field intensity enhancement of only about 4.1×10^2.

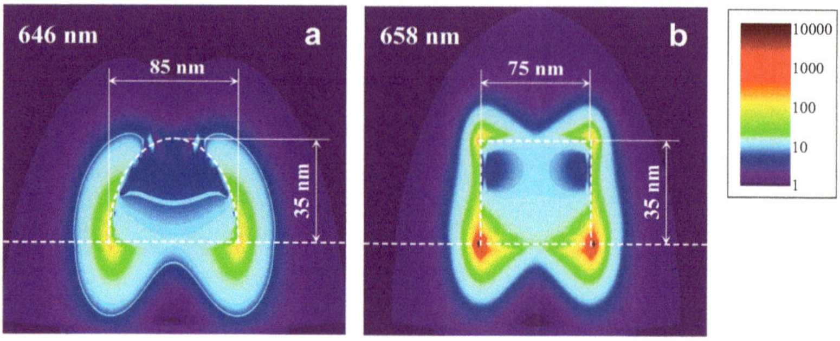

Figure 3.21 Simulated profiles of electric field intensity enhancement for differently shaped Au nanoparticles. Effective geometrical parameters of (a) semi-ellipsoid- and (b) parallelepiped-shaped Au nanoparticles on glass substrates correspond to sample R3 (see inset of Fig. 3.15). Electric field profiles correspond to light wavelengths of 646 and 658 nm for semi-ellipsoid- and parallelepiped-shaped Au nanoparticles, respectively, which provide the maximal intensity enhancement. Electric field polarization and wave vectors lie in the profile plane under the normal incidence conditions. Dashed lines indicate nanoparticle and substrate borders. Different nanoparticle widths are shown due to the exploited model with identical nanoparticle base area for semi-ellipsoid- and parallelepiped-shaped Au nanoparticles. Vertical dimension scale is increased for better visual perception. Reprinted from Ref. [154] under a Creative Commons Attribution 4.0 International License. Figure caption was adapted.

The whole set of maximal electric field enhancement values obtained for each of the nanoparticle geometries is presented in Fig. 3.22. It is evident that the largest enhancement values are exhibited when the equivalent diameters of both types of nanoparticles range from 85 to 143 nm and the height of nanoparticles is 35 nm. These optimized geometrical parameters can be used for manufacturing purposes to fabricate Au NSA providing plasmonic enhancement for different applications. For a defined nanoparticle shape and geometrical parameters, the maximal electric field intensity enhancement is obtained at a specific light wavelength, which was found to range from about 620 nm (for small nanoparticles) to near infrared region (for large nanoparticles). Detailed light wavelength values are contained in labels in Figs. 3.21–3.23.

It is particularly important to discuss the difference between electric field intensity enhancements for semi-ellipsoidal and parallelepiped nanoparticles considering not only the "hot spots" regions but the vicinity of the entire nanoparticle surface where other objects of interest, for example, molecular or biomolecular species, can reside. To accomplish this task, the average electric field intensity enhancements within a 25 nm vicinity (maximal value for a biomolecular monolayer) of both semi-ellipsoidal and parallelepiped Au nanoparticles were calculated (Fig. 3.23). Interestingly, in this case, both nanoparticle shapes have demonstrated comparable average electric field enhancement values that peaked at about 24 and 45 for semi-ellipsoidal and parallelepiped nanoparticles, respectively. This result implies that not only NIL-fabricated nanochips but also NSA samples obtained using island film thermal annealing method can be promising as a possible PE basis for various surface-enhanced spectroscopies. The NIL NSA can be more effective for the detection of small size molecules, which are able to reach "hot spots" in the vicinity of the nanostructure base.

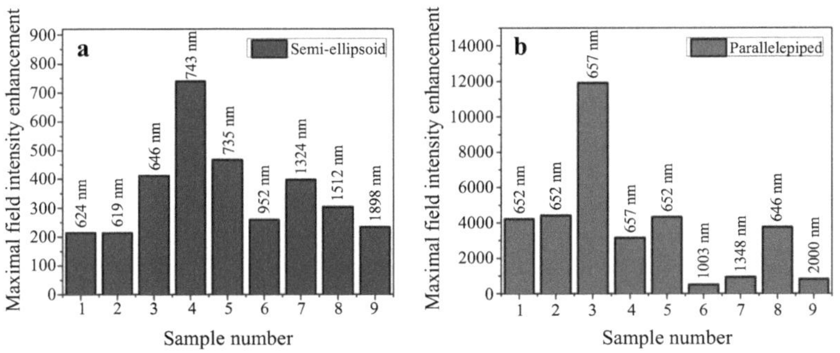

Figure 3.22 Calculated values of maximal electric field intensity enhancement near differently shaped Au nanoparticles. Simulation results correspond to (a) semi-ellipsoid- and (b) parallelepiped-shaped Au nanoparticles on glass substrates. Effective geometrical parameters of Au nanoparticles used in FDTD modeling are the same for semi-ellipsoid and parallelepiped shapes and correspond to the data presented in the inset of Fig. 3.15 for samples R1 to R9, respectively. Labels above the bars indicate the light wavelength, which the maximal electric field intensity enhancement for specific nanoparticle geometry was obtained at. Reprinted from Ref. [154] under a Creative Commons Attribution 4.0 International License. Figure caption was adapted.

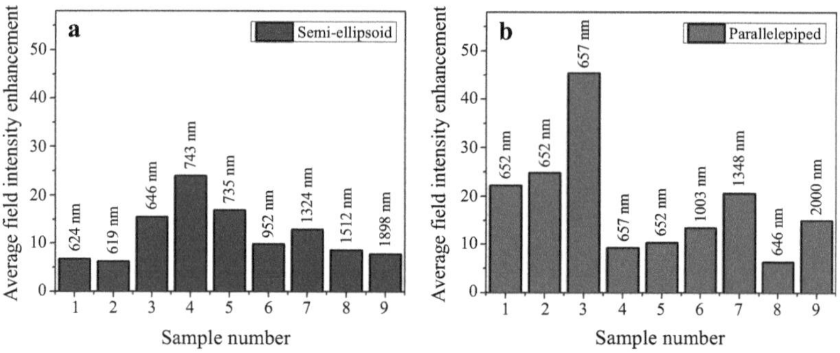

Figure 3.23 Calculated values of average electric field intensity enhancement within a 25 nm vicinity of differently shaped Au nanoparticles. Simulation results correspond to (a) semi-ellipsoid- and (b) parallelepiped-shaped Au nanoparticles on glass substrates. Effective geometrical parameters of Au nanoparticles used in FDTD modeling are the same for semi-ellipsoid and parallelepiped shapes and correspond to the data presented in the inset of Fig. 3.15 for samples R1 to R9, respectively. Labels above the bars indicate the light wavelength, which the maximal electric field intensity enhancement for specific nanoparticle geometry was obtained at. Reprinted from Ref. [154] under a Creative Commons Attribution 4.0 International License. Figure caption was adapted.

3.4 Geometry Factor of Plasmon-Inducing Metal Surface and Its Use in SEIRA Studies

3.4.1 Effect of Plasmonic Enhancement of Infrared Transitions Near the Metal Surface and Its Experimental Application

The enhancement of optical processes of the order of up to 10^2 near the rough metal surface (Au, Ag, Cu, etc.) has been known for the last 30 years for both optical transitions in adsorbed molecules (Raman scattering, luminescence, infrared (IR) absorption) [162–164] and for processes that do not depend on the presence of molecules on the metal surface (for example, second harmonic generation) [165]. The effect of amplification occurs on high-conductive surfaces in the case when the light wavelength far exceeds the value of the radius of curvature at the peaks of surface roughness or the value of the radius of the metal nanoparticle. The phenomenon of optical processes enhancement consists in a significant increase in the intensity of the optical transition (for example, the effective cross section of the Raman scattering increases by 10^2, for IR absorption by $10–10^2$ times) or the efficiency of the process near the metal surface. The explanation of the effect is not straightforward and includes several mechanisms, such as an increase in the intensity of the electromagnetic field near the rough metal surface, island metal films or nanoparticles, the growth of the transition dipole moment of adsorbed molecules, etc. According to Ref. [165], the coefficient of amplification of an electric field is a function of the dielectric constant of a metal. However, it should be noted that the influence of metal under certain conditions and characteristics of the surface and adsorbed particles may, in addition to enhancement, also result in the quenching of optical transitions of molecules near the metal surface [166]. The theoretical interpretation of the amplification effect is given in Ref. [165] in general terms and includes two possible ways: the amplification of the electromagnetic field due to interaction with localized (surface) plasmon oscillations arising on inhomogeneities of the metal surface and the specific growth of the transition dipole moments of the adsorbed molecules on the metal surface.

The electric field generated by a plane electromagnetic wave at the point $\boldsymbol{r}$ of the space near the metal particle can be written as:

$$\boldsymbol{E}(\boldsymbol{r},t) = \exp(-i\omega t)\,[\boldsymbol{E}_0(\omega) + \boldsymbol{E}_1(\boldsymbol{r},\omega)], \tag{3.1}$$

where ω is the angular frequency and t is time. The additive $\boldsymbol{E}_1(\boldsymbol{r},\omega)$ is associated with the excitation of the localized surface plasmon oscillations and has a local character, which is asymptotically fading, then:

$$\boldsymbol{E}_\alpha(\boldsymbol{r},\omega) = g_{\alpha\beta}(\boldsymbol{r},\omega)\boldsymbol{E}_{0\beta}(\omega), \tag{3.2}$$

where $g_{\alpha\beta}(\boldsymbol{r},\omega)$ is the coefficient of the electrical field enhancement.

The effective cross section of the light interaction with a molecule adsorbed on the metal surface can be represented as follows:

$$\sigma_\alpha(r,\omega) = \frac{16\pi^2\omega_{fi}}{cE_0^2}\left|\langle f|H(r,\omega)|i\rangle\right|^2 \delta\left[h\left(\omega-\omega_{fi}\right)\right]$$

$$\sim \sigma_\alpha^{(0)}\left|g(r,\omega)\right|^2\left|h(r)\right|^2, \tag{3.3}$$

where $\sigma_{\alpha}^{(0)}$ is the cross section without metal ($d = d_0$, $E = E_0$), $i \to f$ is the optical transition from state i (with energy E_i) to state f, $cE_0^2/8\pi$ is the incident energy flow density, c is the speed of light, $\langle f|H(r,\omega)|i\rangle$ is the Hamiltonian matrix element of molecule interactions with field, $h(\boldsymbol{r})$ is the dipole moment amplification coefficient of the adsorbed molecule.

In the general case, the amplification of the optical transition is proportional to the field intensity. The calculation of the ratio of the real and imaginary parts of the complex dielectric constant at the corresponding wavelength is also used to determine the magnitude of the surface enhancement [165].

Upon irradiation of the rough metal surface, energy from the photon is transmitted to localized surface plasmon oscillations. Plasmonic oscillations can be in the state of resonances, the frequency of which is determined by the geometric parameters of surface roughness. In this case, energy from plasmonic oscillations is transmitted to the adsorbed molecules, which causes increased light absorption by molecules.

3.4.2 Enhancement Efficiency for Various Experimental Implementations of the SEIRA Technique

There are several experimental implementations of the surface-enhanced infrared absorption method. The main one is the SEIRA technique in reflection geometry, when the amplification takes place due to the increase in the electric field intensity when the localized surface plasmons are excited on the roughness of the metal film. A series of studies [17, 21, 167] connected with the proper use of the amplification effect by the metal surface of infrared absorption of nucleic acids and lipids deposited on the gold substrate were performed in experiments with reflection geometry. Glass plates with a deposited gold layer having thickness of 200–500 Å served as SEIRA substrates (Fig. 3.24). The roughness of the surface of the manufactured samples was different, in the range of 20–100 Å.

Infrared absorption spectra were recorded on the Fourier spectrometer IFS-48 (Bruker, Germany) in the region of 380–5300 cm^{-1} (Fig. 3.25). Spectra processing and the decomposition of complex bands on components were performed using the Opus-2.2 software.

DNA samples were deposited onto the surface of a gold film from an aqueous solution with a concentration of about 1 mg/ml and lyophilized.

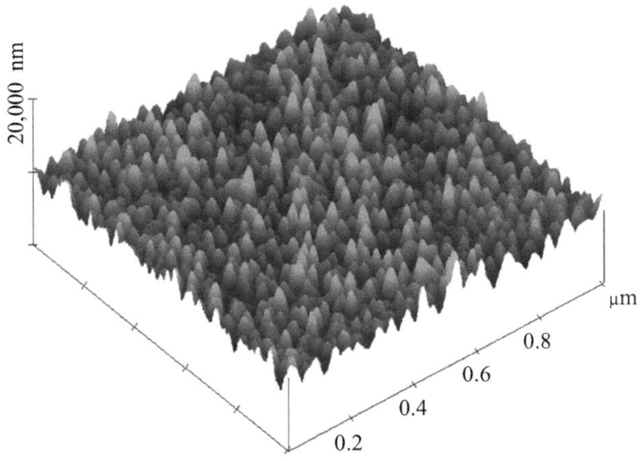

Figure 3.24 AFM 3D image of the gold surface used in SIERA studies.

The obtained results show that due to the use of the SEIRA method for the study of infrared absorption of nucleic acids (analogous to lipids [168]), there is a different amplification of signal magnitude (depending on the geometry of the surface roughness of gold) compared to the spectrum on the neutral CaF_2 substrate (Fig. 3.25). In this case, there was no deformation of the absorption bands contour present in experiments with commonly used CaF_2 substrates. It should be noted that the presented SEIRA technique allows the registration of spectral features of molecules that are not available in usual IR spectroscopy due to the possibility of observing new absorption bands that appear in the enhanced SEIRA spectrum [17].

Studies have shown that signal enhancement in most cases increases with the presence of the characteristic roughness of the gold film surface in the range of 50–100 Å. The IR signal increases by a factor of 3–5 for different vibration frequencies (Fig. 3.25, Table 3.1). The greatest amplification was obtained for the thickness of the gold film of 200 Å, which is probably due to the greater correspondence of the geometry of the relief for film of such a thickness to the 3D geometry of DNA molecule compared with the geometry of the relief for gold films with other thickness. To calculate the possible magnitude of the amplification by a metal surface, the calculation of the ratio of the real and imaginary parts of the complex dielectric permittivity of the gold substrate at the corresponding frequency was used. For example, in the case when the calculated value for this ratio was 16 ($\varepsilon' = 566$, $\varepsilon'' = 35.7$ at 2800 cm^{-1}), in the experiment this value reached only 4.2, indicating the need for a more detailed study of amplification mechanisms due to localized surface plasmons in this case. Comparison of the obtained SEIRA spectra with the ones measured on the CaF_2 substrate shows, in most cases, the coincidence of the wave positions of the bands (Table 3.1). Some frequency shifts are present, but for most bands they are insignificant (1–2 cm^{-1}). This allows concluding that this method can be used for such tasks. The amplification coefficients of integral intensity of absorption bands, which correspond to the vibrations of different functional groups of DNA molecules, were obtained in the experiment, and their value was 3 to 5 depending on the thickness of the deposited gold layer.

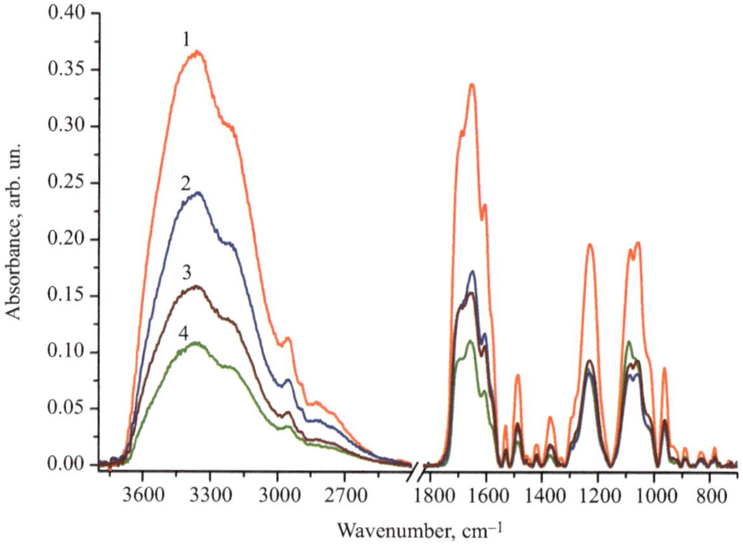

Figure 3.25 SEIRA spectra of DNA: 1 — on a glass substrate with a 200 Å gold film thickness, 2 — 300 Å gold film thickness, 3 — 400 Å gold film thickness, 4 — on a CaF_2 substrate. Adapted with permission from Ref. [17], Copyright 2002, John Wiley and Sons.

Table 3.1 SEIRA enhancement factors of absorption bands integral intensities for different DNA functional groups

DNA on CaF_2	DNA on gold film with thickness 200 Å				DNA on gold film with thickness 300 Å				
		Enhancement factor				Enhancement factor			
			Calc.				Calc.		
Band position, cm^{-1}	Band position, cm^{-1}	Exp. g^2	g	g^2	Band position, cm^{-1}	Exp. g^2	g	g^2	Band assignment
3350	3359	4.3	6.9	47.61	3354	2.0	6.9	47.61	O–H, N–H, C–H
1655	1653	2.9	3.8	14.44	1649	1.4	3.7	13.69	C=O, C=N, N–H
1528	1530	3.0	3.5	12.25	1529	1.1	3.5	12.25	C=N
1419	1418	2.8	3.3	10.89	1418	1.0	3.3	10.89	C–H
1238	1230	2.4	3.0	9.00	1230	1.0	2.9	8.41	PO_2^-
1089	1084	2.1	2.7	7.29	1084	0.8	2.7	7.29	PO_2^-
964	963	2.4	2.4	5.76	964	0.9	2.4	5.76	C–C, C–O
892	889	1.7	2.3	5.29	892	0.9	2.3	5.29	C–C, C–O
833	831	2.0	2.2	4.84	829	0.8	2.2	4.84	C_2

Source: Reproduced with permission from Ref. [17], Copyright 2002, Wiley Periodicals, Inc.

Experimental and estimated amplification coefficients of the absorption bands integral intensity of different DNA functional groups are given in Table 3.1. Experimentally, the amplification coefficients are determined by the ratio of the absorption bands integral intensities of the corresponding DNA functional groups.

The following method implements an enhancement in the attenuated total reflection (ATR) geometry. There is a possibility of resonance amplification of the electric field near the flat surface of the metal by nonradiative surface plasma waves (surface plasmon-polaritons), which are excited by the ATR method. Consideration of the electric field amplification mechanism by the surface plasmon-polariton shows substantially smaller values than the amplification of the electric field by localized plasmon oscillations [162]. Theoretically, the amplification magnitude of an electric field $g(\omega_r) \sim \omega_p/\gamma \sim |\varepsilon'|/\varepsilon''$ at resonance is characterized by competition between the processes of excitation and decay of surface electromagnetic waves (SEW) or modes.

Therefore, the pumping level of SEW (in a given external field E_0) is limited only by the processes of their decay. As shown in Ref. [169], when describing the SEW based on the dielectric function of a metal, the parameter $1/\tau$ describes the complete decay rate of the SEW:

$$\frac{1}{\tau} = \frac{1}{\tau_d} + \frac{1}{\tau_s} + \frac{1}{\tau_r}. \quad (3.4)$$

Here, the following channels of SEW decay are allocated:

- Dissipative (relaxation time τ_d), associated with the transition of SEW energy into heat;
- Radiation (decay into photons per time τ_r);
- Surface (scattering of SEW into other SEW per time τ_s).

According to this, it should be noted that the observed efficiency of the electric field amplification by localized surface plasmon oscillations and surface plasmon electromagnetic waves may be associated with different effects of roughness in relation to the mentioned surface plasmon processes. In fact, when the surface has roughness with a scale smaller than the wavelength of light, the probability of formation of localized surface plasmon oscillations increases, but the lifetime and electric field enhancement by surface electromagnetic waves sharply decreases, determined by the processes of their radiation decay and scattering on surface roughness [165].

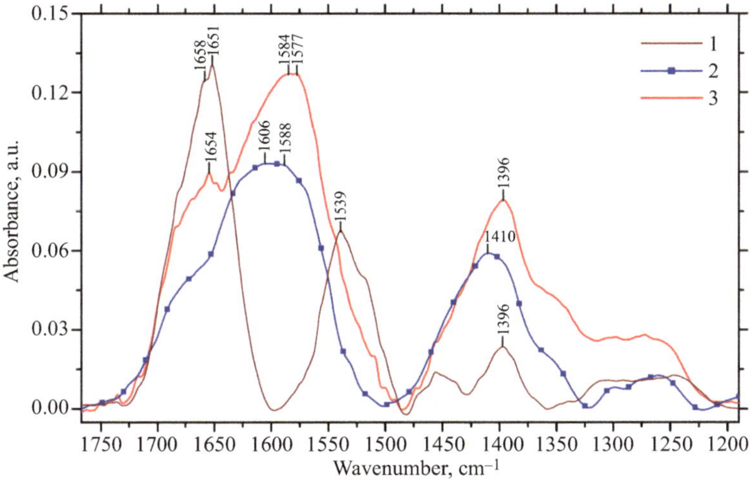

Figure 3.26 SEIRA spectra of (1) DNA, (2) colloidal gold, and (3) DNA-colloidal gold solution on SPR chip. Adapted by permission from Ref. [21], Copyright 2004, Springer Nature.

In the following approach, a study was carried out to investigate the effect of amplification on colloidal gold nanoparticles placed on the surface of the SPR chip. Studies of the spectral features of the condensed state of DNA upon interaction with colloidal gold were conducted using IR spectroscopy. Colloidal gold was added to aqueous DNA solution, which was placed on the surface of the SPR chip. For control, a DNA sample without colloidal gold was used. As a result of the experiment, changes in the spectrum of DNA with colloidal gold were recorded in comparison with spectra without nanoparticles (Fig. 3.26). There was a decrease in the intensity of the bands at certain frequencies, an increase in the half-width of some bands, and other features that are not specific to any of the A, B, Z forms of DNA. These results indicate that colloidal gold probably modified the DNA structure with its new state creation. In fact, the use of arrays of high-conducting structures in the SEIRA method, together with colloidal nanoparticles, requires studies on the effect of the substrate on the processes of amplification of optical transitions.

This makes a comprehensive analysis of the enhancement processes of optical transitions based on simpler models an actual task.

3.5 Plasmonic Nanochips Applications for Surface-Enhanced Fluorescence Studies

3.5.1 Surface-Regulated Fluorometry Using Plasmonic Nanochips: Theoretical Background

The structure of real systems that amplify optical processes is so complex that there is no general model for them that describes the electrodynamic phenomena. However, it is important to take into account the expansion of practical applications of plasmonic nanochips, that is, structures with arrays of high-conductive nanostructures on the surface of the dielectric substrate. In this subsection, the mutual influence of several factors in the system of "high conductive spherical nanoparticle–solid surface" on the local electromagnetic field enhancement near the metal nanostructure and, consequently, on the emission of the organic dye molecules near its surface, was considered for the first time. These factors include the dielectric permittivity of the substrate, the size of the nanostructure and the distance from the dielectric surface. To simulate the relevant parameters, the dipole approximation method was applied using dyadic Green's function for spherical gold nanoparticles with a radius of 20–80 nm at an external excitation wavelength of 650 nm. Calculations of volumetric Lippmann–Schwinger equations showed that in the absence of substrate, the maximum fluorescence excitation rate enhancement near the surface of the spherical gold nanoparticle with a radius of 20 nm in vacuum is 12.54. In the presence of the dielectric substrate, the fluorescence excitation rate always decreases with increase in the value of its dielectric permittivity and equals to 5.59 in the case of glass substrate ($\varepsilon = 2.59$).

3.5.1.1 Simulation of fluorescence rate enhancement near nanostructures

The progress in modern nanotechnology, particularly nanoplasmonics, has initiated the development of a new class of optical sensor devices having sensing elements in the form of a nanochip with an array of high-conductive nanostructures, which is based on a localized surface plasmon resonance phenomenon. Such nanochip presents an optically transparent substrate, which the nanostructured surface of the noble metal, e.g., silver or gold, is somehow created on. If fluorophore molecules are placed at such a surface, then under certain conditions their emission intensity enhancement can be obtained compared to the case in the absence of nanostructures [15, 170, 171]. Sensors of this type make it possible to increase the sensitivity of measurements (e.g., detect the signal even from single molecules [172, 173]), and they are promising for applications in biochemistry and medicine.

To better understand the processes that affect the fluorescence intensity, one must examine in detail the interaction between the fluorophore molecule and the metal nanoparticle. The main characteristic of this interaction is the magnitude of the efficiency of energy exchange between the donor (fluorophore) and acceptor (nanoparticle). The energy exchange between the metal nanoparticle and fluorophore may occur involving the radiative and nonradiative processes that affect the enhancement and quenching of fluorescence. Which processes will dominate depend on the characteristics of the system, such as the distance between donor and acceptor, the mutual orientation of their dipole moments, the fluorophore quantum yield, the size and shape of metal nanoparticles. The possibility to change these characteristics allows regulating fluorescence response of fluorophore molecules, adsorbed on the surface of metal nanostructures, and obtaining the enhancement or quenching of their radiation intensity.

Attention should be paid to the fact that during the nanochip fabrication, metal nanoparticles in most cases are located on the surface of the dielectric substrate, which also makes its impact on the enhancement of the local electromagnetic field near metallic nanostructures and the enhancement of emission intensity of fluorescent molecules that are situated in this field. It should be noted that in the literature on the subject, theoretical approaches are considered that do not account for the presence of a dielectric substrate, which the nanostructures are situated on [174–177]. Therefore, in this study the influence of the substrate dielectric permittivity on the behavior of the local electromagnetic field intensity near the surface of spherical gold nanoparticle was theoretically demonstrated and the effect of the presence of a substrate on the quantum yield and emission of fluorophore molecules, located at the surface of such nanostructure, was determined.

Under the influence of external excitation, the metal nanoparticle in certain conditions induces the local electromagnetic field, which has a much higher strength than the field of external excitation. If a fluorophore molecule is placed at a certain distance from nanostructure, enhancement or quenching of molecule fluorescence can be observed. In order to mathematically simulate this problem, it is rational to divide it into two parts.

The first part of the problem is to find the local field strength near the spherical gold nanoparticle on a dielectric substrate induced by external radiation, and hence the fluorophore molecule excitation rate. The second part involves finding the quantum yield and fluorescence rate of a fluorophore molecule.

Analysis of the electromagnetic properties of nano-objects is closely related to the solution to Maxwell's equations [174]:

$$\begin{aligned} \nabla\times\mathbf{E}(\mathbf{r}) &= i\omega\mathbf{B}(\mathbf{r}), \\ \nabla\times\mathbf{H}(\mathbf{r}) &= -i\omega\mathbf{D}(\mathbf{r})+\mathbf{j}(\mathbf{r}), \\ \nabla\mathbf{D}(\mathbf{r}) &= \rho(\mathbf{r}), \\ \nabla\mathbf{B}(\mathbf{r}) &= 0, \end{aligned} \tag{3.5}$$

where

$$\mathbf{D} = \varepsilon\varepsilon_0\mathbf{E},$$

$$\mathbf{B} = \mu\mu_0\mathbf{H}.$$

The solution to the equations will be electric and magnetic fields as functions of the coordinates of a given distribution of currents **j** that create these fields. In nanophysics, unknown parameters of the electromagnetic field can be found using the analytical Green's function method [175]. This method reduces to integration over the volume, which the currents are distributed in. For convenience, we write the fields **E** and **H** in terms of the vector potential **A** and scalar potential φ [174]:

$$\begin{aligned} \mathbf{E}(\mathbf{r}) &= i\omega\mathbf{A}(\mathbf{r})-\nabla\varphi(\mathbf{r}), \\ \mathbf{H}(\mathbf{r}) &= \frac{1}{\mu\mu_0}\nabla\times\mathbf{A}(\mathbf{r}). \end{aligned} \tag{3.6}$$

Applying the Lorentz gauge to **A**

$$\nabla \mathbf{A}(\mathbf{r}) = i\omega\mu\mu_0\varepsilon\varepsilon_0\varphi(\mathbf{r})$$

and taking into account

$$\nabla \times \nabla \times = -\nabla^2 + \nabla\nabla,$$

we obtain inhomogeneous Helmholtz equations for the vector and scalar potentials, if we substitute these equations into the second equation of (3.5):

$$[\nabla^2 + k^2]\mathbf{A}(\mathbf{r}) = -\mu\mu_0\mathbf{j}(\mathbf{r}),$$
$$[\nabla^2 + k^2]\varphi(\mathbf{r}) = -\frac{\rho(\mathbf{r})}{\varepsilon\varepsilon_0}. \qquad (3.7)$$

The solution to Eqs. (3.7) consists of a sum of homogeneous and particular solutions.

In many works [174–176], the dyadic Green's function defined by the environment is applied to calculate the electromagnetic field, which is created by a source of currents. The electric field **E** at a point **r**, which is created by a point dipole $\boldsymbol{\mu}$ radiation at a point **r**′ (Fig. 3.27), will be determined by the following equation:

$$\mathbf{E}(\mathbf{r}) = \omega^2\mu\mu_0\overleftrightarrow{\mathbf{G}}(\mathbf{r},\mathbf{r}')\boldsymbol{\mu}. \qquad (3.8)$$

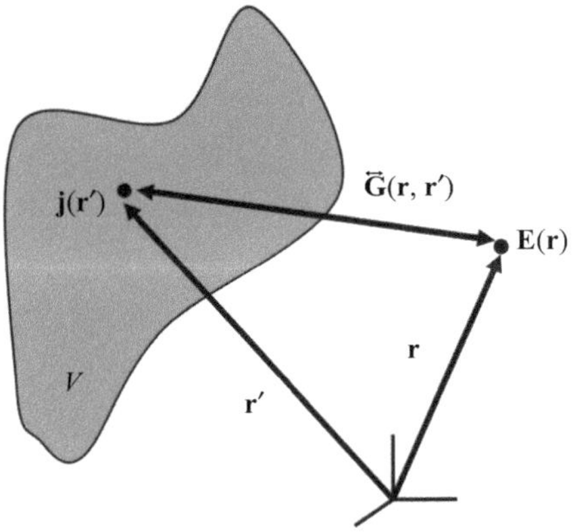

Figure 3.27 Illustration for the determination of the electric field E at a point **r**, generated by the radiation of a current source **j** at a point **r**′. Reprinted with permission from Ref. [178], Copyright 2015, Springer Nature.

For a point current source, Green's function is given by the inhomogeneous equation [174]

$$L\overleftrightarrow{\mathbf{G}}(\mathbf{r},\mathbf{r}') = \overleftrightarrow{\mathbf{I}}\delta(\mathbf{r}-\mathbf{r}'), \tag{3.9}$$

where L is an arbitrary linear operator, $\overleftrightarrow{\mathbf{I}}$ is the unit dyad, and $\delta(\mathbf{r} - \mathbf{r}')$ is the delta function. In deriving the scalar Green's function for the Helmholtz equation, we replace the current source in Eq. (3.7) by the point current source, which is defined by the delta function, and get the following equation:

$$[\nabla^2 + k^2]G_0(\mathbf{r},\mathbf{r}') = -\delta(\mathbf{r}-\mathbf{r}'). \tag{3.10}$$

If a definition for the scalar Green's function is found, one can get a partial solution to Eq. (3.7). For vacuum, the only solution to Eq. (3.10) is

$$G_0(\mathbf{r},\mathbf{r}') = \frac{e^{-ik|\mathbf{r}-\mathbf{r}'|}}{4\pi|\mathbf{r}-\mathbf{r}'|}. \tag{3.11}$$

To determine the dyadic Green's function, we consider the wave equation for the electric field in a homogeneous medium:

$$\nabla\times\nabla\times\mathbf{E}(\mathbf{r}) - k^2\mathbf{E}(\mathbf{r}) = i\omega\mu_0\mu\mathbf{j}(\mathbf{r}). \tag{3.12}$$

Taking into account Eq. (3.8), we obtain

$$\nabla\times\nabla\times\overleftrightarrow{\mathbf{G}}(\mathbf{r},\mathbf{r}') - k^2\overleftrightarrow{\mathbf{G}}(\mathbf{r},\mathbf{r}') = \overleftrightarrow{\mathbf{I}}\delta(\mathbf{r}-\mathbf{r}'). \tag{3.13}$$

Within this approach, the general expressions for the strength of the electromagnetic fields generated by arbitrary-shaped nanoscale objects when irradiated with electromagnetic waves are called volumetric Lippmann–Schwinger equations and are as follows:

$$\begin{aligned} \mathbf{E}(\mathbf{r}) &= \mathbf{E}_0(\mathbf{r}) + i\omega\mu_0\mu \int_V \overleftrightarrow{\mathbf{G}}(\mathbf{r},\mathbf{r}')\mathbf{j}(\mathbf{r}')dV', \mathbf{r}\notin V, \\ \mathbf{H}(\mathbf{r}) &= \mathbf{H}_0(\mathbf{r}) + \int_V [\nabla\times\overleftrightarrow{\mathbf{G}}(\mathbf{r},\mathbf{r}')]\mathbf{j}(\mathbf{r}')dV', \mathbf{r}\notin V, \end{aligned} \tag{3.14}$$

where $\mathbf{E}(\mathbf{r})$ and $\mathbf{H}(\mathbf{r})$ are electric and magnetic fields at a point $\mathbf{r}$, respectively, which are generated by a current source $\mathbf{j}$ at a point $\mathbf{r}'$ (Fig. 3.27), V is a nanostructure volume.

The local field $\mathbf{E}$ can be found from the Lippmann–Schwinger equation (3.14). It is known that the local electromagnetic field enhancement observed near the metallic nanoscale objects of given shape and size occurs due to the surface plasmon resonance phenomenon. Localized surface plasmons can exist only at fixed frequencies for specific conditions, which are called plasmon resonance frequencies. Nanoparticles with defined shapes and sizes have specific resonance frequency values, which the plasmon resonance line is observed at. For the theoretical determination of the plasmon resonance frequency and distribution of the corresponding electromagnetic fields inside and outside the nanoparticle, the method of effective susceptibility [179] is used as one of the main approaches in recent years. For convenience, this method does not operate with the resonant frequencies, but uses the corresponding dielectric permittivities. In this method, the main characteristic of the nanoparticle is its polarizability, which contains information about the dielectric constant of the nanoparticle material and depends on the environment. To calculate the local field strength in the case of spherical nanoparticles, the quasi-static polarizability is used, which is determined by the formula [174]:

$$\overset{\leftrightarrow}{\boldsymbol{\alpha}} = \frac{4\pi\varepsilon_0\varepsilon_1 R^3(\varepsilon - \varepsilon_1)}{\varepsilon + 2\varepsilon_1}\overset{\leftrightarrow}{\mathbf{I}}, \tag{3.15}$$

where R is a nanoparticle radius, ε is the relative dielectric permittivity of the nanoparticle material, ε_1 is the relative dielectric permittivity of the environment, and $\mathbf{I}$ is the unit dyad. Plasmon resonance in this case is observed when approaching the condition $\varepsilon = -2\varepsilon_1$.

The presence of the substrate will affect the polarizability of the nanosphere. According to Ref. [180], the polarizability of a nanoparticle on a dielectric surface can be calculated by the formula

$$\overset{\leftrightarrow}{\boldsymbol{\alpha}}(\varphi) = \left(1 - \frac{\varepsilon_2 - \varepsilon_1}{\varepsilon_1 + \varepsilon_2}\right)\overset{\leftrightarrow}{\boldsymbol{\alpha}}\left[\overset{\leftrightarrow}{\mathbf{I}} - \frac{\overset{\leftrightarrow}{\boldsymbol{\alpha}}\dfrac{\varepsilon_2 - \varepsilon_1}{\varepsilon_1 + \varepsilon_2}}{32\pi\varepsilon_0\varepsilon_1(R+\varphi)^3}\right]^{-1}, \tag{3.16}$$

where ε_2 is the relative dielectric permittivity of the substrate, φ is the surface-to-surface distance between the nanoparticle and the substrate.

So if the polarizability of high-conductive nanoparticle is known and when its size is much smaller than the light wavelength, replacing the spherical nanoparticle with a point dipole with dipole moment

$\alpha(\varphi)\mathbf{E}_0(\mathbf{r}')$ leads to the electric field strength generated by the dipole radiation, which can be calculated by the formula [177, 179]

$$\mathbf{E}(\mathbf{r}) = \frac{k^2}{\varepsilon_0}\overleftrightarrow{\mathbf{G}}(\mathbf{r},\mathbf{r}')\overleftrightarrow{\alpha}(\varphi)\mathbf{E}_0(\mathbf{r}'), \tag{3.17}$$

where, in accordance with Ref. [174], Green's function for the electromagnetic field at a given point will look like

$$\overleftrightarrow{\mathbf{G}}(\mathbf{r},\mathbf{r}') = \left(\overleftrightarrow{\mathbf{I}} + \frac{1}{k^2}\nabla\otimes\nabla\right)G_0(\mathbf{r},\mathbf{r}'). \tag{3.18}$$

Taking into account Eq. (3.11), Eq. (3.18) transforms as follows:

$$\overleftrightarrow{\mathbf{G}}(\mathbf{r},\mathbf{r}') = \frac{e^{ikW}}{4\pi W}\left[\left(1+\frac{ikW-1}{k^2W^2}\right)\overleftrightarrow{\mathbf{I}} + \frac{3-3ikW-k^2W^2}{k^2W^2}\frac{\mathbf{W}\otimes\mathbf{W}}{W^2}\right], \tag{3.19}$$

where $\qquad \mathbf{W} = \mathbf{r} - \mathbf{r}', \quad W = |\mathbf{r} - \mathbf{r}'|.$

In the case of near field, when $W << \lambda$, Eq. (3.19) is dominated by near-field components. In this approximation, Green's function can be found by the following formula [174]:

$$\overleftrightarrow{\mathbf{G}}_{\mathrm{NF}}(\mathbf{r},\mathbf{r}') = \frac{1}{4\pi W}\frac{1}{k^2W^2}\left(-\overleftrightarrow{\mathbf{I}} + 3\frac{\mathbf{W}\otimes\mathbf{W}}{W^2}\right). \tag{3.20}$$

Changes in the studied system configuration due to the nanosphere being placed on a dielectric substrate influence the distribution of electromagnetic field. Green's function in the presence of heterogeneity (in this case, the dielectric substrate) consists of two parts:

$$\overleftrightarrow{\mathbf{G}}(\mathbf{r},\mathbf{r}') = \overleftrightarrow{\mathbf{G}}^{\mathrm{D}}(\mathbf{r},\mathbf{r}') + \overleftrightarrow{\mathbf{G}}^{\mathrm{I}}(\mathbf{r},\mathbf{r}'), \tag{3.21}$$

where $\overleftrightarrow{\mathbf{G}}^{\mathrm{D}}(\mathbf{r},\mathbf{r}')$ is the direct part, where the presence of field heterogeneities is ignored, $\overleftrightarrow{\mathbf{G}}^{\mathrm{I}}(\mathbf{r},\mathbf{r}')$ is the indirect part, which takes into account the existing field heterogeneities and the boundaries between them. When the field source and the point, which the field is observed at, are located in the same half-space with dielectric permittivity ε_1, and the inhomogeneity dielectric permittivity is ε_2, the indirect part in Eq. (3.21) will be as follows:

$$\overleftrightarrow{\mathbf{G}}^{\mathrm{I}}(\mathbf{r},\mathbf{r}') = \frac{\varepsilon_1-\varepsilon_2}{\varepsilon_1+\varepsilon_2}\frac{1}{4\pi k^2W^3}\left(-\overleftrightarrow{\mathbf{I}} + 3\frac{\mathbf{W}\otimes\mathbf{W}}{W^2}\right), \tag{3.22}$$

where $\mathbf{r}'$ is the point, where the field source is situated:

$$\mathbf{r}' = \sqrt{(x-x')^2 + (y-y')^2 + (z+z')^2}.$$

Now the total electric field strength near the nanoparticle can be expressed as follows:

$$\mathbf{E}(\mathbf{r}) = \mathbf{E}_0(\mathbf{r}) + \frac{k^2}{\varepsilon_0} \overleftrightarrow{\mathbf{G}}(\mathbf{r},\mathbf{r}') \overleftrightarrow{\alpha}(\varphi) \mathbf{E}_0(\mathbf{r}'). \tag{3.23}$$

The second part of the problem involves finding the quantum yield and the fluorescence rate of a dye molecule and shows the substrate effect for these values. For this we consider a single molecule that emits light under the influence of external excitation. Let this molecule be located at a point $\mathbf{r}_\mathrm{m}$ and have a dipole moment $\mathbf{p}$ and transition frequency ω (Fig. 3.28). The center of the spherical nanoparticle with a radius of R is at a point $\mathbf{r}_\mathrm{p}$. In the case of weak excitation, molecule fluorescence rate γ_em depends on the molecule excitation rate γ_exc and quantum yield q [177]. Let us consider the excitation rate in more detail. The excitation rate is proportional to the local field strength and dipole moment $\mathbf{p}$ [177]:

$$\gamma_\mathrm{exc} \infty \left|\mathbf{p}\cdot\mathbf{E}\right|^2. \tag{3.24}$$

In the absence of nanoparticle, external radiation $\mathbf{E}_0$ excites the molecule with an excitation rate γ_exc^0. As mentioned earlier, the presence of nanoparticle leads to the occurrence of an additional local electromagnetic field. To assess the impact of the local electromagnetic field intensity near the surface of the metal nanoparticle on the molecule fluorescence rate, we will operate the normalized excitation rate [177]

$$\frac{\gamma_\mathrm{exc}}{\gamma_\mathrm{exc}^0} = \left|\frac{[\mathbf{n}_\mathbf{p}\cdot\mathbf{E}(\mathbf{r}_\mathbf{m})]}{[\mathbf{n}_\mathbf{p}\cdot\mathbf{E}_0(\mathbf{r}_\mathbf{m})]}\right|^2, \tag{3.25}$$

where $\mathbf{n}_\mathrm{p}$ is a unit vector pointing in the direction of $\mathbf{p}$.

Knowing the electric field strength near the nanoparticle surface (3.23), we have everything needed to calculate the normalized excitation rate of a dye molecule (3.25), which is located near the surface of a spherical gold nanoparticle. Taking into account Eqs. (3.15, 3.16, 3.20–3.23), we can write an explicit expression for the normalized excitation rate:

$$\frac{\gamma_{\text{exc}}}{\gamma_{\text{exc}}^0} = \left| 1 + \frac{k^2}{\varepsilon_0} \overset{\leftrightarrow}{\mathbf{G}}(\mathbf{r_m}, \mathbf{r_p}) \overset{\leftrightarrow}{\alpha}(\varphi) \right|^2. \tag{3.26}$$

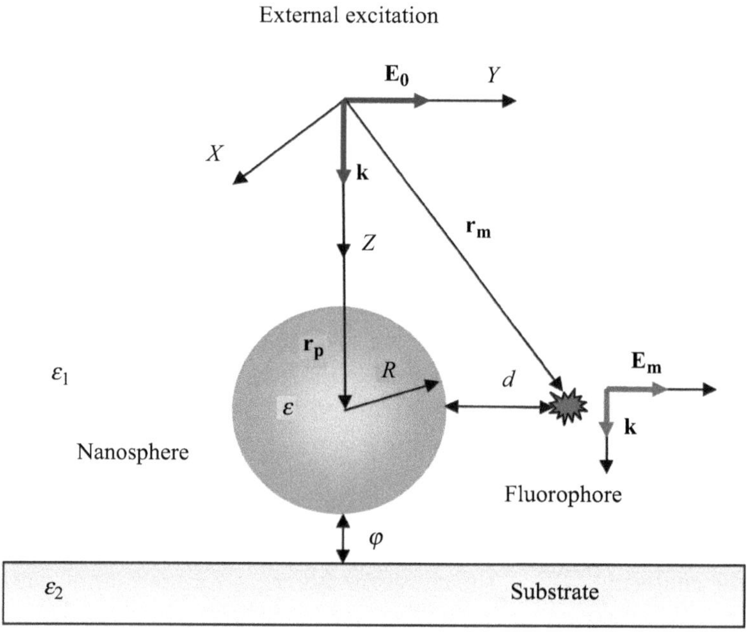

Figure 3.28 Schematic drawing of the system under investigation. Spherical gold nanoparticle with a radius *R* is situated at a certain distance ϕ from the glass dielectric substrate. Fluorophore molecule is located at a distance *d* from the surface of the nanoparticle. Reprinted with permission from Ref. [178], Copyright 2015, Springer Nature.

The fluorescence signal is proportional to the emission probability of molecule (fluorescence rate) γ_{em}:

$$\gamma_{\text{em}} = \gamma_{\text{exc}} \cdot q. \tag{3.27}$$

If γ_{r} is the radiative decay rate and γ is the total decay rate, then the quantum yield is

$$q = \gamma_{\text{r}}/\gamma. \tag{3.28}$$

Then the normalized molecule fluorescence rate near a spherical gold nanoparticle can be written in the following form [177]:

$$\frac{\gamma_{\text{em}}}{\gamma_{\text{em}}^{0}}=\frac{\gamma_{\text{exc}}}{\gamma_{\text{exc}}^{0}}\frac{q}{q^{0}}. \tag{3.29}$$

Here, variables with superscript "0" correspond to the molecule in the absence of nanoparticle and without superscript in its presence. Let us denote the total decay rate of the molecule in the absence of nanoparticle as

$$\gamma^{0}=\gamma_{\text{r}}^{0}+\gamma_{\text{nr}}^{0}, \tag{3.30}$$

where γ_{r}^{0} and γ_{nr}^{0} are the molecule radiative and nonradiative decay rates in the absence of nanoparticle, respectively. The presence of nanoparticle introduces additional losses into the system, such as ohmic losses in the nanoparticle γ_{abs} [177]. Then, in the presence of a nanosphere, the quantum yield of the molecule can be written in the modified form:

$$\begin{gathered}\gamma=\gamma_{\text{r}}+\gamma_{\text{nr}}+\gamma_{\text{abs}},\\ q=\frac{\gamma_{\text{r}}}{\gamma_{\text{r}}+\gamma_{\text{nr}}+\gamma_{\text{abs}}}.\end{gathered} \tag{3.31}$$

If $\gamma_{\text{nr}}^{0}=\gamma_{\text{nr}}$, then the expression for the quantum yield can be presented as

$$q=\frac{\gamma_{\text{r}}/\gamma_{\text{r}}^{0}}{\gamma_{\text{r}}/\gamma_{\text{r}}^{0}+\gamma_{\text{abs}}/\gamma_{\text{r}}^{0}+(1-q^{0})/q^{0}}. \tag{3.32}$$

The rate of energy transfer from the molecule to the nanoparticle γ_{abs} can be calculated via the molecule emission power P_{abs} in the presence of the nanoparticle [174, 177]:

$$\frac{\gamma_{\text{abs}}}{\gamma_{\text{r}}^{0}}=\frac{P_{\text{abs}}}{P^{0}}. \tag{3.33}$$

In general, the emission power of the classical dipole in free space

$$P^{0}=\frac{\omega^{4}\mathbf{p}^{2}}{12\pi\varepsilon_{0}c^{3}},$$

and, according to [174, 177]

$$P_{\text{abs}}=\frac{1}{2}\int_{V}\text{Re}(\mathbf{j}\times\mathbf{E}^{*})d\mathbf{r}^{3}, \tag{3.34}$$

where the integration is performed over the volume of a nanoparticle, **j** is the current density, which is created by the molecule field in the volume of a nanoparticle, and **E** is the electric field strength.

Further, given that γ_r is proportional to the emission power of the classical dipole, we can find the normalized radiative decay rate [174]:

$$\frac{\gamma_r}{\gamma_r^0} = \left| \mathbf{n_p} + \ddot{\alpha}(\varphi)\frac{k^2}{\varepsilon_0} G_0(\mathbf{r_p}, \mathbf{r_m})\mathbf{n_p} \right|^2 . \tag{3.35}$$

Equations (3.29, 3.32–3.34) allow calculating the molecule quantum yield and fluorescence rate near the surface of a spherical metal nanoparticle. The presence of the substrate is taken into account similar to the approach used to determine the field strength in the vicinity of the nanoparticle. That is, Green's function contains the term that depends on the optical characteristics of the substrate, and polarizability of a nanosphere on a substrate is calculated by Eq. (3.16).

3.5.1.2 Influence of dielectric substrate on the plasmon-assisted excitation and fluorescence rates of fluorophore molecule

Simulation shows a significant influence of the dielectric substrate on the local electric field strength in the vicinity of gold nanoparticles and, consequently, on the fluorophore molecule excitation rate. The field strength and, therefore, the excitation rate fall with increasing dielectric permittivity of the substrate material and, conversely, grow with the increasing size of the nanoparticle. Such multi-vector behavior is explained, in the first case, by the increasing energy losses upon the interaction of the nanoparticle field with the substrate material while its dielectric permittivity increases. In the second case, the increase in the excitation rate as a result of increasing size of the nanoparticle prevails over the dielectric losses with a significant decrease in the excitation rate in general. It should be noted that in both cases, the significant dependence of the excitation rate on the distance between the fluorophore molecule and nanoparticle was observed, which exhibits a nonlinear behavior.

In Fig. 3.29, the distribution of electric field strength near the surface of a spherical gold nanoparticle with a radius of 20 nm in the absence of the glass substrate and in the case of its presence (ε_2 = 2.59) is depicted. The maximum field strength for these cases is 3.75 and 2.41, respectively; therefore, in this case the presence of a dielectric environment inhomogeneity reduces the maximum strength of the local electric field generated by a gold nanosphere under the influence of an external excitation. Obviously, the nanoparticle behaves as a classical dipole that emits light, which is consistent with the dipole approximation used in this approach.

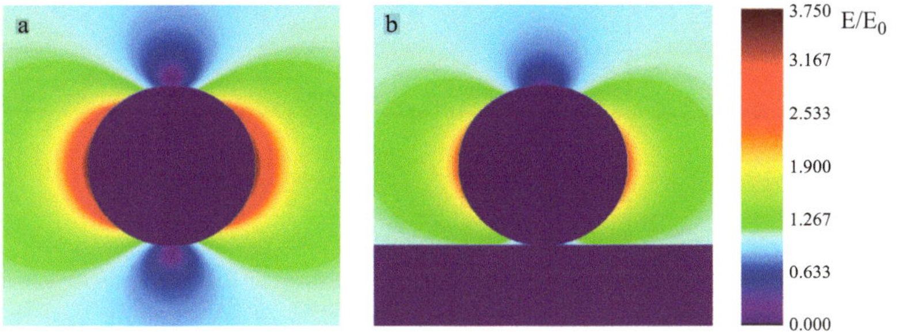

Figure 3.29 Distribution of the electric field strength near the surface of a spherical gold nanoparticle with a radius of 20 nm at an external excitation wavelength of 650 nm (a) without and (b) in the presence of the glass substrate (ε_2 = 2.59). The maximum E/E_0 value in the case (a) is 3.75, and in the case (b) is 2.41. Reprinted with permission from Ref. [178], Copyright 2015, Springer Nature.

Figure 3.30 shows the dependence of the fluorophore molecule excitation rate near spherical gold nanoparticles of different sizes on the distance to the surface of the nanoparticle. It should be noted that with increasing gold nanoparticle size and distance from the surface of the nanoparticle, the excitation rate derivative decreases due to the exponential nature of the field distribution.

The effect of substrate on the nanostructure's local electromagnetic field is determined by the dielectric permittivity of the material from which it is made, and, like in the previous case, there is an obvious dependence of the excitation rate on the distance from the nanoparticle surface to the point of the fluorophore molecule location. The nature of this effect is shown in Fig. 3.31, which shows that in the presence of the glass substrate, the fluorophore excitation rate value is reduced by almost half compared with the case when the substrate is absent. Thus, the best way to obtain the fluorophore excitation rate enhancement is to use the substrate material with the least possible value of the dielectric permittivity.

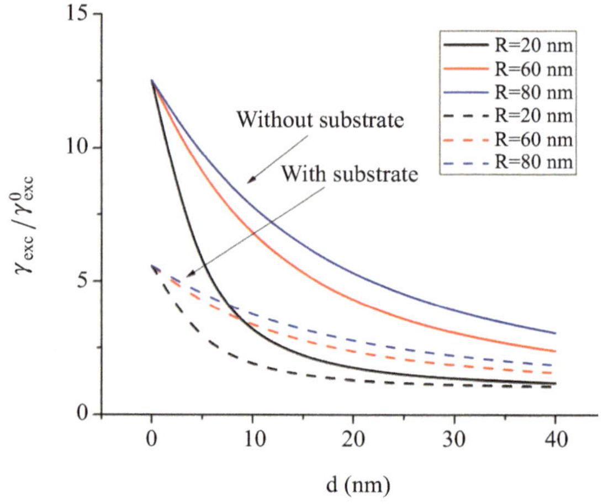

Figure 3.30 Fluorophore molecule excitation rate change near spherical gold nanoparticles of different sizes in the absence and presence of the dielectric glass substrate (ε_2 = 2.59). *R* is the spherical nanoparticle radius, and *d* is the distance from the nanoparticle surface to the point of fluorophore molecule location. Reprinted with permission from Ref. [178], Copyright 2015, Springer Nature.

The decrease in the fluorophore quantum yield, when it approaches the surface of a nanoparticle being in the resonant plasmon oscillations generation mode, is well known. Reduction in the quantum yield shows a gradual nonlinear decline to zero on contact of the fluorophore with the surface of the nanoparticle. Usually, to enhance the fluorescence, a separating layer of dielectric material between the nanoparticle and fluorophore is used [181]. In this study, for modeling generalized system properties, we first use the vacuum as an environment. Subsequently, to agree with experiment, simulation results for a particular case of a dielectric material that was used in the experiment will be presented. Figure 3.32 shows the typical behavior of the molecule quantum yield dependence on the distance to the surface of a gold nanoparticle. It is noticeable that the presence of the substrate generally somewhat reduces the quantum yield. The figure shows the results for a gold nanoparticle with a radius of 40 nm and a model fluorophore with the maximum initial quantum yield q^0 (without the presence of a nanoparticle), which equals to 1. Thus, we can conclude that the resulting reduction in the electric field strength near the nanoparticle in the presence of the dielectric substrate will subsequently reduce the fluorophore molecule quantum yield.

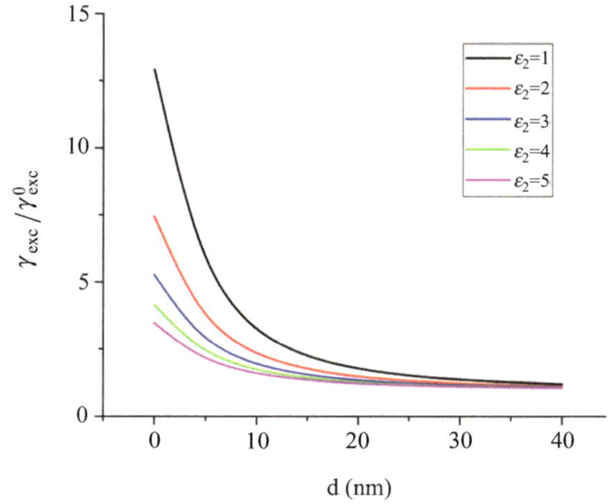

Figure 3.31 Fluorophore molecule excitation rate change near the spherical gold nanoparticle with a radius of 20 nm for different values of the substrate dielectric permittivity. ε_2 is the substrate dielectric permittivity, and d is the distance from the nanoparticle surface to the point of the fluorophore molecule location. Reprinted with permission from Ref. [178], Copyright 2015, Springer Nature.

The theoretical approach using Green's function, which is applied in this study, for the first time allowed us to estimate the fluorescence rate enhancement for fluorophores with different quantum yield in the presence of a spherical gold nanoparticle located on a dielectric substrate (Fig. 3.33). In confirmation of the results of Ref. [177], the dependence of fluorescence rate enhancement factor on the distance between the fluorophore and surface of the nanoparticle exhibits a non-monotonic character and expresses a maximum depending on several factors. Known nonlinear dependence on the initial fluorophore quantum yield, nanoparticle size and the distance between the fluorophore and nanoparticle is supplemented in our case with the influence of the presence of the dielectric substrate. It is evident from Fig. 3.33 that the presence of the dielectric substrate slightly reduces the amount of fluorescence rate enhancement in equidistant point, which increases in proportion to the size of the gold nanoparticle.

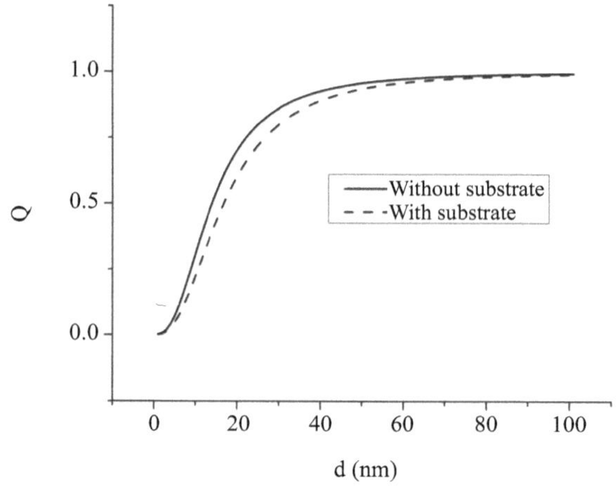

Figure 3.32 Fluorophore molecule ($q^0 = 1$) quantum yield Q depending on the distance to the surface of the spherical gold nanoparticle with a radius of 40 nm in the absence (solid line) and in the presence (dashed line) of the glass substrate with $\varepsilon_2 = 2.59$. d is the distance from the molecule to the nanoparticle surface. Reprinted with permission from Ref. [178], Copyright 2015, Springer Nature.

It is known that the value of the molecule initial quantum yield significantly affects their emission enhancement factor [182, 183]. Since the enhancement factor is inversely proportional to q^0, the smaller the q^0 value, the more significant enhancement of emission can be obtained (Fig. 3.34). Simulation results show that the dependence of the fluorescence rate enhancement factor on the value of the initial quantum yield is pronouncedly nonlinear. In particular, in the case of the fluorophore molecule with $q^0 = 0.1$, located near the gold nanoparticle with a radius of 80 nm, simulation revealed that the maximum $\gamma_{\text{em}}/\gamma_{\text{em}}^0$ value is 22.5 in the absence of substrate and 16.1 in the presence of the glass substrate (ε_2 = 2.59). An important result of the simulation obtained is to find the existence of an optimal distance between the fluorophore and nanoparticle depending on the size of the initial quantum yield. As shown in Fig. 3.34 (inset), the optimal distance for a maximum fluorescence rate enhancement factor is markedly different in the case of low and high values of the fluorophore quantum yield. For example, for q^0 = 0.1 the optimal distance is 12 nm, and for q^0 = 1, it is 18 nm. Significant differences in enhancement values, which are introduced by the presence of the dielectric substrate, indicate the need to consider the impact of this factor in the design of plasmonic nanochips and optimization of dimensional parameters of nanostructures depending on the dielectric characteristics of the substrate. Factor of the substrate becomes even more important in further consideration of the fluorescence enhancement mechanism depending on the wavelength resonance positions of the fluorophore molecule and nanoparticle.

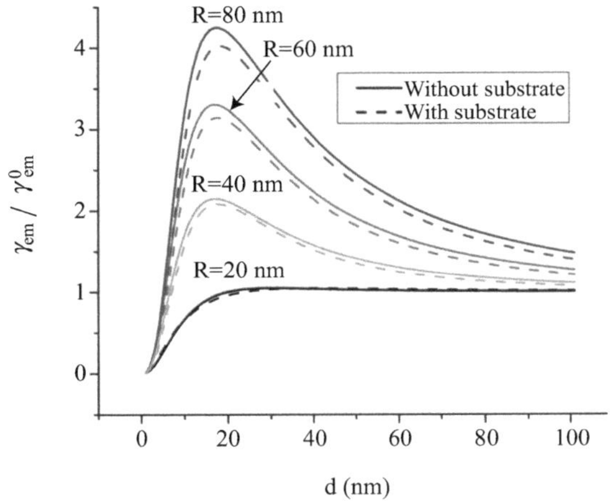

Figure 3.33 Dependence of the fluorescence rate enhancement factor for a fluorophore molecule (q^0 = 1) on the distance from the surface of the spherical gold nanoparticle of different radii. *d* is the distance from the molecule to the nanoparticle surface. Reprinted with permission from Ref. [178], Copyright 2015, Springer Nature.

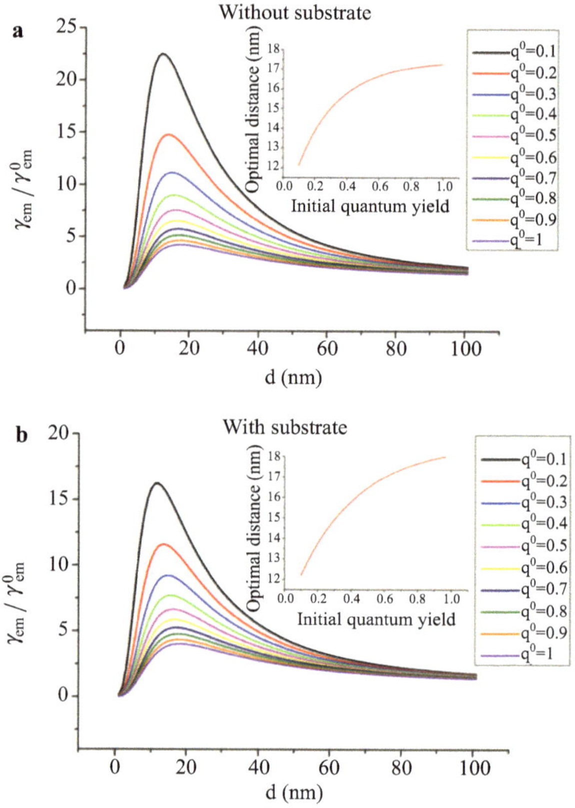

Figure 3.34 Fluorescence rate enhancement factor for fluorophores with different initial quantum yield near the surface of a spherical gold nanoparticle with a radius of 80 nm (a) in the absence and (b) in the presence of the glass substrate with $\varepsilon_2 = 2.59$. *d* is the distance from the molecule to the nanoparticle surface. Insets: dependences of the optimal distance between the fluorophore and nanoparticle on the initial quantum yield value. Reprinted with permission from Ref. [178], Copyright 2015, Springer Nature.

Thus, for the first time, the influence of the dielectric substrate on the amplification of a local electromagnetic field near the metal nanostructure and, consequently, on the emission of fluorophore molecules near its surface, was considered. Calculations have shown that the local electric field strength values and the fluorophore quantum yield decrease with increasing value of the dielectric permittivity of the substrate. Simulation of the fluorophore molecule fluorescence rate enhancement $\gamma_{em}/\gamma_{em}^0$ revealed that the $\gamma_{em}/\gamma_{em}^0$ value increases with the size of the spherical gold nanoparticle and the presence of the substrate slightly reduces its value. The results obtained have also confirmed the known fact that the emission of molecules with low initial quantum yield q^0 is amplified in the presence of gold nanostructures much better than for molecules that are characterized by a high q^0 value. The developed theoretical approach and the obtained results can be used in the development of new plasmonic nanochips based on gold and silver nanostructures for applications as sensor elements of fluorescent sensors and other devices operating on the basis of the localized surface plasmon resonance phenomenon.

3.5.2 Studies of Surface-Enhanced Fluorescence of Dyes Using LSPR Phenomenon in Au and Ag Nanostructures

Fluorescence techniques are widely used in biomedical research due to their high sensitivity, multiplex sensing possibility, compatibility with living organisms, and high response speed [184–186]. It is possible to study living cells and even whole organisms by using fluorescence spectroscopy [185, 187, 188]. Fluorescence is a rapid and sensitive method for studying the structure, kinetics, and functions of biological macromolecules such as nucleic acids and proteins [189–192].

Recently, nanoscale fluorescent emitters have attracted special attention for application in biological studies [193, 194]. In a short time, several applications have been found for these objects in clinical diagnostics [195], environmental monitoring [196], food quality control [197], biological weapons detection [198], and in various related fields. This diversity of applications is provided by bionanointeractions that allow delivery of nanoparticles to the biologically important systems to obtain information about them and even to change their properties [199–201]. Significant progress has been achieved in the application of high-conductive plasmonic nanomaterials for biochemical and diagnostic applications [202]. This subsection describes existing approaches for the development of sensors and biodiagnostic instruments based on surface-enhanced fluorescence [203–205].

3.5.2.1 Fluorescence enhancement by using high-conductive nanostructures

Studying the possibilities of using high-conductive metal nanoparticles (MNPs) has attracted great interest due to the prospects of MNP applications in biophysics and biochemistry, materials science, and fluorescent spectroscopy based on the surface enhancement methods [206]. The optical properties of MNPs are mostly determined by the surface plasmons, which are inherent collective electron oscillations in the metal. LSPR that occurs in metal nanostructures is widely used in highly sensitive sensors and optical devices [170, 207]. To enhance the fluorescence, silver and gold nanoparticles that have unique optical, electronic, and catalytic properties are usually exploited. This is due to their attractive properties, which include not only the simple fabrication, but also such features as the intense plasmon resonance band in the visible spectral region and significant extinction at the LSPR wavelength [154].

Enhancement of fluorescence of the dyes by using silver and gold nanostructures significantly depends on the resonance energy transfer from MNPs generating the plasmon field to the dye molecule, which is located near MNP [208–210], shape and size of MNP [211], distance between the dye molecule and the MNP [170, 207, 208], as well as the characteristics of dye molecule such as intrinsic quantum yield and excited state lifetime [208, 209, 212].

In general, the fluorescence emission enhancement Φ is caused by the resonance energy transfer from plasmon-generating nanostructures to the dye molecule and subsequent change in the quantum yield of the "plasmonic nanostructure–dye molecule" system and can be described by the following equation [213]:

$$\Phi = \left(\frac{E_\mathrm{p}}{E_0} \right)^2 \frac{q}{q^0},$$

where E_p is the plasmon electric field strength in the observation point that is generated by the incident light E_0; q is the quantum yield of the dye molecule induced by the plasmon field E_p; and q^0 is the intrinsic quantum yield of the dye molecule.

3.5.2.2 Factor of optimal distance between fluorophore molecule and plasmonic nanoparticle

The optimal distance between the fluorophore and the metal surface is a very important factor in the mechanism of surface-enhanced fluorescence. This optimal distance depends on the immanent properties of the molecule relevant to energy exchange between the molecule and the surface plasmon. There are three cases for the location of dye molecule near the surface of nanoparticles. First of them is very close to the metal surface when the quenching of fluorescence is observed; second is an optimal distance resulting in the most significant enhancement of fluorescence; and the third is much greater than the distance at which the enhancement occurs [214].

The optimal distance can be determined both theoretically and experimentally [212]. For example, the emission simulation for a layer of fluorescent dye molecules with the thickness of 5 nm placed on the 80 nm spherical gold nanoparticle has shown that the optimal distance between the fluorophores and the metal surface to observe fluorescence enhancement was about 20 nm; in other cases there were weak enhancement or quenching [212]. In Ref. [215], the experimental studies of the dye fluorescence emission near the silver and gold nanostructures were carried out. It was found that the enhancement of fluorescence was observed when the distance between dye molecules and metal surface was 24–25 nm and the quenching was observed at 15 nm. If we consider the model "fluorescent sphere–silver nanoparticle" system [216], it turns out that when the distance between them is equal to 90 nm, the emission quenching by 30% is observed. When the distance is decreased to some optimal value, 1.5-fold enhancement occurs. And if the second silver particle with the same properties is added to the system, 2.7-fold enhancement of fluorescence emission of the fluorescent sphere will be produced.

Consideration of the influence of the distance between the fluorophore and the surface of the metal nanostructures has led to the development of the methods enabling the fluorescence enhancement observation. The most effective technique for distance-controlled surface enhancement is to use the artificially created spacer layer between the fluorescent molecule and the silver or gold nanostructure, which prevents the emission quenching effect [170,

210, 215]. The function of spacers could be provided by organic (e.g., lipid or protein) or inorganic (e.g., silicon dioxide) separation layers [210, 217].

For fluorescence enhancement studies, organic "sandwich" structures are often used [210], which are dye-labeled protein molecules located directly on the surface of the high-conductive nanostructures. Here, protein molecules themselves act as the separation layers between the nanostructure and fluorophore. Such structures allow obtaining up to 18 times enhanced emission of dye molecules. In some experiments, the distance between the nanoparticle and the dye molecule was adjusted by a double-stranded DNA molecule [209] or its replicating form [218]. Another interesting option is to exploit spacer layers fabricated from silicon dioxide. This coating deposited on nanostructures provides strength, chemical inertness, and versatility for binding to biomolecules or any hydrophobic fluorophores, and also allows adjusting the distance between the fluorophores and the nanostructures [205, 217]. For precise spacing of fluorophore molecules from nanostructured surface, the layer-by-layer (LbL) technology is often used [219] to produce oppositely charged polymer "sandwich" that enables fluorescence signal enhancement.

3.5.2.3 Influence of the size, position, and shape of nanostructures on the enhancement effect

As mentioned above, the fluorescence enhancement using high-conductive nanostructures depends on their size and shape. In Ref. [220], a comparative analysis of the fluorescent enhancement of the dye using colloidal gold nanoparticles of different diameters with protein-modified surface was carried out. It was found that nanoparticles with a 40 nm diameter allow obtaining 1.5-fold enhancement of fluorescence signal and nanoparticles of 200 nm diameters provide 2-fold enhancement, respectively. Forty nm cubic and spherical silver nanoparticles were used as the plasmon-generating nanostructures for fluorescence enhancement [221]. The authors experimentally found that the cubic silver nanoparticles provide higher fluorescence enhancement in comparison with the spherical silver nanoparticles of the same size. The effect of the size and the shape of nanoparticles on the enhancement value can be considered through the spectral overlap of the absorption spectra of nanostructure and the emission spectra of the fluorophore [209], because there is a direct relationship of the wavelength position and the half-width of the light extinction spectra with the geometrical parameters of the nanoparticles.

Another important fact is that the emission intensity of the dye placed near the nanostructure exhibits nonlinear dependence on their relative position [173, 218, 222]. In Refs. [170, 173, 223, 224], it was found that the largest enhancement of fluorescence was observed in the gap between closely spaced nanoparticles, where the intensity of the electric field near their surface is greater than at the surface of individual nanostructures. The study in Ref. [173] showed that the enhancement of fluorescence of the dye placed near a plasmon-generating separate silver nanoparticle is about two times smaller than in the case of placing the dye molecule between two silver nanoparticles.

It should be noted that the fluorescence emission enhancement of the dye molecule located between two silver or gold nanoparticles depends on the geometry of this "sandwich" structure. Namely, when the fluorophore is sandwiched in the hot spot region right between the two gold nanoparticles on the axis connecting their centers, it leads to a strong fluorescence enhancement; in the other case, the fluorescence enhancement is less pronounced [216].

3.5.2.4 Mechanisms of surface enhancement

The interaction between the metal surface and fluorescent molecule provides such effects as increasing the radiative or nonradiative decay rates of fluorescent molecule and, accordingly, increasing or decreasing the quantum yield [212]. Enhancement of fluorescence depends on the plasmon resonance energy transfer from the nanostructured metal surface to the dye molecules located near the surface [177, 208, 210, 216, 225]. This transfer is determined by two mechanisms: first of them is the plasmon field generated around the nanoparticle by the incident light that, depending on wavelength, can enhance the excitation of the fluorophore, which, in turn, determines the level of fluorescence emission. The second one is the nanoparticle–fluorophore interaction that reduces the ratio of radiative to nonradiative decay rate and, depending on the presence of the dielectric layer, influences the quantum yield of the fluorophore, resulting in fluorescence quenching [213]. It should be noted that the enhancement of fluorescent emission depends not only on the properties of metal nanoparticle, but also on the properties of the fluorescent molecule. These properties are the quantum yield and the excitation relaxation time, which depend on the probability of quantum transitions involved in the radiative processes. Accordingly, the probability of quantum transitions under the certain conditions increases when fluorescent molecule is located near the metal surface. Quantum yield, which is defined as the ratio of the number of emitted to absorbed photons of the molecule, determines the efficiency of the emission, which, in turn, depends on the distance between the fluorophore and the surface of nanostructures [212, 224]. In Ref. [212], it was theoretically and experimentally shown that the quantum yield of a molecule decreases when the distance between the fluorophore and the surface of metal nanostructure is too small, even in the case of the sample excitation increase, and results in fluorescence quenching.

Studies on fluorescence of the dye solutions containing the high-conductive nanoparticles established that the quantum yield also depends on the pH. In Ref. [206], scientists found that the high pH of the dye solution has no effect on the enhancement; but when pH reduces, the fluorescence enhancement occurs, which is caused by the increasing overlap of the emission spectrum of the dye with the plasmon resonance band of the metal nanoparticles.

The presence of the nanostructured metal surface near the dye molecule affects not only the quantum yield and fluorescence enhancement, but also the radiative lifetime of the excited molecule [172, 226, 227]. For example, a 13-fold emission enhancement of the fluorophore is accompanied by a decrease in the fluorescence lifetime by a factor of 22 [172].

The orientation of the dipole moment of the dye molecules near gold and silver nanostructures and the polarization of excitation light have a considerable effect on the fluorescence enhancement [226, 228]. In the study in Ref. [216], the fluorescence intensity of the "Au nanoparticle–fluorophore–Au nanoparticle" sandwich structure is enhanced when the laser light is polarized parallel to the axis of the sandwich, whereas the fluorescence is decreased when the laser is polarized perpendicular to it. Theoretical calculations showed [177] that the maximum enhancement was observed in the case of perpendicular orientation of the dipole moment of the molecules to the nanostructured surface and when the distance between the molecule and the metal surface was optimal. A single-molecule fluorescence rate as a function of the particle–sample distance for an 80 nm silver particle was studied in the case of perpendicular orientation of the dipole moment of the molecule [225]. Experimentally, it is difficult to identify molecules with such orientation of the dipole moment, but the results of the performed studies indicate that even at the angle of the dipole moment to the normal of the plane in which the nanostructures are placed equal to 15 degrees, 19 times enhancement of fluorescence can be observed [172].

3.5.2.5 Modeling and comparative analysis of gold and silver nanostructures as fluorescent signal amplifiers

Random silver and gold nanostructure arrays were obtained by thermal annealing of silver (mass thickness 8 nm, 250°C, 1 h) and gold (mass thickness 10 nm, 450°C, 2 h) island films [158]. The light extinction spectra of random silver and gold NSA were measured using the LSPR biosensor [136]. Silver and gold NSA exhibited the excitation of localized surface plasmon at 480 nm (Fig. 3.35b) and at 555 nm (Fig. 3.35d), respectively. The average sizes of nanostructures forming the NSA were calculated by matching the wavelength peak positions in the spectra obtained experimentally and theoretically. Thus, for silver nanostructures, the equivalent base diameter was found to be equal to 20 nm and height equal to 10 nm, and for gold nanostructures, the equivalent base diameter is 10 nm and height is 20 nm.

For determined sizes of the gold and silver nanostructures, the profiles of electric field intensity enhancement around the nanoparticles were calculated (Figs. 3.35a,c). These enhancement values correlate with the increase in the fluorescence excitation rate [209]. The calculation was performed at 532 nm, which is an inherent excitation wavelength for the organic dye Rhodamine 6G (R6G). The maximum electric field intensity enhancement value calculated on the 30×30 nm^2 area outside the nanostructure in the plane in which the propagation vector and polarization vector of incident light wave lie was about 100 times for silver nanostructure and about 30 times for gold nanostructure. The broadening in the extinction spectra of silver NSA, observed in the experiment, compared with the simulation results (Fig. 3.35b), can be explained by the growing mismatch between the model semi-ellipsoid and experimental nanostructure shapes and variety of sizes of nanostructures in the NSA. Thus, silver nanostructures produced by thermal annealing of island films are potentially more promising in creating the nanochip. However, the gold nanostructures are mostly used in most scientific researches at the present time due to their higher chemical stability.

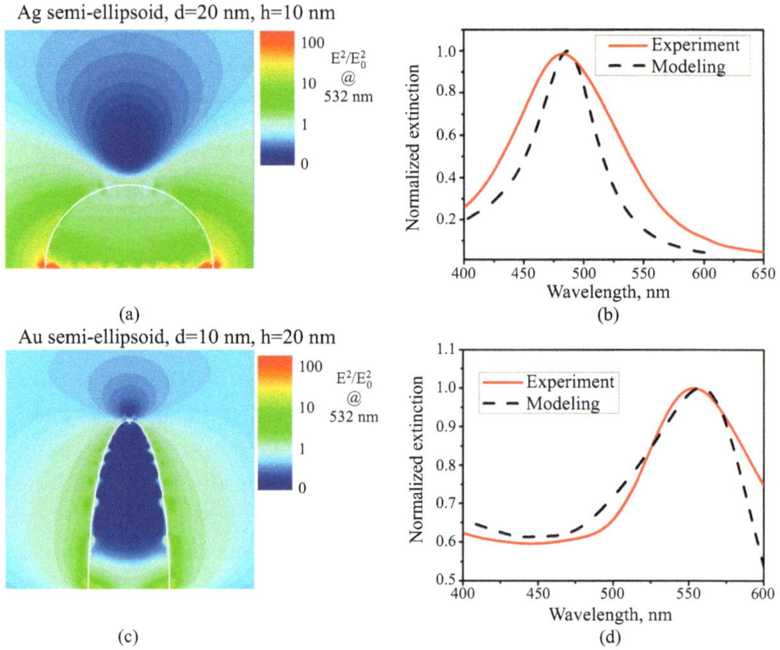

Figure 3.35 Simulated profiles of electric field intensity distributions for (a) Ag and (c) Au semi-ellipsoids on glass substrate; light extinction spectra of unordered (b) Ag and (d) Au nanostructure arrays. Reprinted from Ref. [229] under a Creative Commons Attribution 4.0 International License. Figure and caption were adapted.

3.5.2.6 Studies of Rhodamine 6G dye fluorescence enhancement by using random gold nanostructure arrays

Using the developed method of surface-enhanced fluorometry [205], based on the phenomenon of LSPR in unordered gold nanostructure arrays (GNAs), the fluorescence measurements of R6G were carried out. The dye was placed on the plasmon-generating GNA with various thicknesses of the dielectric SiO_2 coating. The SiO_2 layer deposited on the GNA provides strength, chemical inertness, versatility required to compound the biomolecule or any hydrophobic fluorophore, and also allows adjusting the distance between the fluorophore and the nanoparticle.

The samples of GNA with the thicknesses of dielectric spacer equal to 10, 15, 20, and 25 nm coated with a polymer composite, consisting of an aqueous solution of R6G and polyacrylic acid (PAA), were studied. The R6G concentration in the polymer composite was equal to 10^{-5} mol/L. The fluorescence measurements of R6G on GNA were carried out with the 532 nm green laser used as an excitation light source. For all samples, the enhancement of R6G fluorescence near the GNA was observed compared with the signal on the sample without GNA (Fig. 3.36).

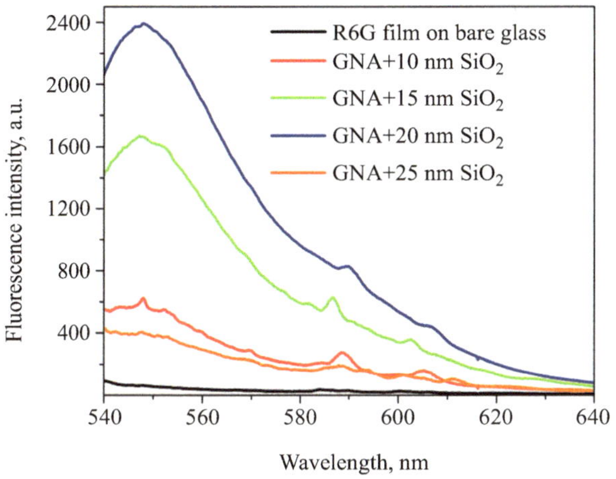

Figure 3.36 Fluorescence spectra of R6G dye for various thicknesses of the dielectric SiO_2 coating placed on the gold nanostructure arrays. Reprinted from Ref. [229] under a Creative Commons Attribution 4.0 International License. Figure was adapted.

The intensity of the fluorescent signal dependence on the SiO_2 coating thickness was nonlinear with an expressed peak. The maximum enhancement of R6G fluorescence was obtained for the sample with 20 nm thickness of SiO_2. Figure 3.37 shows the enhancement factor dependence on the thickness of dielectric

SiO_2 coating. The enhancement factor is defined as the ratio of R6G fluorescence intensities for GNA with SiO_2 shell and bare glass substrate at the maximum emission wavelength. The experimental results of maximum R6G fluorescence enhancement (about 23 times) obtained by using nanochips agree with the calculated value of the electric field enhancement on the gold nanostructures. The disagreement between the calculated and experimental enhancement values is due to the influence of the dielectric substrate and the spread of the size and shape of nanoparticles on the real nanochip.

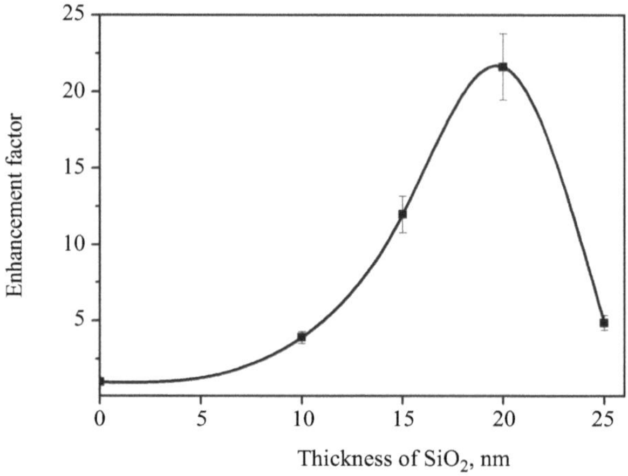

Figure 3.37 Enhancement factor of R6G fluorescence for different thicknesses of SiO_2 coating. Reprinted from Ref. [229] under a Creative Commons Attribution 4.0 International License. Figure was adapted.

3.6 Conclusion

In this chapter, several aspects of plasmonic nanochips development and applications are discussed. First, the approaches for the fabrication of random and ordered noble metal nanostructure arrays and their application for surface-enhanced spectroscopies are reviewed. Second, specific studies described in this chapter demonstrate the interplay between the structural and optical parameters of plasmonic nanochips and their performance in terms of plasmonic enhancement, as well as emphasize the versatility of plasmonic nanochips for different applications ranging from LSPR-based sensing to SEIRA and SEF spectroscopies.

Different techniques used for the fabrication and assembly alignment of Au and Ag nanostructures for preparing plasmonic nanochips are reviewed. The advantages and disadvantages of various nanopatterning techniques are discussed. The evaluation of such technologies for improving biosensor performance in terms of high sensitivity and low detection limits is presented. Several fabrication techniques are highlighted, starting from relatively simple methods, such as preparation of plasmonic nanochips from colloidal solutions of nanoparticles, to more complicated technologies, such as various lithographies.

In a comparative study, samples of random and ordered gold nanoparticle arrays with different morphologies, which were fabricated using thermal annealing of vacuum-evaporated island films and nanoimprint lithography methods, exhibited differences in the maximal level of plasmonic enhancement. However, both nanoparticle shapes have shown comparable average enhancement values in their near vicinity. Thermally annealed island films revealed smooth nanostructure shapes resembling semi-ellipsoids, with easy access to their surfaces by detected molecules possessing a wide range of sizes. In contrast, nanoimprint lithography produced nanostructures with sharp corners that can, on the one hand, generate "hot spots" with strong PE and, on the other hand, hinder the access of larger analyte molecules to these regions, thereby creating more favorable conditions for small molecule detection. A model of LSPR biosensor was developed, which allows evaluating the possibility of registration of molecules with different molecular weights. It was found that the use of nanoparticles with parallelepiped shape leads to greater extinction spectrum shift for adsorption of molecules and expands the range of analyte concentrations registration compared to nanoparticles with the semi-ellipsoids shape. In summary, both studied fabrication approaches can find different implementations to prepare the nanostructured plasmonic arrays for a wide range of applications, such as biosensing and resonant methods using electromagnetic fields of nanostructures.

In SEIRA-related studies, different experimental methods were developed, which give an opportunity to obtain amplification in the IR spectra of molecular absorption. The validity of the SEIRA method for the analysis of nucleic acids was studied. The mechanism of signal amplification includes two possible ways: amplification of the electromagnetic field as a result of interaction with localized surface plasmon oscillations arising on inhomogeneities of the metal surface, and the specific growth of the transition dipole moments of the molecules adsorbed on the metal surface. Under the conditions of the conducted studies, a gain factor of 3–20 times was observed for vibrations of various molecular groups. The amplification is characterized by the structure of inhomogeneities in the system and is determined by the ratio of the real and imaginary parts of the dielectric constant of the metal surface on which the molecules are adsorbed. It should be noted that the given method of SEIRA studies allows registration of the spectral features of molecules that are not available in conventional IR spectroscopy.

In one of the SEF-related studies, the theoretical basis was developed and the dependences of the fluorescence rate enhancement in the "fluorophore molecule–gold nanoparticle–dielectric substrate" system were calculated depending on the quantum yield of the fluorophore, the dielectric constant of the substrate and the geometric configuration of the system elements. Using the dyadic Green's function and volumetric Lippmann–Schwinger equations, computer simulation of fluorescence excitation rate enhancement in the vicinity of nanostructures, which are spherical gold nanoparticles, was conducted in order to study its dependence on the size of the gold nanoparticles and distance from their surface to a fluorophore molecule, as well as on the dielectric permittivity of the substrate, which the nanoparticle is located on. The obtained results are important for the improvement of the fluorescence analysis method using the surface enhancement effect on nanoparticles and are useful for the development of novel nanoscale sensors for registration of low-intensity fluorescence signals in biochemical applications.

In the other SEF-related study, plasmon-generating nanochips for plasmon-enhanced fluorescence based on the gold nanostructure arrays with dielectric SiO_2 coating of different thickness were developed and fabricated. Experimental studies of R6G fluorescence using developed nanochips were carried out. It has been shown that the "gold nanostructure–dielectric coating–dye" system, in which LSPR appears, can be used to obtain the fluorescence enhancement of the R6G dye molecule. For gold nanostructure arrays with defined dimensions, the optimal thickness of the SiO_2 dielectric coating that provides enhancement of R6G fluorescence was found to be 20 nm. This confirms the possibility of using the SEF method with plasmonic nanochips for the detection of low-intensity fluorescence signals in biochemical research. The obtained results are important for improving the fluorescence analysis technique and methods of fluorescent signals registration based on the phenomenon of surface enhancement by using high-conductive nanoparticles, development of the nanoscale SEF-nanochips and optoelectronic LSPR sensors with high sensitivity.

References

1. Kildishev, A. V., Boltasseva, A., and Shalaev, V. M. (2013). Planar photonics with metasurfaces, *Science*, **339**, 1232009.
2. Gjonaj, B., Aulbach, J., Johnson, P. M., Mosk, A. P., Kuipers, L., and Lagendijk, A. (2013). Focusing and scanning microscopy with propagating surface plasmons, *Phys. Rev. Lett.*, **110**, 266804.
3. Bauch, M., Toma, K., Toma, M., Zhang, Q., and Dostalek, J. (2014). Plasmon-enhanced fluorescence biosensors: A review, *Plasmonics*, **9**, pp. 781–799.
4. Ruemmele, J. A., Hall, W. P., Ruvuna, L. K., and Van Duyne, R. P. (2013). A localized surface plasmon resonance imaging instrument for multiplexed biosensing, *Anal. Chem.*, **85**, pp. 4560–4566.
5. Anker, J. N., Hall, W. P., Lambert, M. P., Velasco, P. T., Mrksich, M., Klein, W. L. et al. (2009). Detection and identification of bioanalytes with high resolution LSPR spectroscopy and MALDI mass spectrometry, *J. Phys. Chem. C*, **113**, pp. 5891–5894.
6. Sciacca, B., François, A., Klingler-Hoffmann, M., Brazzatti, J., Penno, M., Hoffmann, P. et al. (2013). Radiative-surface plasmon resonance for the detection of apolipoprotein E in medical diagnostics applications, *Nanomedicine*, **9**, pp. 550–557.
7. Vo-Dinh, T., Fales, A. M., Griffin, G. D., Khoury, C. G., Liu, Y., Ngo, H. et al. (2013). Plasmonic nanoprobes: From chemical sensing to medical diagnostics and therapy, *Nanoscale*, **5**, pp. 10127–10140.
8. Haes, A. J., Hall, W. P., Chang, L., Klein, W. L., and Van Duyne, R. P. (2004). A localized surface plasmon resonance biosensor: First steps toward an assay for Alzheimer's disease, *Nano Lett.*, **4**, pp. 1029–1034.

9. Dantham, V. R., Holler, S., Barbre, C., Keng, D., Kolchenko, V., and Arnold, S. (2013). Label-free detection of single protein using a nanoplasmonic-photonic hybrid microcavity, *Nano Lett.*, **13**, pp. 3347–3351.

10. Henry, A. I., Bingham, J. M., Ringe, E., Marks, L. D., Schatz, G. C., and Van Duyne, R. P. (2011). Correlated structure and optical property studies of plasmonic nanoparticles, *J. Phys. Chem. C*, **115**, pp. 9291–9305.

11. Haes, A. J., Zou, S., Schatz, G. C., and Van Duyne, R. P. (2004). A nanoscale optical biosensor: The long-range distance dependence of the localized surface plasmon resonance of noble metal nanoparticles, *J. Phys. Chem. B*, **108**, pp. 109–116.

12. Liu, K., Bai, Y., Zhang, L., Yang, Z., Fan, Q., Zheng, H. et al. (2016). Porous Au–Ag nanospheres with high-density and highly accessible hotspots for SERS analysis, *Nano Lett.*, **16**, pp. 3675–3681.

13. Deng, W., Xie, F., Baltar, H. T., and Goldys, E. M. (2013). Metal-enhanced fluorescence in the life sciences: Here, now and beyond, *Phys. Chem. Chem. Phys.*, **15**, pp. 15695–15708.

14. Zhang, Z., Yang, P., Xu, H., and Zheng, H. (2013). Surface enhanced fluorescence and Raman scattering by gold nanoparticle dimers and trimers, *J. Appl. Phys.*, **113**, 033102.

15. Xu, S., Cao, Y., Zhou, J., Wang, X., Wang, X., and Xu, W. (2011). Plasmonic enhancement of fluorescence on silver nanoparticle films, *Nanotechnology*, **22**, 275715.

16. Brown, L. V., Zhao, K., King, N., Sobhani, H., Nordlander, P., and Halas, N. J. (2013). Surface-enhanced infrared absorption using individual cross antennas tailored to chemical moieties, *J. Am. Chem. Soc.*, **135**, pp. 3688–3695.

17. Dovbeshko, G. I., Chegel, V. I., Gridina, N. Y., Repnytska, O. P., Shirshov, Y. M., Tryndiak, V. P. et al. (2002). Surface enhanced IR absorption of nucleic acids from tumor cells: FTIR reflectance study, *Biopolymers*, **67**, pp. 470–486.

18. Xue, B., Wang, D., Zuo, J., Kong, X., Zhang, Y., Liu, X. et al. (2015). Towards high quality triangular silver nanoprisms: Improved synthesis, six-tip based hot spots and ultra-high local surface plasmon resonance sensitivity, *Nanoscale*, **7**, pp. 8048–8057.

19. Ryu, Y., Kang, G., Lee, C. W., and Kim, K. (2015). Porous metallic nanocone arrays for high-density SERS hot spots via solvent-assisted nanoimprint lithography of block copolymer, *RSC Adv.*, **5**, pp. 76085–76091.

20. Wang, C., Zhao, X. P., Xu, Q. Y., Nie, X. G., Younis, M. R., Liu, W. Y. et al. (2018). Importance of hot spots in gold nanostructures on direct plasmon-enhanced electrochemistry, *ACS Appl. Nano Mater.*, **1**, pp. 5805–5811.

21. Dovbeshko, G. I., Paschuk, O. P., Fesenko, O. M., Chegel, V. I., Shirshov, Y. M., Nazarova, A. A. et al. (2004) Biological molecule conformations probed and enhanced by metal and carbon nanostructures: SEIRA, AFM and SPR Data, In *Frontiers of Multifunctional Integrated Nanosystems*, Buzaneva E., Scharff P. (eds). NATO Science Series II: Mathematics, Physics and Chemistry, vol 152, (Springer, Dordrecht) pp. 447–466.

22. Snopok, B. A., Kostyukevich, K. V., Lysenko, S. I., Lytvyn, P. M., Lytvyn, O. S., Mamykin, S. V. et al. (2001). Optical biosensors based on the surface plasmon resonance phenomenon: Optimization of the metal layer parameters, *Semicond. Phys. Quantum Electron. Optoelectron.*, **4**, pp. 56–69.

23. Lopatynskyi, A. M., Lopatynska, O. G., Guiver, M. D., Poperenko, L. V., and Chegel, V. I. (2008). Factor of interfacial potential for the surface plasmon-polariton resonance sensor response, *Semicond. Phys. Quantum Electron. Optoelectron.*, **11**, pp. 329–336.

24. Lopatynskyi, A., Guiver, M., and Chegel, V. (2014). Surface plasmon resonance biomolecular recognition nanosystem: Influence of the interfacial electrical potential, *J. Nanosci. Nanotechnol.*, **14**, pp. 6559–6564.

25. Nath, N. and Chilkoti, A. (2002). A colorimetric gold nanoparticle sensor to interrogate biomolecular interactions in real time on a surface, *Anal. Chem.*, **74**, pp. 504–509.

26. Nath, N. and Chilkoti, A. (2002). Immobilized gold nanoparticle sensor for label-free optical detection of biomolecular interactions, *Proc. SPIE*, **4626**, pp. 441–449.

27. Liu, H., Zhang, X., and Gao, Z. (2010). Lithography-free fabrication of large-area plasmonic nanostructures using colloidal gold nanoparticles, *Photonics Nanostructures: Fundam. Appl.*, **8**, pp. 131–139.

28. Malynych, S. and Chumanov, G. (2003). Light-induced coherent interactions between silver nanoparticles in two-dimensional arrays, *J. Am. Chem. Soc.*, **125**, pp. 2896–2898.

29. Hostetler, M. J., Wingate, J. E., Zhong, C. J., Harris, J. E., Vachet, R. W., Clark, M. R. et al. (1998). Alkanethiolate gold cluster molecules with

core diameters from 1.5 to 5.2 nm: Core and monolayer properties as a function of core size, *Langmuir*, **14**, pp. 17–30.

30. Malynych, S., Luzinov, I., and Chumanov, G. (2002). Poly (vinyl pyridine) as a universal surface modifier for immobilization of nanoparticles, *J. Phys. Chem. B*, **106**, pp. 1280–1285.
31. Aslan, K., Leonenko, Z., Lakowicz, J. R., and Geddes, C. D. (2005). Fast and slow deposition of silver nanorods on planar surfaces: Application to metal-enhanced fluorescence, *J. Phys. Chem. B*, **109**, pp. 3157–3162.
32. Jana, N. R., Gearheart, L., and Murphy, C. J. (2001). Wet chemical synthesis of silver nanorods and nanowires of controllable aspect ratio, *Chem. Commun.*, **7**, pp. 617–618.
33. Aslan, K., Lakowicz, J. R., and Geddes, C. D. (2005). Rapid deposition of triangular silver nanoplates on planar surfaces: Application to metal-enhanced fluorescence, *J. Phys. Chem. B*, **109**, pp. 6247–6251.
34. Anker, J. N., Hall, W. P., Lyandres, O., Shah, N. C., Zhao, J., and Van Duyne, R. P. (2008). Biosensing with plasmonic nanosensors, *Nat. Mater.*, **7**, pp. 442–453.
35. Chaney, S. B., Shanmukh, S., Dluhy, R. A., and Zhao, Y. P. (2005). Aligned silver nanorod arrays produce high sensitivity surface-enhanced Raman spectroscopy substrates, *Appl. Phys. Lett.*, **87**, 031908.
36. Zhao, Y. P., Chaney, S. B., Shanmukh, S., and Dluhy, R. A. (2006). Polarized surface enhanced Raman and absorbance spectra of aligned silver nanorod arrays, *J. Phys. Chem. B*, **110**, pp. 3153–3157.
37. Shanmukh, S., Jones, L., Driskell, J., Zhao, Y., Dluhy, R., and Tripp, R. A. (2006). Rapid and sensitive detection of respiratory virus molecular signatures using a silver nanorod array SERS substrate, *Nano Lett.*, **6**, pp. 2630–2636.
38. Hanarp, P., Sutherland, D. S., Gold, J., and Kasemo, B. (2003). Control of nanoparticle film structure for colloidal lithography, *Colloids Surf. A*, **214**, pp. 23–36.
39. Brian, B., Sepúlveda, B., Alaverdyan, Y., Lechuga, L. M., and Käll, M. (2009). Sensitivity enhancement of nanoplasmonic sensors in low refractive index substrates, *Opt. Express*, **17**, pp. 2015–2023.
40. Gao, D., Chen, W., Mulchandani, A., and Schultz, J. S. (2007). Detection of tumor markers based on extinction spectra of visible light passing through gold nanoholes, *Appl. Phys. Lett.*, **90**, 073901.
41. Dahlin, A. B., Tegenfeldt, J. O., and Höök, F. (2006). Improving the instrumental resolution of sensors based on localized surface plasmon resonance, *Anal. Chem.*, **78**, pp. 4416–4423.

42. Dahlin, A., Zäch, M., Rindzevicius, T., Käll, M., Sutherland, D. S., and Höök, F. (2005). Localized surface plasmon resonance sensing of lipid-membrane-mediated biorecognition events, *J. Am. Chem. Soc.*, **127**, pp. 5043–5048.

43. Jonsson, M. P., Jönsson, P., Dahlin, A. B., and Höök, F. (2007). Supported lipid bilayer formation and lipid-membrane-mediated biorecognition reactions studied with a new nanoplasmonic sensor template, *Nano Lett.*, **7**, pp. 3462–3468.

44. Aizpurua, J., Hanarp, P., Sutherland, D. S., Käll, M., Bryant, G. W., and De Abajo, F. G. (2003). Optical properties of gold nanorings, *Phys. Rev. Lett.*, **90**, 057401.

45. Larsson, E. M., Alegret, J., Käll, M., and Sutherland, D. S. (2007). Sensing characteristics of NIR localized surface plasmon resonances in gold nanorings for application as ultrasensitive biosensors, *Nano Lett.*, **7**, pp. 1256–1263.

46. Hanarp, P., Käll, M., and Sutherland, D. S. (2003). Optical properties of short range ordered arrays of nanometer gold disks prepared by colloidal lithography, *J. Phys. Chem. B*, **107**, pp. 5768–5772.

47. Fredriksson, H., Alaverdyan, Y., Dmitriev, A., Langhammer, C., Sutherland, D. S., Zäch, M., and Kasemo, B. (2007). Hole–mask colloidal lithography, *Adv. Mater.*, **19**, pp. 4297–4302.

48. Jonsson, M. P., Dahlin, A. B., Feuz, L., Petronis, S., and Höök, F. (2010). Locally functionalized short-range ordered nanoplasmonic pores for bioanalytical sensing, *Anal. Chem.*, **82**, pp. 2087–2094.

49. Chen, S., Svedendahl, M., Van Duyne, R. P., and Käll, M. (2011). Plasmon-enhanced colorimetric ELISA with single molecule sensitivity, *Nano Lett.*, **11**, pp. 1826–1830.

50. Chen, S., Svedendahl, M., Käll, M., Gunnarsson, L., and Dmitriev, A. (2009). Ultrahigh sensitivity made simple: Nanoplasmonic label-free biosensing with an extremely low limit-of-detection for bacterial and cancer diagnostics, *Nanotechnology*, **20**, 434015.

51. Mazzotta, F., Wang, G., Hägglund, C., Höök, F., and Jonsson, M. P. (2010). Nanoplasmonic biosensing with on-chip electrical detection, *Biosens. Bioelectron.*, **26**, pp. 1131–1136.

52. Dmitriev, A., Pakizeh, T., Käll, M., and Sutherland, D. S. (2007). Gold–silica–gold nanosandwiches: Tunable bimodal plasmonic resonators, *Small*, **3**, pp. 294–299.

53. Pakizeh, T., Dmitriev, A., Abrishamian, M. S., Granpayeh, N., and Käll, M. (2008). Structural asymmetry and induced optical magnetism in plasmonic nanosandwiches, *J. Opt. Soc. Am. B*, **25**, pp. 659–667.

54. Dmitriev, A., Hägglund, C., Chen, S., Fredriksson, H., Pakizeh, T., Käll, M. et al. (2008). Enhanced nanoplasmonic optical sensors with reduced substrate effect, *Nano Lett.*, **8**, pp. 3893–3898.

55. Zhang, X., Sun, B., Friend, R. H., Guo, H., Nau, D., and Giessen, H. (2006). Metallic photonic crystals based on solution-processible gold nanoparticles, *Nano Lett.*, **6**, pp. 651–655.

56. Possin, G. E. (1970). A method for forming very small diameter wires, *Rev. Sci. Instrum.*, **41**, pp. 772–774.

57. Martin, C. R. (1994). Nanomaterials: A membrane-based synthetic approach, *Science*, **266**, pp. 1961–1966.

58. Preston, C. K. and Moskovits, M. (1993). Optical characterization of anodic aluminum oxide films containing electrochemically deposited metal particles. 1. Gold in phosphoric acid anodic aluminum oxide films, *J. Phys. Chem.*, **97**, pp. 8495–8503.

59. Kwon, N., Kim, N., Yeon, J., Yeom, G., and Chung, I. (2011). Fabrication of ordered Au nanodot arrays utilizing anodic aluminum oxide templates formed on Si substrate, *J. Vac. Sci. Technol. B*, **29**, 031805.

60. Iida, C., Sato, M., Nakayama, M., and Sanada, A. (2011). Electrodeposition of Cu_2O nanopyramids using an anodic aluminum oxide template, *Int. J. Electrochem. Sci.*, **6**, pp. 4730–4736.

61. Routkevitch, D., Tager, A. A., Haruyama, J., Almawlawi, D., Moskovits, M., and Xu, J. M. (1996). Nonlithographic nano-wire arrays: Fabrication, physics, and device applications, *IEEE Trans. Electron Devices*, **43**, pp. 1646–1658.

62. Mu, C., Yu, Y. X., Wang, R. M., Wu, K., Xu, D. S., and Guo, G. L. (2004). Uniform metal nanotube arrays by multistep template replication and electrodeposition, *Adv. Mater.*, **16**, pp. 1550–1553.

63. Meng, G., Jung, Y. J., Cao, A., Vajtai, R., and Ajayan, P. M. (2005). Controlled fabrication of hierarchically branched nanopores, nanotubes, and nanowires, *Proc. Natl. Acad. Sci. U. S. A.*, **102**, pp. 7074–7078.

64. Berti, F., Todros, S., Lakshmi, D., Whitcombe, M. J., Chianella, I., Ferroni, M. et al. (2010). Quasi-monodimensional polyaniline nanostructures for enhanced molecularly imprinted polymer-based sensing, *Biosens. Bioelectron.*, **26**, pp. 497–503.

65. Lei, Y., Cai, W., and Wilde, G. (2007). Highly ordered nanostructures with tunable size, shape and properties: A new way to surface nano-patterning using ultra-thin alumina masks, *Prog. Mater. Sci.*, **52**, pp. 465–539.

66. Wurtz, G. A., Evans, P. R., Hendren, W., Atkinson, R., Dickson, W., Pollard, R. J. et al. (2007). Molecular plasmonics with tunable exciton–plasmon coupling strength in J-aggregate hybridized Au nanorod assemblies, *Nano Lett.*, **7**, pp. 1297–1303.

67. Masuda, H. and Satoh, M. (1996). Fabrication of gold nanodot array using anodic porous alumina as an evaporation mask, *Jpn. J. Appl. Phys.*, **35**, L126.

68. Masuda, H., Yasui, K., and Nishio, K. (2000). Fabrication of ordered arrays of multiple nanodots using anodic porous alumina as an evaporation mask, *Adv. Mater.*, **12**, pp. 1031–1033.

69. Lei, Y. and Chim, W. K. (2005). Shape and size control of regularly arrayed nanodots fabricated using ultrathin alumina masks, *Chem. Mater.*, **17**, pp. 580–585.

70. Inguanta, R., Piazza, S., and Sunseri, C. (2008). Novel procedure for the template synthesis of metal nanostructures, *Electrochem. Commun.*, **10**, pp. 506–509.

71. Braunschweig, A. B., Schmucker, A. L., Wei, W. D., and Mirkin, C. A. (2010). Nanostructures enabled by on-wire lithography (OWL), *Chem. Phys. Lett.*, **486**, pp. 89–98.

72. Masuda, H. and Fukuda, K. (1995). Ordered metal nanohole arrays made by a two-step replication of honeycomb structures of anodic alumina, *Science*, **268**, pp. 1466–1468.

73. Lei, Y., Chim, W. K., Zhang, Z., Zhou, T., Zhang, L., Meng, G. et al. (2003). Ordered nanoporous nickel films and their magnetic properties, *Chem. Phys. Lett.*, **380**, pp. 313–318.

74. Yamazaki, K. and Namatsu, H. (2004). 5-nm-order electron-beam lithography for nanodevice fabrication, *Jpn. J. Appl. Phys.*, **43**, 3767.

75. Word, M. J., Adesida, I., and Berger, P. R. (2003). Nanometer-period gratings in hydrogen silsesquioxane fabricated by electron beam lithography, *J. Vac. Sci. Technol. B*, **21**, pp. L12–L15.

76. Hinsberg, W. D., Houle, F. A., Sanchez, M. I., Hoffnagle, J. A., Wallraff, G. M., Medeiros, D. R. et al. (2003). Extendibility of chemically amplified resists: Another brick wall? *Proc. SPIE*, **5039**, pp. 1–15.

77. Haynes, C. L., McFarland, A. D., Zhao, L., Van Duyne, R. P., Schatz, G. C., Gunnarsson, L. et al. (2003). Nanoparticle optics: The importance of radiative dipole coupling in two-dimensional nanoparticle arrays, *J. Phys. Chem. B*, **107**, pp. 7337–7342.

78. Hillenbrand, R., Keilmann, F., Hanarp, P., Sutherland, D. S., and Aizpurua, J. (2003). Coherent imaging of nanoscale plasmon patterns with a carbon nanotube optical probe, *Appl. Phys. Lett.*, **83**, pp. 368–370.

79. Barbillon, G., Bijeon, J. L., Plain, J., and Royer, P. (2009). Sensitive detection of biological species through localized surface-plasmon resonance on gold nanodisks, *Thin Solid Films*, **517**, pp. 2997–3000.

80. Rechberger, W., Hohenau, A., Leitner, A., Krenn, J. R., Lamprecht, B., and Aussenegg, F. R. (2003). Optical properties of two interacting gold nanoparticles, *Opt. Commun.*, **220**, pp. 137–141.

81. Hentschel, M., Saliba, M., Vogelgesang, R., Giessen, H., Alivisatos, A. P., and Liu, N. (2010). Transition from isolated to collective modes in plasmonic oligomers, *Nano Lett.*, **10**, pp. 2721–2726.

82. Bakker, R. M., Yuan, H. K., Liu, Z., Drachev, V. P., Kildishev, A. V., Shalaev, V. M. et al. (2008). Enhanced localized fluorescence in plasmonic nanoantennae, *Appl. Phys. Lett.*, **92**, 043101.

83. Fromm, D. P., Sundaramurthy, A., Schuck, P. J., Kino, G., and Moerner, W. E. (2004). Gap-dependent optical coupling of single "bowtie" nanoantennas resonant in the visible, *Nano Lett.*, **4**, pp. 957–961.

84. Krenn, J. R., Gotschy, W., Somitsch, D., Leitner, A., and Aussenegg, F. R. (1995). Investigation of localized surface plasmons with the photon scanning tunneling microscope, *Appl. Phys. A*, **61**, pp. 541–545.

85. Krenn, J. R., Dereux, A., Weeber, J. C., Bourillot, E., Lacroute, Y., Goudonnet, J. P. et al. (1999). Squeezing the optical near-field zone by plasmon coupling of metallic nanoparticles, *Phys. Rev. Lett.*, **82**, 2590.

86. Auguié, B. and Barnes, W. L. (2008). Collective resonances in gold nanoparticle arrays, *Phys. Rev. Lett.*, **101**, 143902.

87. Clark, A. W., Glidle, A., Cumming, D. R., and Cooper, J. M. (2009). Plasmonic split-ring resonators as dichroic nanophotonic DNA biosensors, *J. Am. Chem. Soc.*, **131**, pp. 17615–17619.

88. Canfield, B. K., Kujala, S., Laiho, K., Jefimovs, K., Turunen, J., and Kauranen, M. (2006). Chirality arising from small defects in gold nanoparticle arrays, *Opt. Express*, **14**, pp. 950–955.

89. Maier, S. A., Kik, P. G., and Atwater, H. A. (2002). Observation of coupled plasmon-polariton modes in Au nanoparticle chain waveguides of different lengths: Estimation of waveguide loss, *Appl. Phys. Lett.*, **81**, pp. 1714–1716.

90. González, M. U., Weeber, J. C., Baudrion, A. L., Dereux, A., Stepanov, A. L., Krenn, J. R. et al. (2006). Design, near-field characterization,

and modeling of 45 surface-plasmon Bragg mirrors, *Phys. Rev. B*, **73**, 155416.

91. Chiu, N. F., Yu, C., Nien, S. Y., Lee, J. H., Kuan, C. H., Wu, K. C. et al. (2007). Enhancement and tunability of active plasmonic by multilayer grating coupled emission, *Opt. Express*, **15**, pp. 11608–11615.
92. Kim, K., Kim, D. J., Moon, S., Kim, D., and Byun, K. M. (2009). Localized surface plasmon resonance detection of layered biointeractions on metallic subwavelength nanogratings, *Nanotechnology*, **20**, 315501.
93. Liu, Y. and Blair, S. (2003). Fluorescence enhancement from an array of subwavelength metal apertures, *Opt. Lett.*, **28**, pp. 507–509.
94. Aksu, S., Yanik, A. A., Adato, R., Artar, A., Huang, M., and Altug, H. (2010). High-throughput nanofabrication of infrared plasmonic nanoantenna arrays for vibrational nanospectroscopy, *Nano Lett.*, **10**, pp. 2511–2518.
95. http://www.nilt.com/
96. Choi, D. G., Jeong, J. H., Sim, Y. S., Lee, E. S., Kim, W. S., and Bae, B. S. (2005). Fluorinated organic–inorganic hybrid mold as a new stamp for nanoimprint and soft lithography, *Langmuir*, **21**, pp. 9390–9392.
97. Chu, J., Meng, F., Han, Z., and Guo, Q. (2010). Large area mold fabrication for the nanoimprint lithography using electron beam lithography, *Sci. China: Technol. Sci.*, **53**, pp. 248–252.
98. Gamo, K. (1996). Nanofabrication by FIB, *Microelectron. Eng.*, **32**, pp. 159–171.
99. Morita, T., Kometani, R., Watanabe, K., Kanda, K., Haruyama, Y., Hoshino, T. et al. (2003). Free-space-wiring fabrication in nano-space by focused-ion-beam chemical vapor deposition, *J. Vac. Sci. Technol. B*, **21**, pp. 2737–2741.
100. Kubena, R. L., Ward, J. W., Stratton, F. P., Joyce, R. J., and Atkinson, G. M. (1991). A low magnification focused ion beam system with 8 nm spot size, *J. Vac. Sci. Technol. B*, **9**, pp. 3079–3083.
101. Gierak, J., Vieu, C., Schneider, M., Launois, H., Ben Assayag, G., and Septier, A. (1997). Optimization of experimental operating parameters for very high-resolution focused ion beam applications, *J. Vac. Sci. Technol. B*, **15**, pp. 2373–2378.
102. Rigneault, H., Capoulade, J., Dintinger, J., Wenger, J., Bonod, N., Popov, E. et al. (2005). Enhancement of single-molecule fluorescence detection in subwavelength apertures, *Phys. Rev. Lett.*, **95**, 117401.

103. Eftekhari, F., Escobedo, C., Ferreira, J., Duan, X., Girotto, E. M., Brolo, A. G. et al. (2009). Nanoholes as nanochannels: Flow-through plasmonic sensing, *Anal. Chem.*, **81**, pp. 4308–4311.

104. Brolo, A. G., Gordon, R., Leathem, B., and Kavanagh, K. L. (2004). Surface plasmon sensor based on the enhanced light transmission through arrays of nanoholes in gold films, *Langmuir*, **20**, pp. 4813–4815.

105. Jeon, H. J., Kim, K. H., Baek, Y. K., Kim, D. W., and Jung, H. T. (2010). New top-down approach for fabricating high-aspect-ratio complex nanostructures with 10 nm scale features, *Nano Lett.*, **10**, pp. 3604–3610.

106. Ferreira, J., Santos, M. J., Rahman, M. M., Brolo, A. G., Gordon, R., Sinton, D., and Girotto, E. M. (2008). Attomolar protein detection using in-hole surface plasmon resonance, *J. Am. Chem. Soc.*, **131**, pp. 436–437.

107. Lesuffleur, A., Im, H., Lindquist, N. C., Lim, K. S., and Oh, S. H. (2008). Laser-illuminated nanohole arrays for multiplex plasmonic microarray sensing, *Opt. Express*, **16**, pp. 219–224.

108. Degiron, A., Lezec, H. J., Yamamoto, N., and Ebbesen, T. W. (2004). Optical transmission properties of a single subwavelength aperture in a real metal, *Opt. Commun.*, **239**, pp. 61–66.

109. Ghaemi, H. F., Thio, T., Grupp, D. E. A., Ebbesen, T. W., and Lezec, H. J. (1998). Surface plasmons enhance optical transmission through subwavelength holes, *Phys. Rev. B*, **58**, 6779.

110. Lesuffleur, A., Im, H., Lindquist, N. C., and Oh, S. H. (2007). Periodic nanohole arrays with shape-enhanced plasmon resonance as real-time biosensors, *Appl. Phys. Lett.*, **90**, 243110.

111. Liu, Z., Wang, Y., Yao, J., Lee, H., Srituravanich, W., and Zhang, X. (2008). Broad band two-dimensional manipulation of surface plasmons, *Nano Lett.*, **9**, pp. 462–466.

112. López-Tejeira, F., Rodrigo, S. G., Martín-Moreno, L., García-Vidal, F. J., Devaux, E., Ebbesen, T. W. et al. (2007). Efficient unidirectional nanoslit couplers for surface plasmons, *Nat. Phys.*, **3**, pp. 324–328.

113. Chang, C. K., Lin, D. Z., Yeh, C. S., Lee, C. K., Chang, Y. C., Lin, M. W. et al. (2007). Experimental analysis of surface plasmon behavior in metallic circular slits, *Appl. Phys. Lett.*, **90**, 061113.

114. Volkov, V. S., Bozhevolnyi, S. I., Devaux, E., Laluet, J. Y., and Ebbesen, T. W. (2007). Wavelength selective nanophotonic components utilizing channel plasmon polaritons, *Nano Lett.*, **7**, pp. 880–884.

115. Hulteen, J. C. and Van Duyne, R. P. (1995). Nanosphere lithography: A materials general fabrication process for periodic particle array surfaces, *J. Vac. Sci. Technol. A*, **13**, pp. 1553–1558.

116. Haynes, C. L. and Van Duyne, R. P. (2001). Nanosphere lithography: A versatile nanofabrication tool for studies of size-dependent nanoparticle optics, *J. Phys. Chem. B*, **105**, pp. 5599–5611.

117. Shah, N. C., Lyandres, O., Yonzon, C. R., Zhang, X., and Van Duyne, R. P. (2007) Surface-enhanced Raman sensing: Glucose and anthrax, Chapter 8, In *New Approaches in Biomedical Spectroscopy, ACS Symposium Series* **963**, Kneipp, K., Aroca, R., Kneipp, H., and Wentrup-Byrne, E., eds., (American Chemical Society, Washington, DC) pp. 107–122.

118. Shah, N. C., Lyandres, O., Walsh, J. T., Glucksberg, M. R., and Van Duyne, R. P. (2007). Lactate and sequential lactate–glucose sensing using surface-enhanced Raman spectroscopy, *Anal. Chem.*, **79**, pp. 6927–6932.

119. Hall, W. P., Modica, J., Anker, J., Lin, Y., Mrksich, M., and Van Duyne, R. P. (2011). A conformation- and ion-sensitive plasmonic biosensor, *Nano Lett.*, **11**, pp. 1098–1105.

120. Zhou, W., Ma, Y., Yang, H., Ding, Y., and Luo, X. (2011). A label-free biosensor based on silver nanoparticles array for clinical detection of serum p53 in head and neck squamous cell carcinoma, *Int. J. Nanomed.*, **6**, pp. 381–386.

121. Haes, A. J. and Van Duyne, R. P. (2002). A nanoscale optical biosensor: Sensitivity and selectivity of an approach based on the localized surface plasmon resonance spectroscopy of triangular silver nanoparticles, *J. Am. Chem. Soc.*, **124**, pp. 10596–10604.

122. Kelf, T. A., Sugawara, Y., Cole, R. M., Baumberg, J. J., Abdelsalam, M. E., Cintra, S. et al. (2006). Localized and delocalized plasmons in metallic nanovoids, *Phys. Rev. B*, **74**, 245415.

123. Mahajan, S., Abdelsalam, M., Suguwara, Y., Cintra, S., Russell, A., Baumberg, J., and Bartlett, P. (2007). Tuning plasmons on nano-structured substrates for NIR-SERS, *Phys. Chem. Chem. Phys.*, **9**, pp. 104–109.

124. Zhu, S., Du, C., and Fu, Y. (2009). Fabrication and characterization of rhombic silver nanoparticles for biosensing, *Opt. Mater.*, **31**, pp. 769–774.

125. Cheung, C. L., Nikolić, R. J., Reinhardt, C. E., and Wang, T. F. (2006). Fabrication of nanopillars by nanosphere lithography, *Nanotechnology*, **17**, 1339.

126. Hicks, E. M., Lyandres, O., Hall, W. P., Zou, S., Glucksberg, M. R., and Van Duyne, R. P. (2007). Plasmonic properties of anchored nanoparticles

fabricated by reactive ion etching and nanosphere lithography, *J. Phys. Chem. C*, **111**, pp. 4116–4124.

127. Hicks, E. M., Zhang, X., Zou, S., Lyandres, O., Spears, K. G., Schatz, G. C., and Van Duyne, R. P. (2005). Plasmonic properties of film over nanowell surfaces fabricated by nanosphere lithography, *J. Phys. Chem. B*, **109**, pp. 22351–22358.
128. Zheng, Y. B. and Huang, T. J. (2008). Surface plasmons of metal nanostructure arrays: From nanoengineering to active plasmonics, *J. Assoc. Lab. Autom.*, **13**, pp. 215–226.
129. Zheng, Y. B., Juluri, B. K., Mao, X., Walker, T. R., and Huang, T. J. (2008). Systematic investigation of localized surface plasmon resonance of long-range ordered Au nanodisk arrays, *J. Appl. Phys.*, **103**, 014308.
130. Gwinner, M. C., Koroknay, E., Fu, L., Patoka, P., Kandulski, W., Giersig, M., and Giessen, H. (2009). Periodic large-area metallic split-ring resonator metamaterial fabrication based on shadow nanosphere lithography, *Small*, **5**, pp. 400–406.
131. Ruan, W., Lu, Z., Zhou, T., Zhao, B., and Niu, L. (2010). Surface micropatterning technique for surface-enhanced Raman scattering analysis, *Anal. Methods*, **2**, pp. 684–687.
132. Chou, S. Y., Krauss, P. R., Zhang, W., Guo, L., and Zhuang, L. (1997). Sub-10 nm imprint lithography and applications, *J. Vac. Sci. Technol. B*, **15**, pp. 2897–2904.
133. Chou, S. Y. and Krauss, P. R. (1997). Imprint lithography with sub-10 nm feature size and high throughput, *Microelectron. Eng.*, **35**, pp. 237–240.
134. Li, M., Chen, L., Zhang, W., and Chou, S. Y. (2002). Pattern transfer fidelity of nanoimprint lithography on six-inch wafers, *Nanotechnology*, **14**, 33.
135. Lucas, B. D., Kim, J. S., Chin, C., and Guo, L. J. (2008). Nanoimprint lithography based approach for the fabrication of large-area, uniformly oriented plasmonic arrays, *Adv. Mater.*, **20**, pp. 1129–1134.
136. Chegel, V., Lucas, B., Guo, J., Lopatynskyi, A., Lopatynska, O., and Poperenko, L. (2009). Detection of biomolecules using optoelectronic biosensor based on localized surface plasmon resonance. Nanoimprint lithography approach, *Semicond. Phys. Quantum Electron. Optoelectron.*, **12**, pp. 91–97.
137. Hu, M., Ou, F. S., Wu, W., Naumov, I., Li, X., Bratkovsky, A. M. et al. (2010). Gold nanofingers for molecule trapping and detection, *J. Am. Chem. Soc.*, **132**, pp. 12820–12822.

138. Malyarchuk, V., Hua, F., Mack, N. H., Velasquez, V. T., White, J. O., Nuzzo, R. G., and Rogers, J. A. (2005). High performance plasmonic crystal sensor formed by soft nanoimprint lithography, *Opt. Express*, **13**, pp. 5669–5675.

139. Stewart, M. E., Mack, N. H., Malyarchuk, V., Soares, J. A., Lee, T. W., Gray, S. K. et al. (2006). Quantitative multispectral biosensing and 1D imaging using quasi-3D plasmonic crystals, *Proc. Natl. Acad. Sci. U. S. A.*, **103**, pp. 17143–17148.

140. Malyarchuk, V., Stewart, M. E., Nuzzo, R. G., and Rogers, J. A. (2007). Spatially resolved biosensing with a molded plasmonic crystal, *Appl. Phys. Lett.*, **90**, 203113.

141. Yao, J., Stewart, M. E., Maria, J., Lee, T. W., Gray, S. K., Rogers, J. A. et al. (2008). Seeing molecules by eye: Surface plasmon resonance imaging at visible wavelengths with high spatial resolution and submonolayer sensitivity, *Angew. Chem. Int. Ed.*, **120**, pp. 5091–5095.

142. Stewart, M. E., Yao, J., Maria, J., Gray, S. K., Rogers, J. A., and Nuzzo, R. G. (2009). Multispectral thin film biosensing and quantitative imaging using 3D plasmonic crystals, *Anal. Chem.*, **81**, pp. 5980–5989.

143. Kim, S., Jung, J. M., Choi, D. G., Jung, H. T., and Yang, S. M. (2006). Patterned arrays of Au rings for localized surface plasmon resonance, *Langmuir*, **22**, pp. 7109–7112.

144. McPhillips, J., McClatchey, C., Kelly, T., Murphy, A., Jonsson, M. P., Wurtz, G. A. et al. (2011). Plasmonic sensing using nanodome arrays fabricated by soft nanoimprint lithography, *J. Phys. Chem. C*, **115**, pp. 15234–15239.

145. Yan, X. M., Kwon, S., Contreras, A. M., Bokor, J., and Somorjai, G. A. (2005). Fabrication of large number density platinum nanowire arrays by size reduction lithography and nanoimprint lithography, *Nano Lett.*, **5**, pp. 745–748.

146. Alvarez-Puebla, R., Cui, B., Bravo-Vasquez, J. P., Veres, T., and Fenniri, H. (2007). Nanoimprinted SERS-active substrates with tunable surface plasmon resonances, *J. Phys. Chem. C*, **111**, pp. 6720–6723.

147. Lee, S. W., Lee, K. S., Ahn, J., Lee, J. J., Kim, M. G., and Shin, Y. B. (2011). Highly sensitive biosensing using arrays of plasmonic Au nanodisks realized by nanoimprint lithography, *ACS Nano*, **5**, pp. 897–904.

148. Ko, S. H., Park, I., Pan, H., Grigoropoulos, C. P., Pisano, A. P., Luscombe, C. K. et al. (2007). Direct nanoimprinting of metal nanoparticles for nanoscale electronics fabrication, *Nano Lett.*, **7**, pp. 1869–1877.

149. Park, I., Ko, S. H., Pan, H., Grigoropoulos, C. P., Pisano, A. P., Fréchet, J. M. et al. (2008). Nanoscale patterning and electronics on flexible substrate by direct nanoimprinting of metallic nanoparticles, *Adv. Mater.*, **20**, pp. 489–496.

150. Torres, C. S., Zankovych, S., Seekamp, J., Kam, A. P., Cedeno, C. C., Hoffmann, T. et al. (2003). Nanoimprint lithography: An alternative nanofabrication approach, *Mater. Sci. Eng. C*, **23**, pp. 23–31.

151. Khang, D. Y. and Lee, H. H. (2000). Room-temperature imprint lithography by solvent vapor treatment, *Appl. Phys. Lett.*, **76**, pp. 870–872.

152. Khang, D. Y., Yoon, H., and Lee, H. H. (2001). Room-temperature imprint lithography, *Adv. Mater.*, **13**, pp. 749–752.

153. Kedem, O., Tesler, A. B., Vaskevich, A., and Rubinstein, I. (2011). Sensitivity and optimization of localized surface plasmon resonance transducers, *ACS Nano*, **5**, pp. 748–760.

154. Lopatynskyi, A. M., Lytvyn, V. K., Nazarenko, V. I., Guo, L. J., Lucas, B. D., and Chegel, V. I. (2015). Au nanostructure arrays for plasmonic applications: Annealed island films versus nanoimprint lithography, *Nanoscale Res. Lett.*, **10**, 99.

155. Chien, F. C. and Chen, S. J. (2006). Direct determination of the refractive index and thickness of a biolayer based on coupled waveguide-surface plasmon resonance mode, *Opt. Lett.*, **31**, pp. 187–189.

156. Wright, A. K. and Thompson, M. R. (1975). Hydrodynamic structure of bovine serum albumin determined by transient electric birefringence, *Biophys. J.*, **15**, pp. 137–141.

157. Barrios, C. A., Banuls, M. J., Gonzalez-Pedro, V., Gylfason, K. B., Sanchez, B., Griol, A. et al. (2008). Label-free optical biosensing with slot-waveguides, *Opt. Lett.*, **33**, pp. 708–710.

158. Chegel, V. I. (2012) Nanostructured materials for biosensor applications: Comparative review of preparation methods, Chapter 13, In *Manipulation of Nanoscale Materials: An Introduction to Nanoarchitectonics, RSC Nanoscience & Nanotechnology* **24**, Ariga, K., ed., (The Royal Society of Chemistry, Cambridge) pp. 318–355.

159. Kinnan, M. K. and Chumanov, G. (2010). Plasmon coupling in two-dimensional arrays of silver nanoparticles: II. Effect of the particle size and interparticle distance, *J. Phys. Chem. C*, **114**, pp. 7496–7501.

160. Noguez, C. (2005). Optical properties of isolated and supported metal nanoparticles, *Opt. Mater.*, **27**, pp. 1204–1211.

161. Hu, M., Novo, C., Funston, A., Wang, H., Staleva, H., Zou, S. et al. (2008). Dark-field microscopy studies of single metal nanoparticles:

Understanding the factors that influence the linewidth of the localized surface plasmon resonance, *J. Mater. Chem.*, **18**, pp. 1949–1960.

162. Chang, R. K. and Furtak, T. E. (1982). *Surface Enhanced Raman Scattering* (Plenum Press, New York).

163. Hartstein, A., Kirtley, J. R., and Tsang, J. C. (1980). Enhancement of the infrared absorption from molecular monolayers with thin metal overlayers, *Phys. Rev. Lett.*, **45**, pp. 201–204.

164. Osawa, M. and Ikeda, M. (1991). Surface-enhanced infrared absorption of p-nitrobenzoic acid deposited on silver island films: Contributions of electromagnetic and chemical mechanisms, *J. Phys. Chem.*, **95**, pp. 9914–9919.

165. Kosobukin, V. A. (1983). Effect of enhancement of external electric field near metal surface and its manifestation in spectroscopy, *Phys. Chem. Mech. Surfaces*, **12**, pp. 5–20.

166. Bondar, V. V. and Kurik, M. V. (1980). Exciton luminescence on anthracene-gold boundary, *J. Exp. Theor. Phys.*, **51**, pp. 46–49.

167. Dovbeshko, G. I., Shirshov, Y. M., Chegel, V. I., and Fesenko, O. M. (2004). Experimental and calculated enhancement factor in the SEIRA method, *Proc. SPIE*, **5507**, pp. 386–396.

168. Kuhne, C., Steiner, G., Fischer, W. B., and Salzer, R. (1998). Surface enhanced FTIR spectroscopy on membranes, *Fresenius J. Anal. Chem.*, **360**, pp. 750–754.

169. Kosobukin, V. A. (1985). Collective modes in enhancement of external electric field on the surface of the metals, *Proc. USSR Acad. Sci. Phys.*, **49**, pp. 1111–1120 (in Russian).

170. Chumanov, G., Sokolov, K., Gregory, B. W., and Cotton, T. M. (1995). Colloidal metal films as a substrate for surface-enhanced spectroscopy, *J. Phys. Chem.*, **99**, pp. 9466–9471.

171. Sun, J., Li, Z., Sun, Y., Zhong, L., Huang, J., Zhang, J. et al. (2018). Uniform and reproducible plasmon-enhanced fluorescence substrate based on PMMA-coated, large-area Au@Ag nanorod arrays, *Nano Res.*, **11**, pp. 953–965.

172. Kühn, S., Håkanson, U., Rogobete, L., and Sandoghdar, V. (2006). Enhancement of single-molecule fluorescence using a gold nanoparticle as an optical nanoantenna, *Phys. Rev. Lett.*, **97**, 017402.

173. Zhang, J., Fu, Y., Chowdhury, M. H., and Lakowicz, J. R. (2007). Metal-enhanced single-molecule fluorescence on silver particle monomer and dimer: Coupling effect between metal particles, *Nano Lett.*, **7**, pp. 2101–2107.

174. Novotny, L. and Hecht, B. (2006) *Principles of Nano-Optics* (Cambridge University Press, New York).

175. Jackson, J. D. (1998). *Classical Electrodynamics*, 3rd Ed. (John Wiley & Sons, New York).

176. Carminati, R., Greffet, J. J., Henkel, C., and Vigoureux, J. M. (2006). Radiative and non-radiative decay of a single molecule close to a metallic nanoparticle, *Opt. Commun.*, **261**, pp. 368–375.

177. Anger, P., Bharadwaj, P., and Novotny, L. (2006). Enhancement and quenching of single-molecule fluorescence, *Phys. Rev. Lett.*, **96**, 113002.

178. Chegel, V. I., Naum, O. M., Lopatynskyi, A. M., and Lytvyn, V. K. (2015). Plasmonic nanochips for application in surface-enhanced fluorescence spectroscopy: Factor of dielectric substrate, Chapter 27, In *Nanoplasmonics, Nano-Optics, Nanocomposites, and Surface Studies*, Fesenko, O., Yatsenko, L., eds., Springer Proceedings in Physics 167, (Springer: Switzerland) pp. 395–412.

179. Lozovski, V. (2010). The effective susceptibility concept in the electrodynamics of nano-systems, *J. Comput. Theor. Nanosci.*, **7**, pp. 2077–2093.

180. Myroshnychenko, V., Rodríguez-Fernández, J., Pastoriza-Santos, I., Funston, A. M., Novo, C., Mulvaney, P. et al. (2008). Modelling the optical response of gold nanoparticles, *Chem. Soc. Rev.*, **37**, pp. 1792–1805.

181. Bardhan, R., Grady, N. K., Cole, J. R., Joshi, A., and Halas, N. J. (2009). Fluorescence enhancement by Au nanostructures: Nanoshells and nanorods, *ACS Nano*, **3**, pp. 744–752.

182. Kim, J., Dantelle, G., Revaux, A., Bérard, M., Huignard, A., Gacoin, T., and Boilot, J. P. (2010). Plasmon-induced modification of fluorescent thin film emission nearby gold nanoparticle monolayers, *Langmuir*, **26**, pp. 8842–8849.

183. Derom, S., Berthelot, A., Pillonnet, A., Benamara, O., Jurdyc, A. M., Girard, C. et al. (2013). Metal enhanced fluorescence in rare earth doped plasmonic core–shell nanoparticles, *Nanotechnology*, **24**, 495704.

184. Lakowicz, J. R. (2006). *Principles of Fluorescence Spectroscopy*, 3rd Ed. (Springer US, New York).

185. Drummen, G. P. (2012). Fluorescent probes and fluorescence (microscopy) techniques—illuminating biological and biomedical research, *Molecules*, **17**, pp. 14067–14090.

186. Demchenko, A. P. (2009). *Introduction to Fluorescence Sensing* (Springer, Dordrecht).

187. Valeur, B. and Berberan-Santos, M. N. (2012). *Molecular Fluorescence: Principles and Applications*, 2nd Ed. (Wiley-VCH Verlag, Weinheim).

188. Demchenko, A. P., Mély, Y., Duportail, G., and Klymchenko, A. S. (2009). Monitoring biophysical properties of lipid membranes by environment-sensitive fluorescent probes, *Biophys. J.*, **96**, pp. 3461–3470.

189. Millar, D. P. (1996). Fluorescence studies of DNA and RNA structure and dynamics, *Curr. Opin. Struct. Biol.*, **6**, pp. 322–326.

190. Royer, C. A. (2006). Probing protein folding and conformational transitions with fluorescence, *Chem. Rev.*, **106**, pp. 1769–1784.

191. Shendure, J. and Ji, H. (2008). Next-generation DNA sequencing, *Nat. Biotechnol.*, **26**, pp. 1135–1145.

192. McGinn, S. and Gut, I. G. (2013). DNA sequencing: Spanning the generations, *New Biotechnol.*, **30**, pp. 366–372.

193. Liu, S. and Tang, Z. (2010). Nanoparticle assemblies for biological and chemical sensing, *J. Mater. Chem.*, **20**, pp. 24–35.

194. Hötzer, B., Medintz, I. L., and Hildebrandt, N. (2012). Fluorescence in nanobiotechnology: Sophisticated fluorophores for novel applications, *Small*, **8**, pp. 2297–2326.

195. Santos, G. M., Zhao, F., Zeng, J., Li, M., and Shih, W. C. (2015). Label-free, zeptomole cancer biomarker detection by surface-enhanced fluorescence on nanoporous gold disk plasmonic nanoparticles, *J. Biophotonics*, **8**, pp. 855–863.

196. Riu, J., Maroto, A., and Rius, F. X. (2006). Nanosensors in environmental analysis, *Talanta*, **69**, pp. 288–301.

197. Jahn, M., Patze, S., Hidi, I. J., Knipper, R., Radu, A. I., Mühlig, A. et al. (2016). Plasmonic nanostructures for surface enhanced spectroscopic methods, *Analyst*, **141**, pp. 756–793.

198. Gooding, J. J. (2006). Biosensor technology for detecting biological warfare agents: Recent progress and future trends, *Anal. Chim. Acta*, **559**, pp. 137–151.

199. Amin, R., Hwang, S., and Park, S. H. (2011). Nanobiotechnology: An interface between nanotechnology and biotechnology, *Nano*, **6**, pp. 101–111.

200. Wang, J. (2009). Biomolecule-functionalized nanowires: From nanosensors to nanocarriers, *ChemPhysChem*, **10**, pp. 1748–1755.

201. Suh, W. H., Suh, Y. H., and Stucky, G. D. (2009). Multifunctional nanosystems at the interface of physical and life sciences, *Nano Today*, **4**, pp. 27–36.

202. Howes, P. D., Rana, S., and Stevens, M. M. (2014). Plasmonic nanomaterials for biodiagnostics, *Chem. Soc. Rev.*, **43**, pp. 3835–3853.

203. Karolin, J. O. and Geddes, C. D. (2012). Reduced lifetimes are directly correlated with excitation irradiance in metal-enhanced fluorescence (MEF), *J. Fluoresc.*, **22**, pp. 1659–1662.

204. Motl, N. E., Smith, A. F., DeSantis, C. J., and Skrabalak, S. E. (2014). Engineering plasmonic metal colloids through composition and structural design, *Chem. Soc. Rev.*, **43**, pp. 3823–3834.

205. Chegel, V. I., Lytvyn, V. K., Lopatynskyi, A. M., Shepeliavyi, P. E., Lytvyn, O. S., and Goltvyanskyi, Y. V. (2015). Plasmon-enhanced fluorometry based on gold nanostructure arrays. Method and device, *Semicond. Phys. Quantum Electron. Optoelectron.*, **18**, pp. 272–278.

206. Tovmachenko, O. G., Graf, C., van den Heuvel, D. J., van Blaaderen, A., and Gerritsen, H. C. (2006). Fluorescence enhancement by metal-core/silica-shell nanoparticles, *Adv. Mater.*, **18**, pp. 91–95.

207. Drozdowicz-Tomsia, K. and Goldys, E. M. (2011). Gold and silver nanowires for fluorescence enhancement, Chapter 14, In *Nanowires: Fundamental Research*, Hashim, A. (ed.), (InTech, Rijeka, Croatia) pp. 309–332.

208. Krutyakov, Y. A., Kudrinskiy, A. A., Olenin, A. Y., and Lisichkin, G. V. (2008). Synthesis and properties of silver nanoparticles: Advances and prospects, *Russ. Chem. Rev.*, **77**, pp. 233–257.

209. Chen, Y., Munechika, K., and Ginger, D. S. (2007). Dependence of fluorescence intensity on the spectral overlap between fluorophores and plasmon resonant single silver nanoparticles, *Nano Lett.*, **7**, pp. 690–696.

210. Geddes, C. D., Parfenov, A., Roll, D., Fang, J., and Lakowicz, J. R. (2003). Electrochemical and laser deposition of silver for use in metal-enhanced fluorescence, *Langmuir*, **19**, pp. 6236–6241.

211. Gartia, M. R., Hsiao, A., Sivaguru, M., Chen, Y., and Liu, G. L. (2011). Enhanced 3D fluorescence live cell imaging on nanoplasmonic substrate, *Nanotechnology*, **22**, 365203.

212. Kruszewski, S., Wybranowski, T., Cyrankiewicz, M., Ziomkowska, B., and Pawlaczyk, A. (2008). Enhancement of FITC fluorescence by silver colloids and silver island films, *Acta Phys. Pol. A*, **113**, pp. 1599–1608.

213. Kang, K. A., Wang, J., Jasinski, J. B., and Achilefu, S. (2011). Fluorescence manipulation by gold nanoparticles: From complete quenching to extensive enhancement, *J. Nanobiotechnol.*, **9**, 16.

214. Feng, A. L., You, M. L., Tian, L., Singamaneni, S., Liu, M., Duan, Z. et al. (2015). Distance-dependent plasmon-enhanced fluorescence of upconversion nanoparticles using polyelectrolyte multilayers as tunable spacers, *Sci. Rep.*, **5**, 7779.

215. Dragan, A. I. and Geddes, C. D. (2011). Excitation volumetric effects (EVE) in metal-enhanced fluorescence, *Phys. Chem. Chem. Phys.*, **13**, pp. 3831–3838.

216. Bek, A., Jansen, R., Ringler, M., Mayilo, S., Klar, T. A., and Feldmann, J. (2008). Fluorescence enhancement in hot spots of AFM-designed gold nanoparticle sandwiches, *Nano Lett.*, **8**, pp. 485–490.

217. Aslan, K., Wu, M., Lakowicz, J. R., and Geddes, C. D. (2007). Fluorescent core–shell Ag@SiO_2 nanocomposites for metal-enhanced fluorescence and single nanoparticle sensing platforms, *J. Am. Chem. Soc.*, **129**, pp. 1524–1525.

218. Aslan, K., Gryczynski, I., Malicka, J., Matveeva, E., Lakowicz, J. R., and Geddes, C. D. (2005). Metal-enhanced fluorescence: An emerging tool in biotechnology, *Curr. Opin. Biotechnol.*, **16**, pp. 55–62.

219. Ray, K., Badugu, R., and Lakowicz, J. R. (2007). Polyelectrolyte layer-by-layer assembly to control the distance between fluorophores and plasmonic nanostructures, *Chem. Mater.*, **19**, pp. 5902–5909.

220. Aslan, K., Malyn, S. N., and Geddes, C. D. (2007). Metal-enhanced fluorescence from gold surfaces: Angular dependent emission, *J. Fluoresc.*, **17**, pp. 7–13.

221. Leonard, K., You, J., Takahashi, Y., Yonemura, H., Kurawaki, J., and Yamada, S. (2015). Enhanced photoelectrochemical response of polythiophene photoelectrodes with controlled arrays of silver nanocubes, *J. Phys. Chem. C*, **119**, pp. 8829–8837.

222. Camposeo, A., Persano, L., Manco, R., Wang, Y., Del Carro, P., Zhang, C. et al. (2015). Metal-enhanced near-infrared fluorescence by micropatterned gold nanocages, *ACS Nano*, **9**, pp. 10047–10054.

223. Xu, S., Cao, Y., Zhou, J., Wang, X., Wang, X., and Xu, W. (2011). Plasmonic enhancement of fluorescence on silver nanoparticle films, *Nanotechnology*, **22**, 275715.

224. Rainò, G., Stöferle, T., Park, C., Kim, H. C., Topuria, T., Rice, P. M. et al. (2011). Plasmonic nanohybrid with ultrasmall Ag nanoparticles and fluorescent dyes, *ACS Nano*, **5**, pp. 3536–3541.

225. Bharadwaj, P., Anger, P., and Novotny, L. (2006). Nanoplasmonic enhancement of single-molecule fluorescence, *Nanotechnology*, **18**, 044017.

226. Kinkhabwala, A., Yu, Z., Fan, S., Avlasevich, Y., Müllen, K., and Moerner, W. E. (2009). Large single-molecule fluorescence enhancements produced by a bowtie nanoantenna, *Nat. Photonics*, **3**, pp. 654–657.

227. Acuna, G. P., Möller, F. M., Holzmeister, P., Beater, S., Lalkens, B., and Tinnefeld, P. (2012). Fluorescence enhancement at docking sites of DNA-directed self-assembled nanoantennas, *Science*, **338**, pp. 506–510.

228. Ming, T., Zhao, L., Yang, Z., Chen, H., Sun, L., Wang, J., and Yan, C. (2009). Strong polarization dependence of plasmon-enhanced fluorescence on single gold nanorods, *Nano Lett.*, **9**, pp. 3896–3903.

229. Chegel, V. I., Lopatynskyi, A. M., Lytvyn, V. K., Naum, O. M., Nazarenko, V. I., and Demchenko, A. P. (2015). High-conductive nanostructures in biochemical studies: Fluorescence enhancing, *Biotechnol. Acta*, **8**, pp. 9–18.

Chapter 4

Peculiarities of Surface Plasmon Resonance Method Application for the Investigation of Biomolecules and Biomolecular Interactions

4.1 Introduction

Biosensing is a growing field with main applications for registration and analysis of (bio)molecular interactions in biochemical, biomedical, pharmaceutical, diagnostic, and environmental industries [1, 2]. Biosensors based on surface plasmon resonance (SPR) as representatives of surface optical affinity biosensors possess several unique features, among which are label-free detection, real-time operation, and high sensitivity [3]. SPR and localized SPR (LSPR) biosensors have been applied for characterization of various (bio) molecular systems, including small molecules [4, 5], nucleotides [6, 7], viruses [8, 9], peptides [10, 11], and proteins [12, 13]. Substantial scientific efforts are concentrated on developing novel sensor elements and improving the sensitivity of SPR and LSPR biosensors [14, 15].

This chapter presents the results of specific practical studies of biomolecular interactions with SPR and LSPR techniques. The first study deals with SPR measurement of IgG–anti-IgG immunological reaction and calculation of surface molecular concentration and

Molecular Plasmonics: Theory and Applications
Volodymyr I. Chegel and Andrii M. Lopatynskyi

ISBN 978-981-4800-65-5 (Hardcover), 978-0-429-29511-9 (eBook)
www.jennystanford.com

mass of respective biomolecular coverings. The second study is focused on developing an LSPR sensor based on gold nanoparticle arrays fabricated using nanoimprint lithography and reports sensor capability to monitor antigen–antibody specific reactions between bovine serum albumin (BSA) and anti-BSA immunoglobulin.

4.2 Experimental Procedure of Surface Plasmon Resonance Technique

The SPR detection of biomolecules or their interactions with quantitative determination of mass of the substance (that is present at the sensitive surface of the sensing instrument) involves the following operations:

- The sensitive surface of a metal (gold, silver) layer of the sensor (Fig. 4.1) is brought into contact with an ambience (say, air or water) whose refractive index is known.

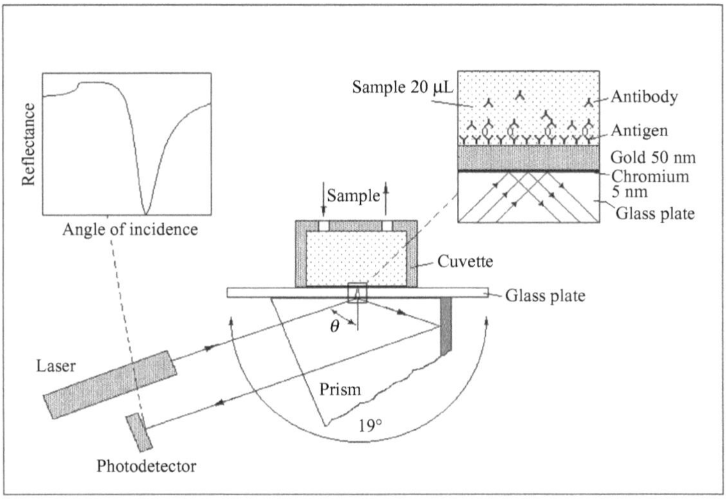

Figure 4.1 Simplified scheme of SPR sensor.

- The angular dependence of the intensity of reflected light is measured over the range of angles of incidence $\theta_{min} < \theta < \theta_{max}$ (here θ_{min} and θ_{max} are the minimal and maximal angles, respectively; they are chosen in such a way that the above range includes the plasmon resonance angle).

- The optical parameters (the refractive index $n + ik$ and thickness d of conducting film) of the layered structure in the absence of the substance to be analyzed are determined from the results of measurements using the procedure of fitting the experimental and theoretical SPR curves. Usually, the parameters $n + ik$ and d of the conducting film are known.
- The conducting film surface is brought into contact with a sample of the substance to be analyzed in such a way that the latter could be adsorbed or, in some other way, be bound to the surface.
- The angular dependence of the intensity of reflected light is measured.
- The refractive index of the adsorbed substance n_x is set. It may be chosen in different ways, so that the condition $n_x > n_0$ (where n_0 is the refractive index of the ambience) is fulfilled. The optimal n_x value is that close to the refractive index of the substance to be analyzed. For instance, if the substance to be analyzed is protein molecules, and their study is performed in a water solution, then one may take $n_x = 1.45$.
- The thickness d_x of the layer of the substance to be analyzed is determined from the results of measurements using the procedure of fitting the experimental and theoretical SPR curves.
- Knowing the optical parameters n and d of the layer studied, one can determine the mass of the adsorbed substance using the following equation obtained from the expression for molecular refraction of the substance analyzed [16]:

$$\Gamma = d(n - n_0)/(dn/dc), \tag{4.1}$$

where Γ [g/cm^2] is the mass of the adsorbed substance; $d(n)$ is the thickness (refractive index) of the layer studied; and n_0 is the refractive index of the ambience. The typical value of dn/dc (i.e., the ratio between the refractive index and solution concentration increments) for different proteins is 0.188 cm^3/g [17]. It should be noted that the only true physical quantity that is invariant under variation of n and d values and characterizes the actual amount of adsorbed substance is the Γ value, while n_x and d_x are not physically meaningful at the above measurement procedure. If the optical characteristics of a multilayer sensor structure vary with time, then the angular position of SPR varies accordingly. To register the kinetics of variation of the angular position of SPR, one should measure both the total SPR curve and the angular position of SPR at set intervals. In such way, one can register the kinetics of additional mass (say, biomolecules) inflow to the area of damping electric field of surface plasmon, as well as kinetics of mass outflow from that area. Then the kinetic curves $\theta_{\text{sp}}(t)$ (SPR sensograms) can be analyzed using special-purpose computer software to obtain the kinetic constants of association, dissociation, etc. If biomolecules have expressed eccentricity (like immunoglobulin, etc.), the method of SPR response estimation (Chapter 2, Section 2.2.2) could be used to improve the accuracy of measurements, as shown below in the original experimental research [18].

4.3 Surface Plasmon Resonance Study of IgG–Anti-IgG Biospecific Reaction

Biospecific reactions were investigated between the complementary (biospecific) pairs of IgG–anti-IgG immunoglobulins and between protein G and IgG in the flow mode using phosphate-buffered saline (PBS) as a buffer solution (pH 7.3). Concentrations of both IgG and anti-IgG solutions were 200 μg/mL. Glycine buffer solution (pH 2.2) was used to evaluate the reaction specificity. Experimental kinetic dependences of the SPR angular position for IgG–anti-IgG and protein G-IgG specific reactions are presented in Figs. 4.2a and 4.2b.

Immunoglobulins are described as molecules with an expressed eccentricity (size 3 nm × 4 nm × 25 nm) [19], so the theoretical approach described in Subsection 2.2.2.3 can be applied to this biostructure. As mentioned above, the theoretical approximation presented relates to monolayers. From this point of view, parts A–B and A–C of the immunological response (Fig. 4.2a) can be theoretically considered as monolayers with different shape parameters ζ for IgG molecules and IgG–anti-IgG complexes. Here $\zeta = h_{||}/h_{\perp}$ (where $h_{||}$ and $h_{\perp}$ are the semiaxes of ellipsoid, parallel (||) and perpendicular (⊥) to the substrate surface plane), which defines the shape of the molecules of similar mass (see Subsection 2.2.2.3).

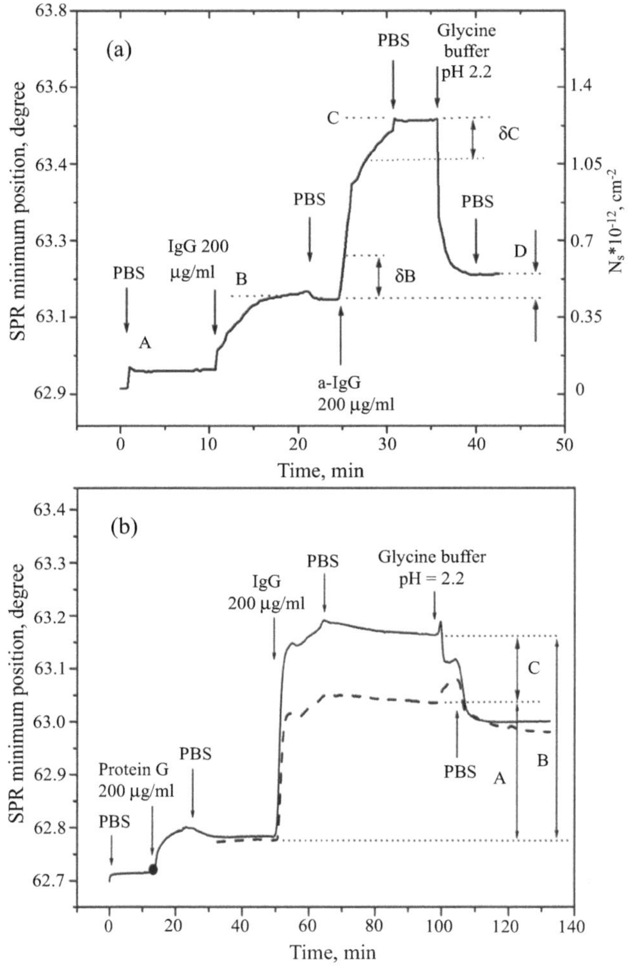

Figure 4.2 (a) Kinetic dependence of the SPR angular position for the IgG–anti-IgG reaction. The right axis reflects the surface concentration of the IgG monolayer (to level B) or the IgG–anti-IgG complex (to level C); *δB* and *δC* are deviations of the SPR angular position (directions of deviations are indicated by arrows); level D is associated with the specificity of reaction (if difference between levels B and D is close to zero, specificity is close to 100%). (b) Typical comparative kinetic dependence of the SPR angular position upon adsorption of IgG molecules on the bare gold surface (dashed line) and the surface covered with protein G molecules (solid line): A is the SPR response for adsorption of IgG onto the bare gold surface after washing with PBS; B is the same response for a surface modified by protein G; C is the difference associated with the modification of the surface of the sensor. Reprinted from Ref. [18], Copyright 2008, with permission from Elsevier.

The SPR angular position after rinsing with low-pH glycine buffer indicates that the effect of non-specific adsorption of anti-IgG on the SPR response in this case is small and the gradient between levels B and C reflects preferably the binding of anti-IgG molecules to specific sites of IgG molecules and complex formation. The value of ζ for the IgG–anti-IgG complex can be considered to be much smaller than the value for one IgG molecule due to the increased eccentricity of the complex relative to one molecule and, accordingly, the shift in the SPR angular position after IgG–anti-IgG binding has to be bigger. This assumption correlates with the calculated values of the SPR shift (see Fig. 2.11a) and with values observed in Fig. 4.2a for kinetic dependence, where the SPR angular position for the binding of anti-IgG is significantly greater than that for the adsorption of the IgG monolayer despite the same concentration of IgG and anti-IgG (200 μg/mL) and approximately identical IgG and anti-IgG molecular weights. It is worth noting that the result shown in Fig. 4.2a represents the maximum reported response to the anti-IgG binding, since when the experiments were repeated, there was a deviation in the SPR response within certain limits, denoted as δB and δC. This strong deviation is often observed in SPR experiments; it can be easily explained by the change in the parameter ζ for the monolayers under consideration during IgG adsorption onto the gold surface and for the IgG-anti-IgG binding process. This assumption is confirmed by SPR experiments with the oriented immobilization of IgG molecules using protein G (see Fig. 4.2b). It is known that the protein G molecules, immobilized as the first layer, promote the position of immunoglobulin normally oriented relative to the surface of sensor [20] (see Fig. 4.3a).

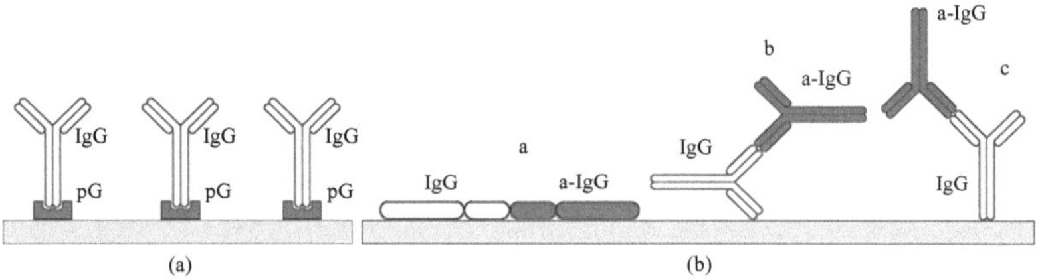

Figure 4.3 (a) Scheme of possible orientation of IgG molecules relative to the gold surface modified by protein G. (b) Scheme of possible orientation of the IgG–anti-IgG complex relative to the gold surface. Here $\zeta_a > \zeta_b > \zeta_c$. Reprinted from Ref. [18], Copyright 2008, with permission from Elsevier.

The sequence of comparative SPR experiments shows a clear difference in the SPR response for IgG immobilization onto the gold surface that was initially modified by protein G. The rational explanation for the observation of this difference lies in the presence of an initial layer of protein G, which may affect the normal orientation of IgG molecules relative to the gold surface. This explanation is acceptable, since SPR response after the removal of specifically immobilized IgG molecules (using glycine buffer, pH 2.2) is the same as the response to the randomly adsorbed IgG monolayer on the bare surface of gold, so the SPR response value in this case depends on the part of the normally oriented IgG molecules for both comparative experiments.

Following the theoretical approach (Chapter 2, Section 2.2.2) for the results presented here, the SPR response in this study depends on the spatial position of the individual biomolecule or biomolecular complex on the Au surface, if the shape of these objects is elongated, as in the case of immunoglobulin. Clearly, the difference in the orientation of such biomolecules is not the same as the difference between prolate and oblate ellipsoids; rather, it is a strict approximation. However, there is a high probability that a maximum SPR response will be observed when individual molecules or complexes are oriented perpendicular to the gold surface when immobilized (position c in Fig. 4.3b). For the intermediate position b, the response decreases and becomes minimal when the complex or molecule is oriented parallel to the surface (position a). In addition, it can be assumed that the presence of molecular complexes of type c at sufficiently high concentrations provides a clear SPR angle shift since for the positions a and b, which are simulated by oblate ellipsoids, the SPR angle shift weakly depends on the shape of the particle, as the calculations showed.

The surface roughness of Au is important in the disordered nature of the IgG adsorption and increases the values of δB and δC. The influence of the shape of biomolecules on the deviation of the SPR response becomes smaller with decreasing eccentricity of the biomolecule shape and is minimal for spherical objects.

The above-mentioned consideration of the spatial shape of biomolecules makes it possible to calculate nomograms for monolayers of a certain type of biomolecules or biomolecular complexes, taking into account their geometric shape (or position). Based on the assumption that the monolayers of biomolecules in the described experiments are located on a surface with average distances between molecules in the order of their linear size, nomograms were calculated for determining the surface concentration of IgG molecules or IgG–anti-IgG complexes (see Fig. 4.4).

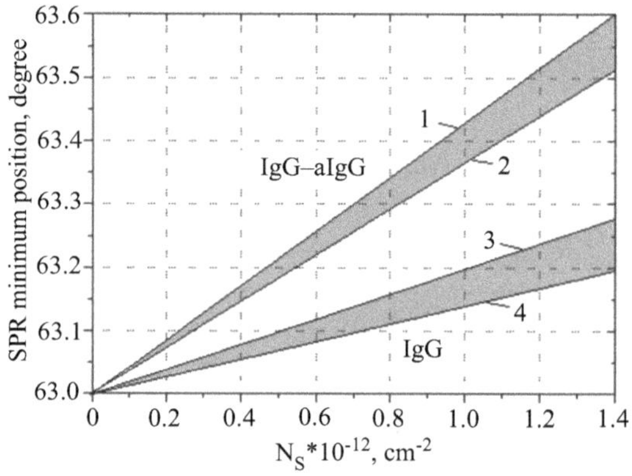

Figure 4.4 Calculated dependences of the angular SPR response on the surface concentration of IgG molecules and IgG–anti-IgG complexes. For positions 1 and 3, $\zeta = 0.15$; for positions 2 and 4, $\zeta = 2$. Reprinted from Ref. [18], Copyright 2008, with permission from Elsevier.

The deviation of the observed surface concentration N_S (shaded area) is due to the value of the parameter ζ, which describes the spatial position of the IgG molecule or the IgG–anti-IgG complex. It is worth to note that nomograms consider only specifically constructed complexes, without taking into account the possible non-specific adsorption of anti-IgG on the gold surface. Thus, with the help of the SPR method, an approximate estimate of molecular adsorption can be made directly in units of the surface concentration N_S, which in the experiments under consideration was about $(1.1 \pm 0.1) \times 10^{12}$ cm^{-2}. In addition, Eq. (4.2) can be also used to calculate the biomolecular layer mass in mg/m^2 [21]:

$$\Gamma = \frac{N_S M}{A_m}, \tag{4.2}$$

where M is the molecular weight of the biomolecule or biomolecular complex and $A_m = (\tilde{\chi}_{\perp} + \tilde{\chi}_{||})/2$ (where the indices $\perp$ and $||$ indicate that the molecules are perpendicular or parallel to the surface) considering that the number of prolate and oblate particles is the same. The mass of the IgG layer was determined as equal to 3.66 ± 0.5 mg/m^2, while the mass of the complex layer was 6.5 ± 1.0 mg/m^2, which is approximately twice as much, taking into account the observed weak non-specific adsorption.

4.4 Biomolecules Registration Using an Optoelectronic Biosensor Based on LSPR

4.4.1 Peculiarities of LSPR Technique Biosensing Applications

The phenomenon of localized surface plasmon resonance, as a collective oscillation of free electrons in nanosized structures of high-conductive metals [22, 23], has become widely attractive to researchers since the development of various methods of production of nanostructures that made it possible to fabricate relatively large arrays of nanoparticles and nanostructured films [24, 25]. In recent decades, numerous important applications of localized surface plasmon resonance have been demonstrated. In 1995, Kreibig [22] showed that the optical density of the immobilized monolayer of colloidal gold nanoparticles is a parameter sensitive to the refractive index of the surrounding solvent. Several experimental works related to the influence of the parameters of nanoparticle systems (e.g., shape, size, and interparticle distance) on their light extinction properties and observed optical dichroism were published [26–32].

Several basic methods for fabrication of nanostructure arrays to be used as sensitive elements in LSPR biosensors were demonstrated. Two of the most accepted methods—deposition of nanoparticles from a colloidal solution and nanosphere lithography—were widely used to study the fundamental properties of these systems [27, 33–35]. Van Duyne et al. [36] and Nathan et al. [37, 38] showed that silver or gold nanostructures on the surface of mica or glass could be fabricated using nanosphere lithography with monolayer formation, which could be optically detected and allowing the registration of the biomolecular interactions. G. Chumanov et al. [39] showed the possibility of creating stable monolayers of silver nanoparticles obtained from colloidal solutions using a transition polymer layer, which, by partially fixing a nanoparticle in its volume, does not prevent the appearance of LSPR on a polymer-free surface. The arrays of silver nanostructures manufactured by this method exhibit extremely narrow peaks in the absorption spectra, which can be used to create highly sensitive biosensors. The main disadvantages of these technologies are their limitations connected with the shape of nanoparticles, monodispersity, and reproduction, which negatively affect the possibility of their widespread use.

The nanoimprint lithography (NIL) technology was used to make gold nanoparticle arrays (NPAs) on glass substrates. It allows the creation of uniformly oriented and homogeneous NPA with controlled sizes, shapes, and interparticle distances, which can be the basis for the development of sensors for the detection of biomolecules. The most prominent advantage of this method is the use of matrix templates for the fabrication of nanostructures, making NIL a highly efficient and inexpensive process that allows the creation of templates with relatively large linear sizes and a resolution of less than 10 nm [40, 41]. This method can represent the general basis for the rapid prototyping of different types of NPAs (shapes, sizes, compositions, and interparticle distances) and enables simultaneous production of more than a billion monodisperse nanoparticles directly on a variety of substrates in a consistent, single process that is fully reproducible. The use of one-dimensional diffraction grating for producing two-dimensional templates allows obtaining three-dimensional nanostructures of a given shape, which form an NPA with the desired optical properties.

Among many areas of possible use of highly ordered NPAs, application in biosensors looks like one of the most promising due to the ability to detect interactions between biomolecules in real time [42, 43]. In order to develop a sensor structure for the registration of biomolecules, the NIL method was used to produce arrays of gold nanostructures with parallelepiped shape obtained from one-dimensional gratings (see Subsection 3.2.3.5 and Section 3.3.1). As a result, arrays of nanostructures with strictly defined dimensions (120 × 130 × 50 nm), shape, and interparticle distances were obtained.

Using this approach, a method for measuring biomolecular interactions in real time using a UV-visible range spectrophotometer [44] is presented below. The developed approach has minimal technological requirements, namely, the presence of UV-visible spectrophotometer and peristaltic pump. The ability to biomolecular recognition depends on the change in the extinction spectrum of the NPA as a function of biomolecular adsorption onto the gold surface. Although NPAs are a completely non-selective sensor platform, a high degree of selectivity to the analyte can be achieved by the specificity of the attached ligands and the passivation of the sensor surface to non-specific binding [45, 46].

Atomic force microscopy (AFM) images and a section of 120 nm × 130 nm × 50 nm gold nanostructures with a distance between the nanoparticles of 237 nm are shown in Fig. 4.5. It should be noted that the AFM technology does not allow for a correct estimation of the distance between the NPA due to the limitations concerning the geometry of the AFM tip.

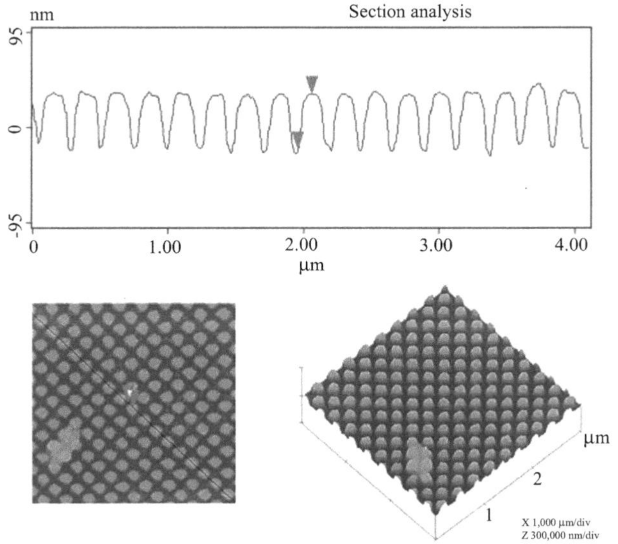

Figure 4.5 AFM image and section of Au NPA. Reprinted from Ref. [44] under a Creative Commons Attribution-NoDerivatives 4.0 International License. Figure caption was adapted.

The nanostructured biosensors based on LSPR spectroscopy operate in the same way as their SPR analogues by transducing the refractive index changes on the surface of a high-conductive metal into a plasmonic response (a shift in wavelength on which an SPR is observed). Biosensors are real-time liquid sensors, so initial testing of the NPA sensitivity to the change in the refractive index of the surrounding liquid medium was performed. For this purpose, aqueous solutions of glycerin with different refractive indices were used. Measurements for registering changes in the refractive index were carried out in the flow mode, with the registration of real-time changes in the extinction of light. The simplified scheme of LSPR biosensor setup is shown in Fig. 4.6.

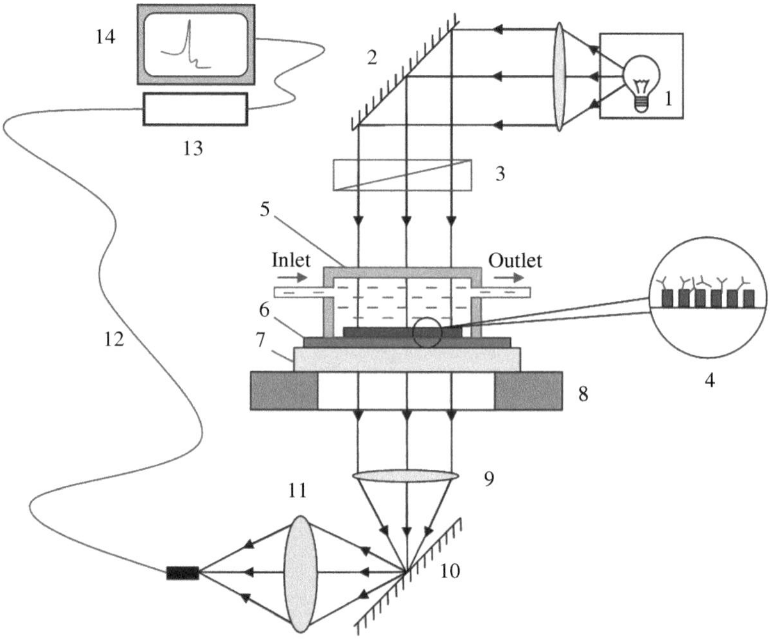

Figure 4.6 Simplified scheme of LSPR biosensor setup: 1 — halogen lamp; 2, 10 — mirrors; 3 — polarizer; 4 — gold nanoparticles array; 5 — flow cell; 6 — glass substrate; 7, 8 — sample holders; 9, 11 — lenses; 12 — fiber-optic cable; 13 — spectrometer; 14 — computer. Reprinted from Ref. [44] under a Creative Commons Attribution-NoDerivatives 4.0 International License. Figure caption was adapted.

The results obtained for calibrated glycerin solutions clearly show that the optical extinction of the NPA structures is sensitive to the change in the refractive index of the surrounding liquid medium (Fig. 4.7). A near-linear dependence of extinction on the right slope of the extinction peak (λ = 660 nm) on the refractive index was observed, which is related to the LSPR wavelength shift (Fig. 4.7, inset). From the point of view of maximizing the LSPR response, it was found for most cases that it is more efficient to use the wavelengths of the right side of the LSPR peak.

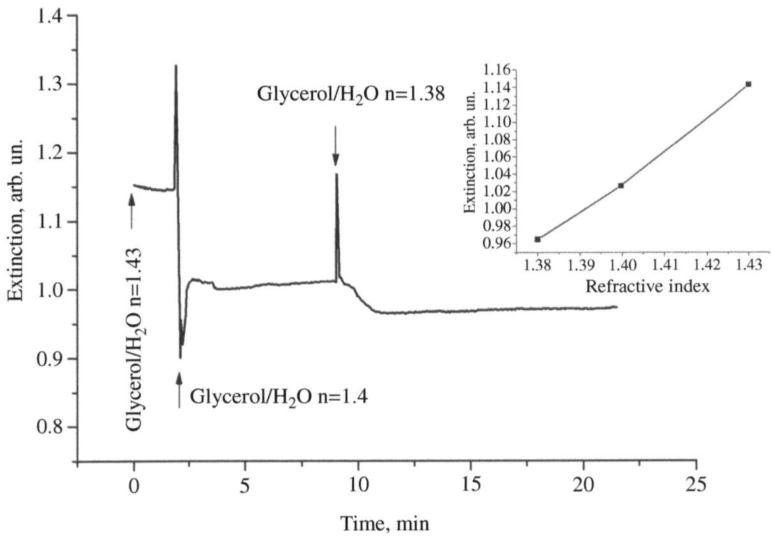

Figure 4.7 The real-time kinetic dependence of light extinction by NPA when the refractive index of a liquid medium changes. Inset: dependence of extinction on refractive index. Reprinted from Ref. [44] under a Creative Commons Attribution-NoDerivatives 4.0 International License. Figure caption and inset were adapted.

4.4.2 Study of a Biomolecular Antigen–Antibody Reaction and Molecular Recognition Using LSPR Biosensors

To evaluate the kinetic response of the LSPR biosensor to the adsorption of real biological materials, NPAs were used to detect a biospecific reaction between the protein BSA (molecular weight

= 67 kDa) and anti-BSA immunoglobulin (molecular weight = 150 kDa). Experimental results are shown in Fig. 4.8, which shows the real-time kinetic dependence of light extinction (sensogram) for a specific reaction between BSA and anti-BSA, dissolved in a PBS buffer solution, with different concentrations of the anti-BSA solution. Figure 4.9 shows respective changes in the light extinction spectra.

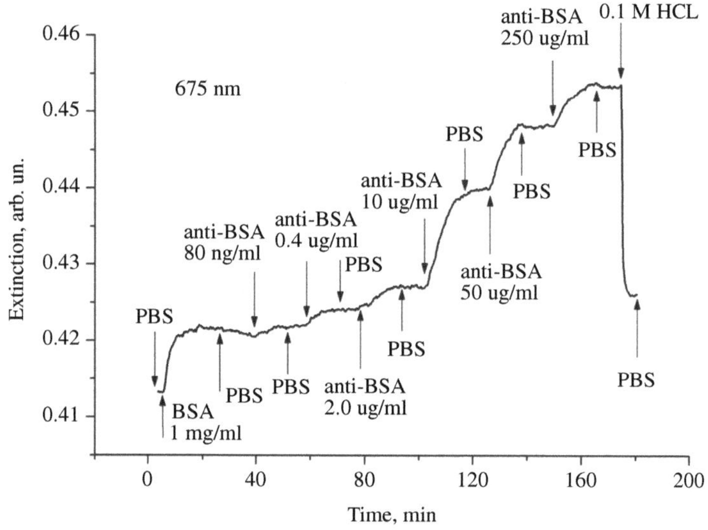

Figure 4.8 Real-time kinetic dependence of light extinction by nanorectangle NPA for the BSA–anti-BSA specific reaction with different concentrations of anti-BSA solution.

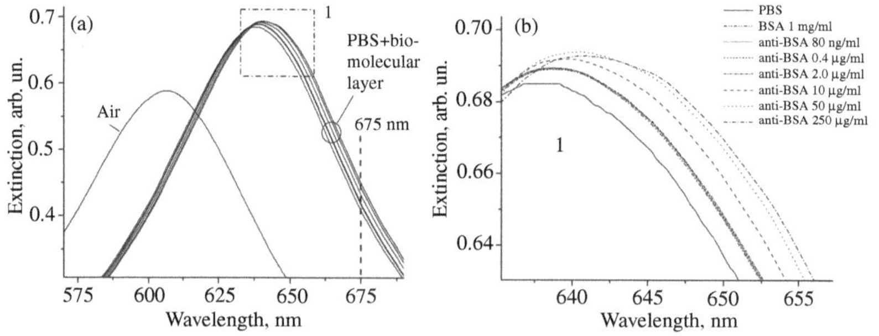

Figure 4.9 (a) Light extinction spectra for nanorectangle NPA in air, in a buffer solution and for BSA–anti-BSA specific reaction with different concentrations of the anti-BSA solution. (b) Zoomed in portion of the spectra detailing the extinction shifts.

As it can be seen from Figs. 4.8 and 4.9, a clear response to the specific BSA–anti-BSA reaction occurs when an anti-BSA solution is injected at a concentration of 80 ng/mL. In addition, a high degree of reaction specificity was observed during the experiment after washing the IgG layer with a 0.1 M HCl solution when about 90% of the antibodies were eliminated, while the BSA layer completely remained on the surface of the sensor. As it can be seen from the sensogram, the developed technique allows the regeneration of a sensitive element.

4.5 Comparative Study of SPR and LSPR Techniques for Small Molecule Detection

To recognize the ability of SPR sensors for small molecule detection, SPR experiment with thiourea (molecular weight = 76 Da) was performed using the saline-sodium citrate (1xSSC) buffer as the base solution. The results of experiment (Fig. 4.10) reveal low recognition limit of about 0.8 μM, whereas the expressed SPR response starts

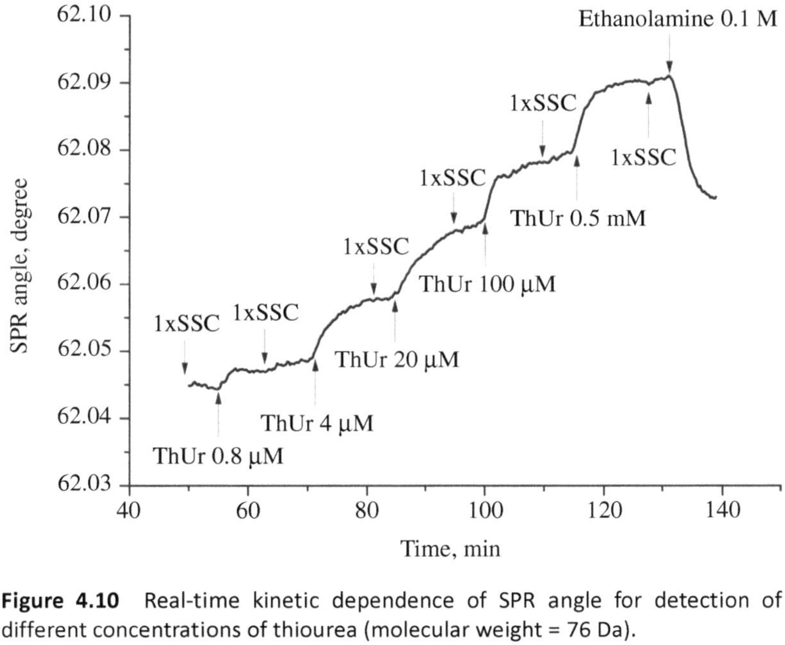

Figure 4.10 Real-time kinetic dependence of SPR angle for detection of different concentrations of thiourea (molecular weight = 76 Da).

from the thiourea concentration of 4 μM. Due to the presence of the thio group in the structure of this compound, the removal of molecules was not revealed after washing with buffer solution due to the strong chemical binding with gold surface. Moreover, no saturation of SPR response was observed even for the highest concentration of 0.5 mM used in the experiment, which indicates the high binding capacity for the surface of SPR slides.

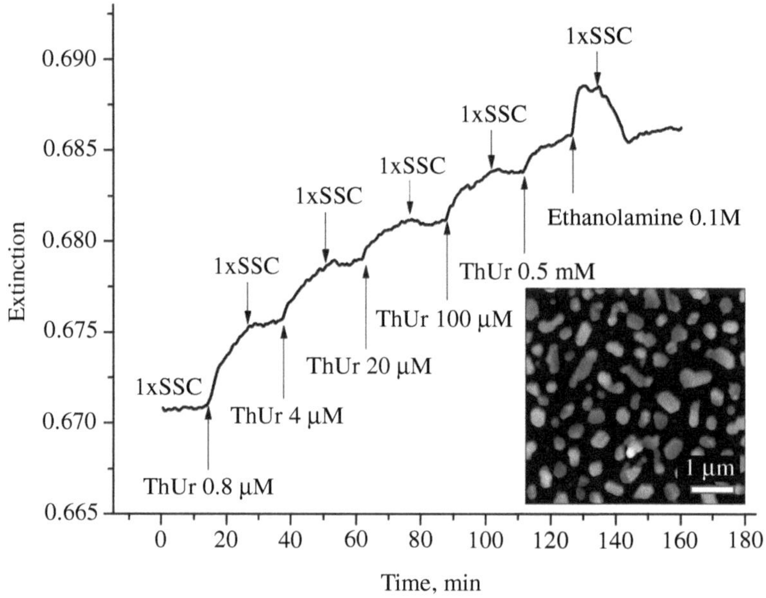

Figure 4.11 Real-time kinetic dependence of light extinction of random NPA for detection of different concentration of thiourea (molecular weight = 76 Da). Inset: AFM image of a random array of gold nanostructures on a glass substrate fabricated by thermal annealing.

It should be noted that LSPR biosensor may be more suitable for the registration of biomolecules with small molecular weight due to more developed surface of gold compared with conventional SPR sensors and ability to fill the volume between nanostructures with small biomolecules that can easily penetrate and adsorb on their side faces, where distribution of local electric field is still strong. The experiment with thiourea detection was repeated under the similar conditions (Fig. 4.11) using a random array of gold nanostructures on a glass substrate (Fig. 4.11, inset) fabricated by thermal annealing (see Section 3.3.1). The sensogram of experiment expressed noticeable difference in the behavior of LSPR response in comparison with SPR. The LSPR sensor clearly recognized the low limit concentration of thiourea molecules (0.8 μM); however, the saturation of response was observed with higher concentration of molecules.

The result of the experiment proves the above assumption and shows that LSPR biosensors can register the nanomolar concentration of molecules with small molecular weight even using random NPA arrays. The obtained results suggest the promising use of random NPA for the construction of LSPR biosensors, primarily due to the simpler manufacturing technology compared with NIL. At the same time, it is necessary to keep in mind the possible deviation of parameters between the samples of random NPAs due to the existing dispersion of nanostructures in size and shape as well as the lower long-time stability of parameters due to the absence of adhesive chromium layer (see Section 3.3.1).

4.6 Conclusion

Specific studies described in this chapter demonstrate the potential of the SPR method as a basis for building high-sensitive SPR and LSPR biosensors capable of real-time registration of biomolecular interactions. Both sensor platforms can register biospecific interactions and exhibit concentration-dependent response, which can be calibrated for obtaining various molecular absorption parameters, for example, biomolecular layer mass.

In the first study, the experimental approach for the 3D quantification of the surface concentration of biological molecules from SPR data was proposed. SPR measurements of biospecific reaction between IgG and anti-IgG molecules on the Au surface were performed and the surface concentration of biomolecular layer was estimated. This approach may be useful to determine or estimate the orientation of biomolecules relative to the surface of SPR sensors for a more comprehensive understanding and quantitative estimation of biomolecular layers.

The second study presented developed LSPR biosensors based on the approach to fabricate noble metal nanoparticle arrays using nanoimprint lithography and thermal annealing. The capability of ordered and random Au nanostructure arrays for real-time detection of adsorption and interaction of biomolecules with different molecular weight was demonstrated. The performed experiments showed promising prospects of the LSPR method for the exploitation of NPA structures as the basis for biomolecular detection and monitoring of specific biomolecular interactions.

References

1. Ramanavièius, A., Herberg, F. W., Hutschenreiter, S., Zimmermann, B., Lapënaitë, I., Kaušaitë, A., Finkelšteinas A., and Ramanavièienë, A. (2005). Biomedical application of surface plasmon resonance biosensors, *Acta Med. Litu.*, **12**, pp. 1–9.
2. Olaru, A., Bala, C., Jaffrezic-Renault, N., and Aboul-Enein, H. Y. (2015). Surface plasmon resonance (SPR) biosensors in pharmaceutical analysis, *Crit. Rev. Anal. Chem.*, **45**, pp. 97–105.
3. Yadav, S. P., Bergqvist, S., Doyle, M. L., Neubert, T. A., and Yamniuk, A. P. (2012). MIRG Survey 2011: Snapshot of rapidly evolving label-free technologies used for characterizing molecular interactions, *J. Biomol. Tech.*, **23**, pp. 94–100.
4. Wartchow, C. A., Podlaski, F., Li, S., Rowan, K., Zhang, X., Mark, D., and Huang, K. S. (2011). Biosensor-based small molecule fragment screening with biolayer interferometry, *J. Comput. Aided Mol. Des.*, **25**, pp. 669–676.
5. Park, J. H., Byun, J. Y., Jang, H., Hong, D., and Kim, M. G. (2017). A highly sensitive and widely adaptable plasmonic aptasensor using berberine for small-molecule detection, *Biosens. Bioelectron.*, **97**, pp. 292–298.
6. Ananthanawat, C., Hoven, V. P., Vilaivan, T., and Su, X. (2010). Surface plasmon resonance study of PNA interactions with double-stranded DNA, *Biosens. Bioelectron.*, **26**, pp. 1918–1923.
7. Wang, G., Akiyama, Y., Takarada, T., and Maeda, M. (2016). Rapid non-crosslinking aggregation of DNA-functionalized gold nanorods and nanotriangles for colorimetric single-nucleotide discrimination, *Chem. Eur. J.*, **22**, pp. 258–263.
8. Nilsson, C. E., Abbas, S., Bennemo, M., Larsson, A., Hamalainen, M. D., and Frostell-Karlsson, A. (2010). A novel assay for influenza virus

quantification using surface plasmon resonance, *Vaccine*, **28**, pp. 759–766.

9. Kim, J., Oh, S. Y., Shukla, S., Hong, S. B., Heo, N. S., Bajpai, V. K., Chun, H. S., Jo, C.-H., Choi, B. G., Huh, Y. S., and Han, Y. K. (2018). Heteroassembled gold nanoparticles with sandwich-immunoassay LSPR chip format for rapid and sensitive detection of hepatitis B virus surface antigen (HBsAg), *Biosens. Bioelectron.*, **107**, pp. 118–122.
10. Olaru, A., Gheorghiu, M., David, S., Wohland, T., and Gheorghiu, E. (2009). Assessment of the multiphase interaction between a membrane disrupting peptide and a lipid membrane, *J. Phys. Chem. B*, **113**, pp. 14369–14380.
11. Tadepalli, S., Kuang, Z., Jiang, Q., Liu, K. K., Fisher, M. A., Morrissey, J. J., Kharasch, E. D., Slocik, J. M., Naik, R. R., and Singamaneni, S. (2015). Peptide functionalized gold nanorods for the sensitive detection of a cardiac biomarker using plasmonic paper devices, *Sci. Rep.*, **5**, 16206.
12. Vachali, P. P., Li, B., Bartschi, A., and Bernstein, P. S. (2015). Surface plasmon resonance (SPR)-based biosensor technology for the quantitative characterization of protein–carotenoid interactions, *Arch. Biochem. Biophys.*, **572**, pp. 66–72.
13. Sanders, M., Lin, Y., Wei, J., Bono, T., and Lindquist, R. G. (2014). An enhanced LSPR fiber-optic nanoprobe for ultrasensitive detection of protein biomarkers, *Biosens. Bioelectron.*, **61**, pp. 95–101.
14. Mauriz, E., García-Fernández, M. C., and Lechuga, L. M. (2016). Towards the design of universal immunosurfaces for SPR-based assays: A review, *Trends Anal. Chem.*, **79**, pp. 191–198.
15. Satija, J., Bharadwaj, R., Sai, V. V. R., and Mukherji, S. (2010). Emerging use of nanostructure films containing capped gold nanoparticles in biosensors, *Nanotechnol. Sci. Appl.*, **3**, pp. 171–188.
16. de Bruijn, H. E., Minor, M., Kooyman, R. P. H., and Greve, J. (1993). Thickness and dielectric constant determination of thin dielectric layers, *Opt. Commun.*, **95**, pp. 183–188.
17. Malmsten, M. (1994). Ellipsometry studies of protein layers adsorbed at hydrophobic surfaces, *J. Colloid Interface Sci.*, **166**, pp. 333–342.
18. Chegel, V., Chegel, Y., Guiver, M. D., Lopatynskyi, A., Lopatynska, O., and Lozovski, V. (2008). 3D-quantification of biomolecular covers using surface plasmon-polariton resonance experiment, *Sens. Actuators B Chem.*, **134**, pp. 66–71.
19. Hall, C. E., Nisonoff, A., and Slayter, H. S. (1959). Electron microscopic observations of rabbit antibodies, *J. Cell Biol.*, **6**, pp. 407–412.

20. Bae, Y. M., Oh, B. K., Lee, W., Lee, W. H., and Choi, J. W. (2005). Study on orientation of immunoglobulin G on protein G layer, *Biosens. Bioelectron.*, **21**, pp. 103–110.
21. Shirshov, Y. M., Chegel, V. I., Subbota, Y. V., Matsas, E. P., Rachcov, A. E., and Sergeeva, T. A. (1995). Determination of dielectric constant and thickness of thin biological layers using surface plasmon resonance, *Proc. SPIE*, **2648**, pp. 118–123.
22. Kreibig, U. and Vollmer, M. (1995). *Optical Properties of Metal Clusters* (Springer-Verlag, Berlin).
23. Dmitruk, N. L., Goncharenko, A. V., and Venger, E. F. (2009). *Optics of Small Particles and Composite Media* (Naukova Dumka, Kyiv).
24. Maier, S. A. (2005). Plasmonics: Towards subwavelength optical devices, *Curr. Nanosci.*, **1**, pp. 17–22.
25. Atwater, H. A., Maier, S., Polman, A., Dionne, J. A., and Sweatlock, L. (2005). The new "p–n junction": Plasmonics enables photonic access to the nanoworld, *MRS Bull.*, **30**, pp. 385–389.
26. Mock, J. J., Smith, D. R., and Schultz, S. (2003). Local refractive index dependence of plasmon resonance spectra from individual nanoparticles, *Nano Lett.*, **3**, pp. 485–491.
27. Mock, J. J., Barbic, M., Smith, D. R., Schultz, D. A., and Schultz, S. (2002). Shape effects in plasmon resonance of individual colloidal silver nanoparticles, *J. Chem. Phys.*, **116**, pp. 6755–6759.
28. Su, K. H., Wei, Q. H., Zhang, X., Mock, J. J., Smith, D. R., and Schultz, S. (2003). Interparticle coupling effects on plasmon resonances of nanogold particles, *Nano Lett.*, **3**, pp. 1087–1090.
29. Duval Malinsky, M., Kelly, K. L., Schatz, G. C., and Van Duyne, R. P. (2001). Nanosphere lithography: Effect of substrate on the localized surface plasmon resonance spectrum of silver nanoparticles, *J. Phys. Chem. B*, **105**, pp. 2343–2350.
30. Dirix, Y., Bastiaansen, C., Caseri, W., and Smith, P. (1999). Oriented pearl-necklace arrays of metallic nanoparticles in polymers: A new route toward polarization-dependent color filters, *Adv. Mater.*, **11**, pp. 223–227.
31. Haynes, C. L. and Van Duyne, R. P. (2003). Dichroic optical properties of extended nanostructures fabricated using angle-resolved nanosphere lithography, *Nano Lett.*, **3**, pp. 939–943.
32. Canfield, B. K., Kujala, S., Kauranen, M., Jefimovs, K., Vallius, T., and Turunen, J. (2005). Remarkable polarization sensitivity of gold nanoparticle arrays, *Appl. Phys. Lett.*, **86**, 183109.

33. Oldenburg, S. J., Hale, G. D., Jackson, J. B., and Halas, N. J. (1999). Light scattering from dipole and quadrupole nanoshell antennas, *Appl. Phys. Lett.*, **75**, 1063.

34. Sherry, L. J., Chang, S. H., Schatz, G. C., Van Duyne, R. P., Wiley, B. J., and Xia, Y. (2005). Localized surface plasmon resonance spectroscopy of single silver nanocubes, *Nano Lett.*, **5**, pp. 2034–2038.

35. Bouhelier, A., Bachelot, R., Im, J. S., Wiederrecht, G. P., Lerondel, G., Kostcheev, S., and Royer, P. (2005). Electromagnetic interactions in plasmonic nanoparticle arrays, *J. Phys. Chem. B*, **109**, pp. 3195–3198.

36. Riboh, J. C., Haes, A. J., McFarland, A. D., Ranjit Yonzon, C., and Van Duyne, R. P. (2003). A nanoscale optical biosensor: Real-time immunoassay in physiological buffer enabled by improved nanoparticle adhesion, *J. Phys. Chem. B*, **107**, pp. 1772–1780.

37. Grabar, K. C., Freeman, R. G., Hommer, M. B., and Natan, M. J. (1995). Preparation and characterization of Au colloid monolayers, *Anal. Chem.*, **67**, pp. 735–743.

38. Freeman, R. G., Grabar, K. C., Allison, K. J., Bright, R. M., Davis, J. A., Guthrie, A. P., Hommer, M. B., Jackson, M. A., Smith, P. C., Walter, D. G., and Natan, M. J. (1995). Self-assembled metal colloid monolayers: An approach to SERS substrates, *Science*, **267**, pp. 1629–1632.

39. Malynych, S. and Chumanov, G. (2006). Coupled planar silver nanoparticle arrays as refractive index sensor, *J. Opt. A Pure Appl. Opt.*, **8**, pp. S144–S147.

40. Chou, S. Y. and Krauss, P. R. (1997). Imprint lithography with sub-10 nm feature size and high throughput, *Microelectron. Eng.*, **35**, pp. 237–240.

41. Li, M., Chen, L., Zhang, W., and Chou, S. Y. (2002). Pattern transfer fidelity of nanoimprint lithography on six-inch wafers, *Nanotechnology*, **14**, pp. 33–36.

42. McFarland, A. D. and Van Duyne, R. P. (2003). Single silver nanoparticles as real-time optical sensors with zeptomole sensitivity, *Nano Lett.*, **3**, pp. 1057–1062.

43. Nath, N. and Chilkoti, A. (2002). A colorimetric gold nanoparticle sensor to interrogate biomolecular interactions in real time on a surface, *Anal. Chem.*, **74**, pp. 504–509.

44. Chegel, V., Lucas, B., Guo, J., Lopatynskyi, A., Lopatynska, O., and Poperenko, L. (2009). Detection of biomolecules using optoelectronic biosensor based on localized surface plasmon resonance. Nanoimprint

lithography approach, *Semicond. Phys. Quantum Electron. Optoelectron.*, **12**, pp. 91–97.

45. Schuck, P. (1997). Use of surface plasmon resonance to probe the equilibrium and dynamic aspects of interactions between biological macromolecules, *Annu. Rev. Biophys. Biomol. Struct.*, **26**, pp. 541–566.
46. Knoll, W. (1998). Interfaces and thin films as seen by bound electromagnetic waves, *Annu. Rev. Phys. Chem.*, **49**, pp. 569–638.

Chapter 5

Application of Molecular Imprinting for Development of Plasmonic Bio- and Chemosensors

5.1 Introduction

Imprinting of molecular recognition sites in organic [1] or inorganic polymers [2, 3] has been the subject of extensive research efforts [4] (Fig. 5.1). There are two general methods to generate imprinted sites in polymer membranes. One approach [5] involves the polymerization of monomers that include a complementary function to the imprinted substrate such as H-bonds, electrostatic interactions, and π-donor–acceptor interactions. Polymerization of the monomer–substrate complex, followed by the removal of the substrate molecules acting as a template for the polymerization, yields the imprinted sites in the polymer. The second approach involves the covalent attachment [6] or coordination [7] of the substrate to polymerizable monomer units, followed by the copolymerization of the functional monomers with other monomers, to yield rigidified polymer matrixes. Cleavage of the polymer-linked substrate units leads to the formation of the polymer with imprinted sites. Polymers with imprinted sites revealing structural [8] and chiral [9, 10] selectivities have been prepared. Molecularly imprinted polymers (MIPs) have been used as specific sensing interfaces [11], functional

Molecular Plasmonics: Theory and Applications
Volodymyr I. Chegel and Andrii M. Lopatynskyi

ISBN 978-981-4800-65-5 (Hardcover), 978-0-429-29511-9 (eBook)
www.jennystanford.com

materials for chromatographic separations [12], and matrixes for selective and catalyzed chemical transformations [13]. This artificial molecular recognition system has significant advantages such as mechanical/chemical stability, low cost, and ease of preparation, thus have increasingly attracted considerable attention [14, 15].

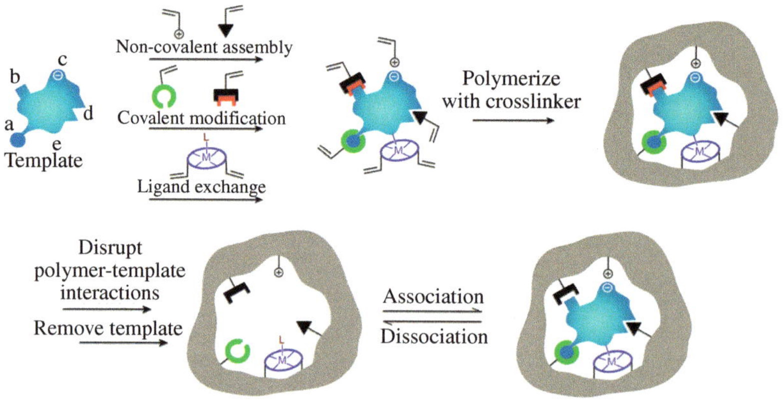

Figure 5.1 Highly schematic representation of the molecular imprinting process. The formation of reversible interactions between the template and polymerizable functionality may involve one or more of the following interactions: (a) reversible covalent bond(s), (b) covalently attached polymerizable binding groups that are activated for non-covalent interaction by template cleavage, (c) electrostatic interactions, (d) hydrophobic or van der Waals interactions, or (e) coordination with a metal centre; each formed with complementary functional groups or structural elements of the template, (a–e) respectively. A subsequent polymerization in the presence of crosslinker(s), a crosslinking reaction, or other process results in the formation of an insoluble matrix (which itself can contribute to recognition through steric, van der Waals and even electrostatic interactions) in which the template sites reside. Template is then removed from the polymer through disruption of polymer—template interactions, and extraction from the matrix. The template, or analogues thereof, may then be selectively rebound by the polymer in the sites vacated by the template, the "imprints." Reproduced with permission from Ref. [4], Copyright 2014, John Wiley & Sons.

The use of imprinted polymers is particularly tempting for sensing applications since membranes with tailored recognition functions can be generated. The major difficulty encountered in the use of imprinted polymers as active components in sensor devices is, however, the coupling of the sensing membrane with an electronic transducer. The imprinted organic polymer is usually relatively thick, and the recognition sites lack direct electrical contact with the transducer. Indeed, most of the sensor devices based on imprinted polymers are either optical, and include chromogenic markers [16], or involve the microgravimetric analysis of the bound substrate using piezoelectric crystals (quartz crystal microbalance) [17]. Only a few reports have addressed the use of surface plasmon resonance (SPR) to follow the association of substrates to imprinted polymers [18–20]. In these studies, however, the SPR spectra followed mass changes associated with the binding of the substrate to the polymer. These mass changes (for low molecular-weight substrates) are small, resulting in low signals and limited analytical performance of the respective sensors.

Addressing the limitations of conventional molecular imprinting techniques, several research groups have begun to explore alternative approaches for developing new imprinting methodologies. An effective approach is to control templates to locate at the surface of imprinted materials, typically exampled by surface imprinting, which is carried out by immobilizing template molecules at the surface of suitable substrates, forming thin imprinted films [21, 22]. Surface molecular imprinting is especially valuable as it solves the problems of limited mass transfer and template removal to some extent, as compared to the conventional molecular imprinting technique. Another attempt to address the limitations of the conventional MIPs is the development of molecularly imprinted nanomaterials [23]. Nanostructured imprinted substrates have extremely high surface-to-volume ratio, so that most of the template molecules can be situated at the surface of the materials, resulting in a large amount of effective imprinted sites, a high binding capacity, and good site accessibility for the molecularly imprinted nanomaterials [24, 25].

An important practical chemosensing field, which has significant prospects for improvement by application of MIPs, is the selective and highly sensitive detection of explosives, which attracts substantial efforts in homeland security research [26–30]. Different analytical procedures for the detection of explosives were reported, and these include optical [31–33], electrochemical [34–36], surface acoustic wave [37–39], or competitive immunoassay [40] methods. Recently, imprinted Au nanoparticle (Au NP) composites as ultrasensitive and selective matrixes for the electrochemical or SPR analyses of the trinitrotoluene (TNT) or RDX explosives [41–43] were introduced. The key element of this approach involves the generation of molecularly imprinted sites in Au NP matrixes. In contrast to the well-established principles of preparation of molecularly imprinted matrixes in organic or inorganic polymers (e.g., TiO_2, SiO_2) [44–46], the synthesis of molecularly imprinted matrixes of metal nanoparticles, for example, Au NPs, demonstrates significant advantages emerging from the nanoscale dimensions of the building blocks of the composite-sensing interface. According to this principle, Au NPs modified with electropolymerizable thioaniline units are electropolymerized onto Au-coated surfaces in the presence of a molecular template that acts as a structural analogue for the respective substrate analyte and exhibits affinity interactions with the thioaniline-modifying groups, or with the electrogenerated bis(aniline) bridging units. The electropolymerization of the functionalized NPs onto a thioaniline-modified surface, and the subsequent removal of the template molecules from the resulting bis(aniline)-crosslinked Au NP matrixes, resulted in the formation of molecularly imprinted contours that bind the analyte by complementary affinity interactions. For example, by the electropolymerization of the functionalized NPs in the presence of picric acid or Kemp's acid, imprinted Au NP composites for the ultrasensitive detection of TNT or RDX were, respectively, demonstrated. The binding of the substrates to the imprinted sites was monitored by SPR spectroscopy.

This chapter presents the results of specific studies related to the development of SPR bio- and chemosensors based on the exploitation of molecular imprinting with two different approaches for SPR signal enhancement. The first study applies imprinted polymers that undergo a swelling process upon binding of the substrates, resulting in a substantial change in the refractive index of the polymer-sensing interface. This enables the improved application of SPR spectroscopy to follow the association of substrates to the imprinted polymers. As a result, the possibility to detect nicotinamide adenine dinucleotide cofactors $NAD(P)^+$ and NAD(P)H using the imprinted polymer films by means of SPR spectroscopy was demonstrated. Additionally, these functionalized membranes have been used to analyze biocatalytic transformations involving NAD^+-dependent enzyme lactate dehydrogenase. The second study is based on the application of molecularly imprinted composite polymer-Au NP matrix for the detection of ultralow concentrations of pentaerythritol tetranitrate

(PETN), nitroglycerin (NG), and ethylene glycol dinitrate (EGDN) explosives.

5.2 Application of Macromolecules Conformational Changes as a Signal Parameter for Studying the Biospecific Reactions Using SPR and Molecular Imprinting

5.2.1 Investigation of Enzymatic Reactions Involving NAD(P)$^+$ and NAD(P)H

Investigation of enzymatic reactions involving β-nicotinamide adenine dinucleotide (NAD^+) and β-nicotinamide adenine dinucleotide phosphate ($NADP^+$), as well as their reduced forms NAD(P)H (see Fig. 5.2), occupies one of the leading places in biochemistry and bioelectrochemistry [47, 48]. Due to the control of the content of these cofactors, it becomes possible to significantly expand the range of enzymes that can be used for the development of intelligent biosensors and monitoring of biocatalytic transformations involving these cofactors. However, the role of the NADH/NAD^+ and NADPH/$NADP^+$ pairs is not limited to the promotion of enzymatic reactions. In many cases, it is much more convenient to determine the amount of the reduced (or oxidized) form of the cofactor than any product of the reaction. Often, the quantitative analysis of one of the forms of these cofactors is carried out electrochemically. However, the amperometric analysis of NAD(P)$^+$ reveals irreversible electrochemical reduction accompanied by high overpotentials [49] resulting in a non-enzymatically active dimer [50]. The NAD(P)$^+$ cofactors were reduced electrocatalytically or bioelectrocatalytically in the presence of Rh complexes [51], or in the presence of NAD(P)$^+$-dependent enzymes (e.g., ferredoxin-$NADP^+$ reductase, lipoamide dehydrogenase, formate dehydrogenase) [52]. Direct, non-mediated electrochemical reduction of NAD(P)$^+$ was also achieved at modified electrodes, for example, in the presence of an L-histidine-modified Ag electrode [53]. Similarly, the direct oxidation of NAD(P)H is electrochemically irreversible and involves high overpotentials

[54]. The electrocatalyzed two-electron oxidation of NAD(P)H was extensively studied in the presence of different electrocatalysts [55], for example *o*-quinones [56], phenazine, phenoxazine, and phenothiazine derivatives [57]. There is no method, however, to selectively analyze NAD^+ in the presence of $NADP^+$ or alternatively to selectively analyze NADH and NADPH. The proposed alternative non-electrochemical method for the determination of $NAD(P)^+$ and NAD(P)H [58] is based on the principle of molecular imprinting, which in recent years has begun to be used for the identification and quantification of various substances.

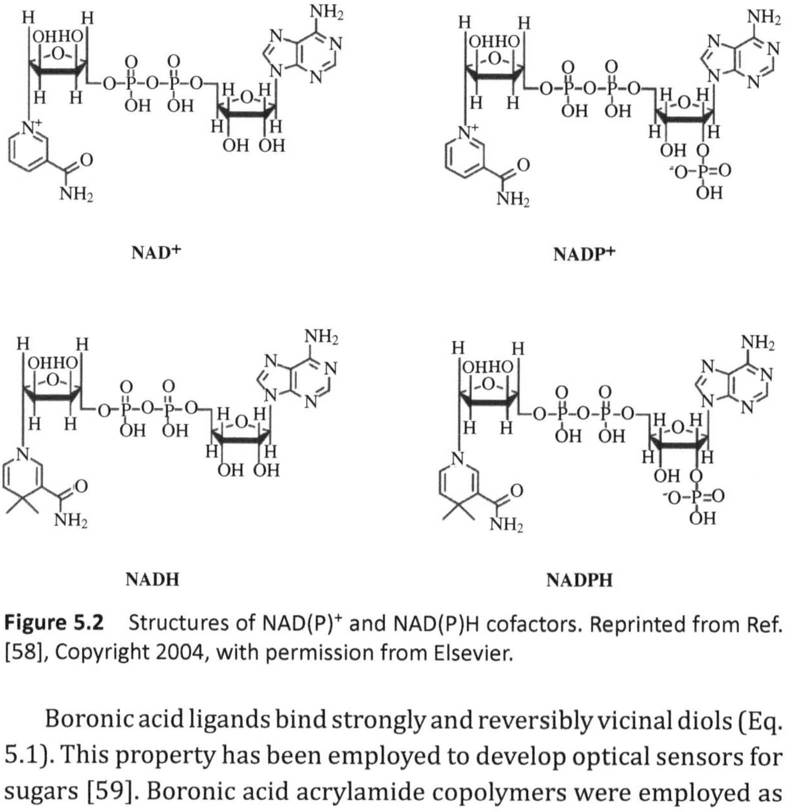

Figure 5.2 Structures of $NAD(P)^+$ and NAD(P)H cofactors. Reprinted from Ref. [58], Copyright 2004, with permission from Elsevier.

Boronic acid ligands bind strongly and reversibly vicinal diols (Eq. 5.1). This property has been employed to develop optical sensors for sugars [59]. Boronic acid acrylamide copolymers were employed as active matrixes for the sensing of glucose [60] or nucleotides [61]. The boronic acid ligand is often used to imprint molecular recognition sites in polymers for specific binding [62] and separation [63] of sugars. The selective analysis of the NAD^+, $NADP^+$, NADH, and NADPH cofactors in acrylamide–acrylamidophenylboronic acid copolymer membranes associated with ISFET devices was reported [61]. In complementary microgravimetric quartz crystal microbalance experiments, it was found that the association of the NAD(P)$^+$ or the NAD(P)H cofactors to the imprinted polymer matrixes linked to the Au-quartz crystals involved the uptake of water and the swelling of the polymer films.

(5.1)

5.2.2 SPR Spectroscopy as a Research Tool for Molecular Imprinting of NAD(P)$^+$ and NAD(P)H Cofactors

5.2.2.1 Preparation of molecularly imprinted polymer

As a polymer base, a copolymer based on acrylamide and 3-acrylamidophenylboronic acid with a moderate content of a crosslinker, *N,N′*-methylene bisacrylamide, has been used, which makes the polymer base rather durable and at the same time allows its swelling and shrinking. Each of the oxidized cofactors NAD$^+$ or NADP$^+$, and the reduced cofactors NADH or NADPH (Fig. 5.2), were imprinted in a crosslinked acrylamide–acrylamidophenylboronic acid copolymer associated with an Au-coated glass support. The method for the generation of the imprinted polymer is schematically depicted in Fig. 5.3.

As the NAD(P)$^+$ or NAD(P)H cofactors include ribose units, the cofactors yield a complex with the 3-acrylamidophenylboronic acid. This linkage, together with complementary H-bonds between the different functional units of the cofactors and the acrylamide and 3-acrylamidophenylboronic acid monomers, leads, upon polymerization accompanied by crosslinking in the presence of *N,N′*-methylene bisacrylamide, to the cofactor-embedded polymer matrixes. To improve the adhesion of the imprinted polymer film on the Au-coated glass support, a primary cystamine monolayer is assembled on the Au surface, and acrylic acid is covalently linked to the monolayer interface. Thus, upon the radical-induced polymerization of the film, the solution-solubilized monomers are copolymerized with the surface-confined monomer units, leading to a surface-bound crosslinked film. The elimination of the template cofactors with 1% NH_3 generates the imprinted sites for the respective cofactors. The existence of functional groups at the periphery of the imprinted molecular contours (boronic acid residues, amide-functionalities, etc.) is expected to yield selective sites for the accommodation of the respective oxidized or reduced cofactors.

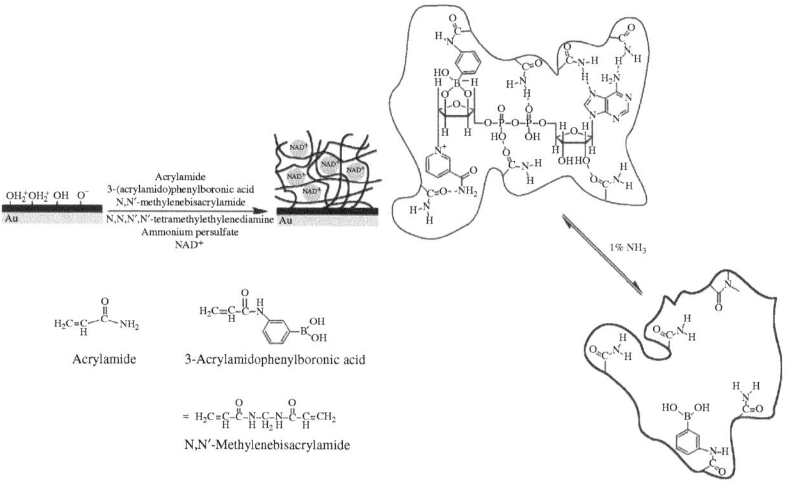

Figure 5.3 Schematic preparation of the $NAD(P)^+$/NAD(P)H cofactors-imprinted acrylamide–acrylamidophenylboronic acid copolymer sensing film on an Au-coated glass support. The support was primarily modified with cystamine monolayer followed by its coupling to acrylic acid. The scheme exemplifies the imprinting of NAD^+. Reprinted from Ref. [58], Copyright 2004, with permission from Elsevier.

5.2.2.2 Detection of NAD^+, $NADP^+$, NADH, and NADPH cofactors

Mathematical modeling of the SPR spectra for bare and thin-film-modified gold surfaces (Fig. 5.4) allowed estimating the NAD^+-imprinted polymer thin film thickness, which was found to be 22 ± 3 nm. The shift in the minimum reflectivity angle observed, before (curve b) and after the imprinted polymer rinsing (curve c), corresponds to the elimination of the imprinted NAD^+ from the polymer. To estimate the analyte content in the imprinted polymer, Eq. (5.2) [64] was applied:

$$\Gamma = d(n_1 - n_2)/(dn/dc), \quad (5.2)$$

where Γ is the surface coverage, *d* is the thickness of the polymer layer, n_1 and n_2 correspond to the refractive indexes of media near the gold surface before and after polymer interaction with an aqueous solution of NAD^+, respectively, and *dn/dc* reflects the influence of the analyte concentration (*c*) on the refractive index of the surrounding medium. Using the derived values of n_1 and n_2 (1.45 and 1.40, respectively), and dn/dc = 0.188 cm^3/g [65], the loading of the imprinted NAD^+ sites in the polymer was estimated as 22.6 ng/mm^2.

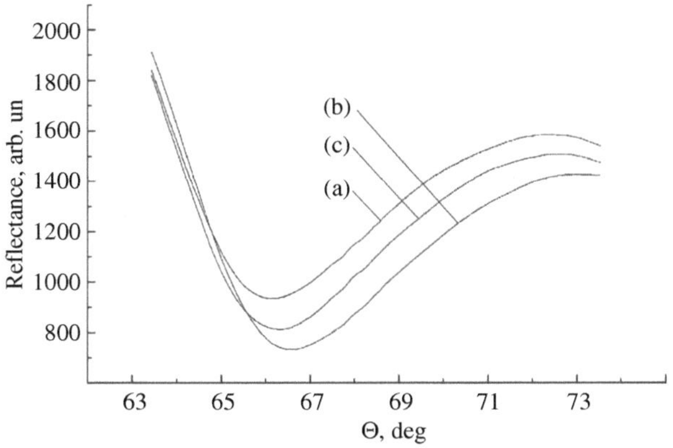

Figure 5.4 The SPR spectra of (a) the bare gold electrode, (b) the gold electrode after the assembly of NAD^+-imprinted polymer film, and (c) the gold electrode with NAD^+-imprinted polymer film after rinsing with 1% NH_3 solution. Adapted from Ref. [58], Copyright 2004, with permission from Elsevier.

Figure 5.5 shows the SPR spectra of the NAD^+-imprinted polymer film in the absence of NAD^+ (curve a), and after treatment with NAD^+, 1×10^{-3} M (curve b). The minimum reflectivity angle is shifted to lower values. In a control experiment, an acrylamide–acrylamidophenylboronic acid copolymer film was assembled without the imprint of the NAD^+ cofactor. Treatment of the nonimprinted film with NAD^+, 1×10^{-3} M, does not yield any noticeable change in the SPR spectrum of the film. This control experiment clearly indicates that no non-specific binding of NAD^+

to the polymer takes place. It also reveals that minute changes in the refractive index of the solution resulting upon solubilization of NAD^+ do not affect the SPR spectrum. Thus, the changes in the SPR spectrum upon interaction with NAD^+ originate from the specific association of the oxidized cofactor to the imprinted sites. The mathematical modeling of SPR spectrum shown in curve b indicates a decrease in the refractive index of the polymer upon the association of NAD^+ from 1.45 to 1.40 and an increase in the polymer thickness of 3.0 ± 0.2 nm, as a result of the swelling of the polymer film. The changes in the SPR spectrum of the NAD^+-imprinted polymer film are controlled by the bulk concentration of NAD^+ in solution.

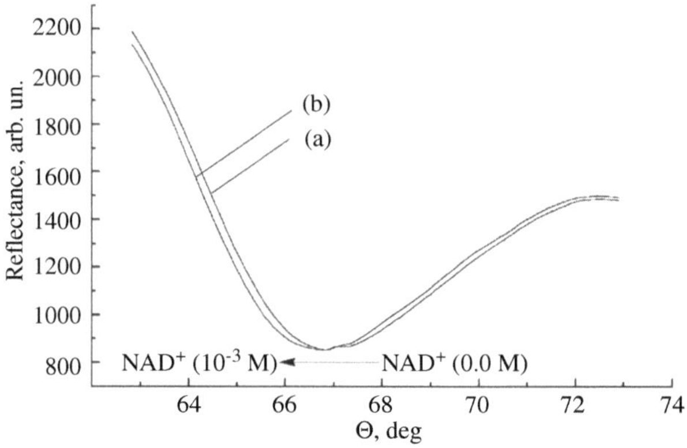

Figure 5.5 The SPR spectra of the NAD^+-imprinted polymer recorded (a) after the removal of the NAD^+ template molecules and (b) after its treatment with NAD^+ (1×10^{-3} M) solution. Adapted from Ref. [58], Copyright 2004, with permission from Elsevier.

The NAD^+-imprinted polymer film reveals selectivity upon analysis of the imprinted cofactor. Figure 5.6 shows the sensogram corresponding to the changes in the minimum reflectivity angles of the SPR spectra in the course of analyzing a range of concentrations of the different oxidized and reduced cofactors solutions by the NAD^+-imprinted film. In this experiment, the NAD^+-imprinted film was initially treated with solutions containing different concentrations of NAD^+. Clearly, as the bulk concentration of NAD^+ increases, the minimum reflectivity angle decreases, and bulk concentrations of NAD^+ as low as 1×10^{-6} M can be sensed by the system. Upon binding of a substrate to an imprinted polymer film, two processes will affect the SPR spectrum change: a mass increase due to the association of NAD^+ with the sensitive sensor surface and a decrease in the refractive index of the polymer layer due to the saturation of the polymer with water (the refractive index of water, $n = 1.33$, is essentially lower than the refractive index of polymer, $n_1 = 1.45$). The increase in the mass of the polymer–cofactor complex should be reflected by an increase in the minimum reflectivity angle of the SPR spectrum, whereas the decrease in the refractive index of the film should lower the minimum reflectivity angle of the SPR spectrum. Experimentally, it was observed that the swelling of the polymer upon the binding of NAD^+ to the imprinted sites results in a decrease in the minimum reflectivity angle, implying that the changes in the refractive index of the polymer, as a result of the swelling, contribute mostly to the overall features of the experimental SPR spectra.

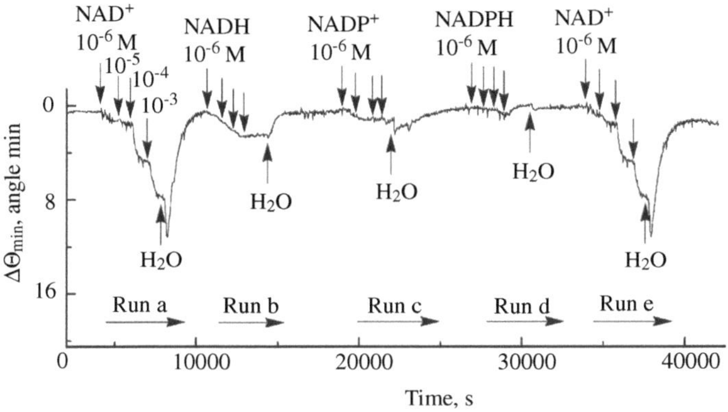

Figure 5.6 The SPR sensogram corresponding to the analyses of 1×10^{-6} M to 1×10^{-3} M solutions of (a) NAD^+, (b) NADH, (c) $NADP^+$, and (d) NADPH by the NAD^+-imprinted polymer film. Run (e) corresponds to repetitive analyses of NAD^+ by the imprinted film. Adapted from Ref. [58], Copyright 2004, with permission from Elsevier.

After the completion of the analysis cycle of NAD^+, the film was washed with water. The bound NAD^+ is washed off as evident by the fact that the original minimum reflectivity angle of the imprinted film that lacks bound NAD^+ is achieved. The resulting film was then interacted with the NADH cofactor using different concentrations (1×10^{-6} M to 1×10^{-3} M). Only minute changes in the minimum reflectivity angle were observed. Similarly, washing off the bound NADH and treatment of the NAD^+-imprinted film with $NADP^+$ or NADPH, respectively, do not yield any changes in the minimum reflectivity angles of the SPR spectra of the film. To reveal that the NAD^+-imprinted film remained functionally intact toward the sensing of NAD^+, the system that was employed for the analysis of NAD^+, NADH, $NADP^+$, and NADPH was applied again for the analysis of variable concentrations of NAD^+. Clearly, the changes in the minimum reflectivity angle of the SPR spectra characteristic to the film in the presence of variable bulk concentrations of NAD^+ were reproduced, similarly to the first analysis cycle of NAD^+. These experiments clearly indicate that the NAD^+-imprinted polymer film reveals selectivity toward the imprinted substrate, and that the oxidized cofactor $NADP^+$ or the reduced cofactors NADH or NADPH exhibit low affinity, or no affinity, for the imprinted sites.

Figure 5.7 depicts the calibration curves corresponding to the changes in the minimum reflectivity angles of the NAD^+-imprinted film upon analysis of the different cofactors. It is obvious that in the concentration range of the cofactor, 1×10^{-5} to 1×10^{-3} M, the determination of NAD^+ by the proposed method can be carried out selectively. The NAD^+-imprinted film revealed stability for at least 10 days upon daily operation (with only 10% decay in the outcome signal but with no lack in the selectivity).

Similar results of SPR registration were also obtained for polymer films imprinted with $NADP^+$, NADPH, and NADH cofactors before and after interaction with 1×10^{-3} M solution of respective cofactors. As in the case of NAD^+, binding of other cofactors to the imprinted polymer results in a shift in the SPR minimum reflectivity angle. Theoretical fitting of the SPR spectra after the association of imprinted polymer films with cofactors indicates a decrease in the refractive index of the polymer and an increase in the film thickness due to its swelling. Control experiments revealed that treatment with solutions containing different concentrations of cofactors does not yield any change in the SPR spectrum of a nonimprinted acrylamide–acrylamidophenylboronic acid copolymer film, implying that no non-specific binding of cofactors to the polymer film occurs.

Minute changes in the minimum reflectivity angles of the film are observed in the presence of nonimprinted cofactors, indicating that these cofactors exhibit very little affinity to the imprinted sites. For the reduced cofactor NADH, some decrease in the minimum reflectivity angles was observed upon interaction with the NAD^+-imprinted polymer. This may be explained by the fact that the reduced cofactors NADH or NADPH always include, as impurities, the oxidized cofactors NAD^+ and $NADP^+$, and thus the observed responses may originate from these components present during polymerization. It is interesting to note that the NAD^+-imprinted polymer is unaffected in the presence of $NADP^+$, and the $NADP^+$-imprinted film is not affected by the NAD^+ cofactor. Thus, the imprinting procedure leads to impressive selectivity where the single phosphate substituent differentiating $NADP^+$ from NAD^+ is sufficient to generate specific molecular contours for the respective cofactors.

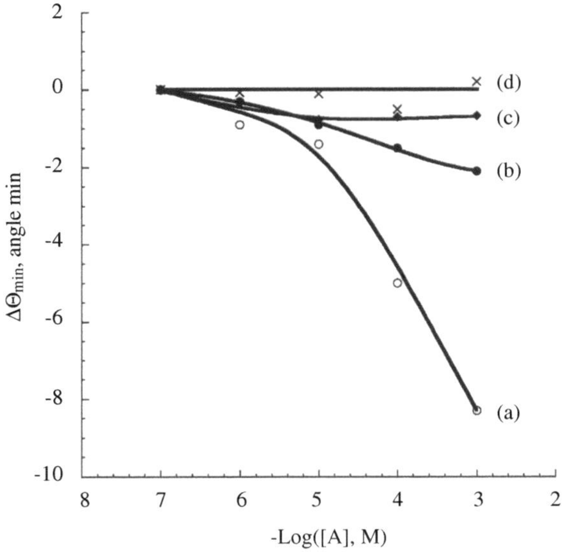

Figure 5.7 Calibration plots corresponding to the changes in the minimum reflectivity angles at variable concentrations of the $NAD(P)^+$/NAD(P)H cofactors: (a) for NAD^+, (b) for NADH, (c) for $NADP^+$, and (d) for NADPH. Adapted from Ref. [58], Copyright 2004, with permission from Elsevier.

These results demonstrate that SPR sensors based on polymers imprinted with NADH or NADPH cofactors can be used to selectively detect the NADH or NADPH molecules, respectively. NADH- and NADPH-imprinted films exhibit cyclic sensing activities, under continuous operation, for at least several days, with no observable degradation of the SPR signals.

5.2.3 SPR Monitoring of Biocatalytic Oxidation of Lactate with NAD$^+$ Cofactor Using NADH-Imprinted Polymer

The NADH-imprinted film was also utilized for SPR analysis of a biocatalytic process that involves the oxidation of lactate to pyruvate by NAD$^+$ in the presence of lactate dehydrogenase (LDH) (Fig. 5.8).

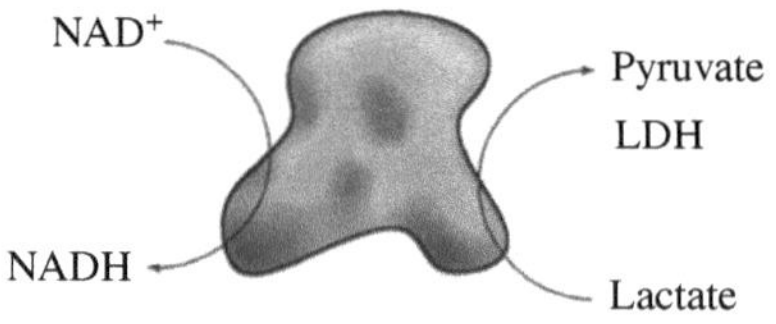

Figure 5.8 Biocatalytic oxidation of lactate to pyruvate in the presence of NAD$^+$ and lactate dehydrogenase (LDH). Reprinted from Ref. [58], Copyright 2004, with permission from Elsevier.

The NADH being formed in the biocatalytic process binds to the NADH-imprinted film and results in its swelling by the uptake of water. As discussed earlier, the binding of NADH to the imprinted polymer should lead to the increase in the polymer mass and an expected increase in the minimum reflectivity angle. The uptake of water and subsequent swelling, however, decreased the refractive index of the polymer, and concomitantly a decrease in the minimum reflectivity angle was observed similar to the above-mentioned cases.

The effectiveness of the imprinted polymer film toward the sensing of NADH is demonstrated in Fig. 5.9, run (a). Upon interaction with 1×10^{-3} M NADH, the minimum reflectivity angle of the NADH-imprinted polymer changed by $\Delta\theta_{\rm min}$ = 8.2 min. Polymer wash-off with water resulted in its complete regeneration (Fig. 5.9, point (x)). At point (y), the analysis of the biocatalyzed oxidation of lactate was initiated, run (b). At this point, the sensor was brought into contact with a solution containing 0.1 μg/ml LDH and 1×10^{-3} M NAD$^+$. It is evident that only a minute change in the minimum reflectivity angle is observed in this case. At point (z) of run (b), after the signal stabilization, lactate, 1×10^{-2} M, was added to the solution. The drastic change in the signal indicates the binding of biocatalytically formed NADH to the NADH-imprinted polymer. Control experiments revealed that addition of lactate alone to the system had no effect on the minimum reflectivity angle. This implies that the observed changes in the minimum reflectivity angle originate from the biocatalyzed generation of NADH. As noted above, the change in the minimum reflectivity angle is due to the uptake of water, and, accordingly, a decrease in the refractive index of the polymer. Here, the minimum reflectivity angle of the polymer film stabilized at a value of $\Delta\theta_{min}$ = 7.9 min, which is similar to the signal change registered upon the detection of pure NADH (Fig. 5.9, run (a)). This result is consistent with the fact that the concentration of NAD^+ in the system was 1×10^{-3} M, whereas the concentration of lactate was 10-fold higher, 1×10^{-2} M. Thus, the oxidized cofactor NAD^+ was essentially fully transformed to the reduced cofactor NADH, which was being sensed by the imprinted polymer. By analyzing the kinetic curve (Fig. 5.9, inset), the pseudo first-order rate constant $k = 3.2\times10^{-3}\ s^{-1}$ was calculated for the LDH-biocatalyzed oxidation reaction of lactate by NAD^+ cofactor.

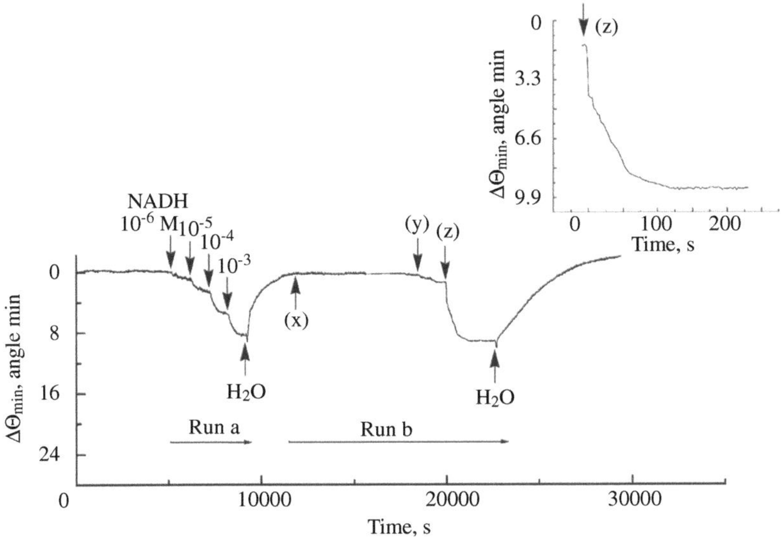

Figure 5.9 The SPR sensogram corresponding to (run a) the analysis of NADH, and (run b) analysis of the LDH-mediated oxidation of lactate in the presence of NAD^+, using the NADH-imprinted polymer film as sensing interface. Inset shows enlarged time-dependent SPR response of the sensing interface in the presence of lactate (1×10^{-2} M), LDH (1×10^{-7} g/ml), and NAD^+ (1×10^{-3} M). Adapted from Ref. [58], Copyright 2004, with permission from Elsevier.

5.3 SPR Detection of Low-Molecular-Weight Species Using Molecular Imprinting of Gold Nanoparticles Matrix

5.3.1 Analytical Approaches for Detection of Small Molecules

It is known that upon the association of low-molecular-weight substrates, the dielectric changes at the Au surface are too small to allow detectable SPR spectral changes. Different methods to enhance the dielectric changes upon binding the analyte to the surface were reported by implementing latex particles [66], liposomes [67], or proteins [68] as amplifying labels for the recognition events. One effective method to amplify the recognition process has involved the conjugation of Au NPs to the SPR chip surface. The coupling between the localized plasmon of the NPs and the surface plasmon wave was found to significantly affect the SPR spectrum [69, 70]. Due to the proximity of gold nanoparticles to the binding sites, even small dielectric changes caused by a small amount of analyte close to the sites change its equilibrium. Thus, changes in the dielectric properties of the imprinted matrix as a result of the binding of the analyte could be amplified by the conjugation of metallic NPs ensemble to the sensing complex.

5.3.2 Molecular Imprinting in Polymer-Gold Nanoparticles Composite Matrix

As the carboxylic acid residues exhibit similar dimensions and structural features to the nitro groups in the explosives, and since ionic and/or H-bonds between the carboxylate and the anilinium residues could provide affinity interactions between the carboxylic acids and the electrogenerated bis(aniline)-crosslinked Au NP composite, it was argued that the structurally related carboxylic acids could act as templates for the generation of imprinted sites for structurally analogous nitrosubstituted explosives [71]. The scheme in Fig. 5.10 outlines the method to synthesize an imprinted Au NP composite for the detection of PETN (1) or NG (2).

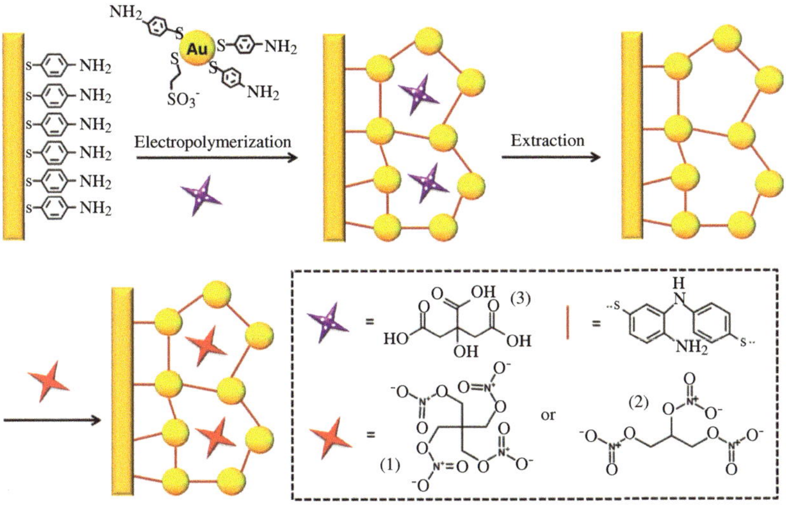

Figure 5.10 Schematic presentation for the electropolymerization of a composite of bis(aniline)-crosslinked Au NPs for the sensing of PETN or NG using citric acid as an imprinting template. Reprinted with permission from Ref. [71]. Copyright (2011) American Chemical Society.

The thioaniline-functionalized Au NPs were electropolymerized on a thioaniline-modified Au-coated glass surface in the presence of citric acid (3) as an imprinting template. The resulting bis(aniline)-crosslinked matrix was then rinsed to remove the imprinting template molecule. The electropolymerization of the Au NPs and the elimination of the imprinting template were followed by SPR spectroscopy.

A nonimprinted Au NP composite was similarly prepared by the electropolymerization of the modified Au NPs in the absence of the imprinting template. Figure 5.11A exemplifies the SPR spectra observed upon the treatment of the citrate-imprinted Au NP matrix (panel I) and the nonimprinted Au NP composite (panel II) with 2 pM PETN. Whereas a noticeable shift in the SPR curve is observed for this low PETN concentration on the citrate-imprinted composite, no shift is evident for the nonimprinted matrix. Figure 5.11B, curve I, shows the reflectance changes of the citric-acid-imprinted Au NP composite upon interaction with different concentrations of PETN. As the concentration of PETN increases, the reflectance changes are intensified, consistent with a higher content of PETN that binds to the composite. For comparison, Fig. 5.11B, curve II, depicts the reflectance values observed upon the treatment of the nonimprinted Au NP composite within the same PETN concentration range. Evidently, only minute reflectance changes are observed, implying that PETN does not bind to the nonimprinted Au NP matrix. The calibration curve corresponding to the reflectance changes of the citrate-imprinted composite upon treatment with different concentrations of PETN is shown in Fig. 5.11C.

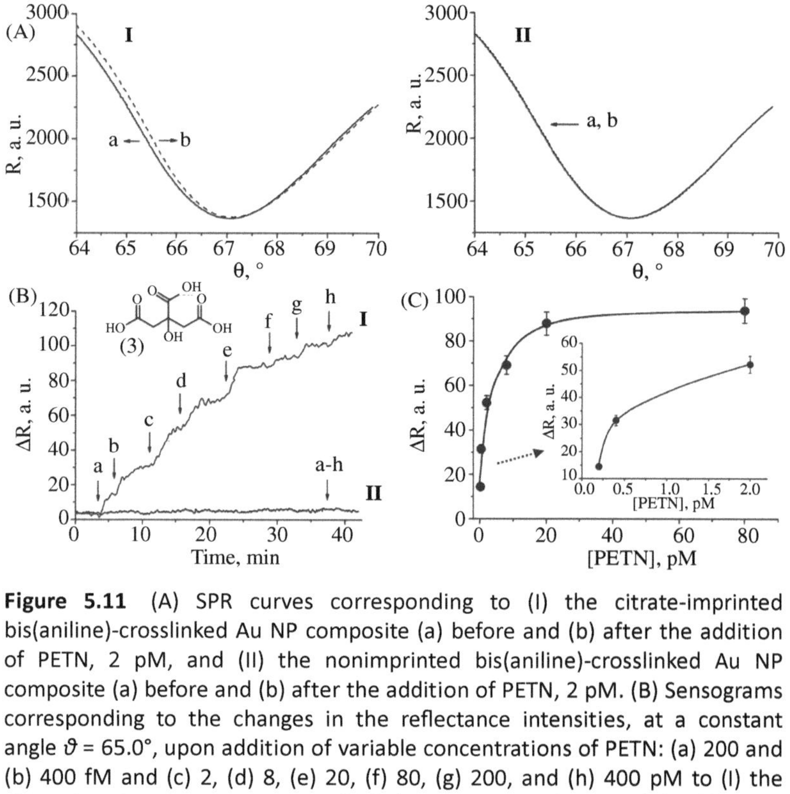

Figure 5.11 (A) SPR curves corresponding to (I) the citrate-imprinted bis(aniline)-crosslinked Au NP composite (a) before and (b) after the addition of PETN, 2 pM, and (II) the nonimprinted bis(aniline)-crosslinked Au NP composite (a) before and (b) after the addition of PETN, 2 pM. (B) Sensograms corresponding to the changes in the reflectance intensities, at a constant angle $\vartheta = 65.0°$, upon addition of variable concentrations of PETN: (a) 200 and (b) 400 fM and (c) 2, (d) 8, (e) 20, (f) 80, (g) 200, and (h) 400 pM to (I) the citrate-imprinted Au NP matrix and (II) the nonimprinted matrix. (C) Calibration curve relating the reflectance changes to the concentrations of PETN on the citrate-imprinted matrix. The inset shows the lower concentration region of the calibration curve. Error bars correspond to a set of $N = 5$ measurements. All measurements were performed in ethanol. Reprinted with permission from Ref. [71]. Copyright (2011) American Chemical Society.

The detection limit achieved for analyzing PETN by the citrate-imprinted matrix is 200 fM. The matrix interface can be used repeatedly, regenerating it after an analysis by rinsing, although a slight degradation of the matrix with each subsequent analysis has been observed.

5.3.3 Peculiarities of SPR Detection of Explosives Using LSPR Nanoantenna

Upon the application of dicarboxylic acids, such as succinic acid or fumaric acid, as imprinting templates, sensing abilities toward PETN decreased (not shown). These results emphasize the significance in the selection of an appropriate template analogue for PETN detection. Nitroglycerin exhibits, however, structural similarity to the citric acid imprinting template, and thus it was anticipated that the citrate-imprinted matrix would also sense the NG explosive. Figure 5.12A, curve I, shows the reflectance changes of the citrate-imprinted matrix upon interaction with different concentrations of NG. For comparison, Fig. 5.12A, curve II, depicts the reflectance changes in the nonimprinted Au NP composite upon interaction with different concentrations of NG. Figure 5.12B shows the calibration curves for both citrate-imprinted and nonimprinted matrixes.

The detection limit for analyzing NG by the imprinted composite corresponds to 20 pM, while reflectance changes level off to a saturation value at a concentration of ca. 2 nM, consistent with the saturation of the imprinted sites. Evidently, the detection limit for analyzing PETN by the citrate-imprinted Au NP matrix is ca. 100-fold lower as compared to the analysis of NG on the same matrix. Calibration curve in Fig. 5.12B relates the reflectance changes to the concentrations of NG and allows deriving the association constant of NG to the citrate-imprinted sites ($K_a = 2.5\times10^{10}$ M^{-1}). In comparison, the enhanced sensitivity to PETN was reflected by the significantly higher association constant (ca. 40-fold higher) of PETN to the citrate-imprinted sites. The enhanced association of PETN to the citric-acid-imprinted sites, as compared to the NG binding to these sites, may be attributed to the fact that, in addition to the three carboxylic acid residues that mimic the $-ONO_2$ functionalities of PETN, the additional OH group present in the imprint molecule (citric acid) acts cooperatively in generating a binding domain for the fourth $-ONO_2$ group present in PETN. That is, the imprinted sites are structurally optimized to accommodate the substrate with four

$-ONO_2$ groups. Figure 5.12B, curve b, shows the calibration curve for analyzing NG by the nonimprinted matrix. The reflectance changes are minute, and they do not allow the detection of NG. It was also found that the imprint of other carboxylic acids, such as isocitric, succinic, or maleic acids, does not lead to Au NP matrixes that allow effective sensing of NG at the low-concentration range.

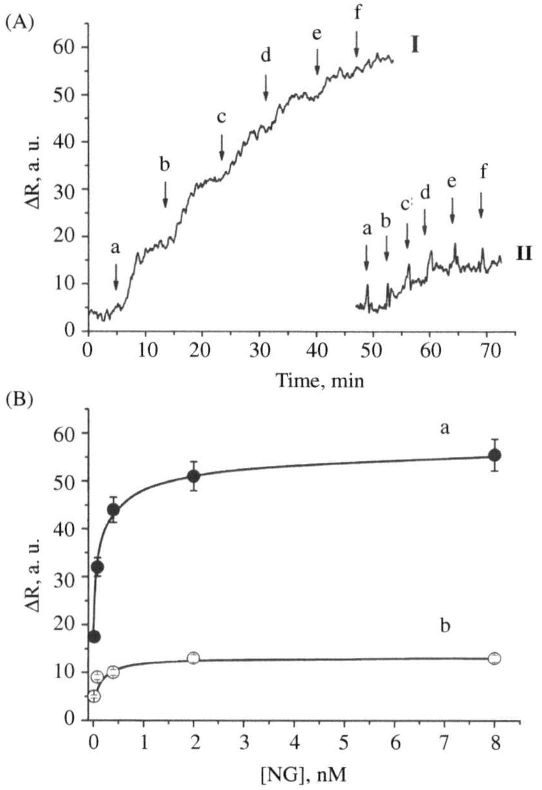

Figure 5.12 (A) SPR sensograms corresponding to the changes in the reflectance intensities, at a constant angle $\theta = 65.0°$, upon addition of variable concentrations of NG: (a) 20, (b) 80, and (c) 400 pM and (d) 2, (e) 8, and (f) 40 nM to (I) the citrate-imprinted Au NP matrix and (II) the nonimprinted matrix. (B) Calibration curves relating the reflectance changes to the concentrations of NG on (a) the citrate-imprinted and (b) the nonimprinted matrixes. Error bars correspond to a set of $N = 5$ measurements. All measurements were performed in ethanol. Reprinted with permission from Ref. [71]. Copyright (2011) American Chemical Society.

An attempt was also performed to develop an imprinted matrix for the sensing of EGDN (4) explosive. Based on the imprinting rationale described for the generation of imprinted matrixes for PETN or NG using the structurally related carboxylic acids, it was argued that fumaric acid (5), succinic acid (6), or maleic acid (7) may act as imprinting templates for EGDN. Accordingly, each of these dicarboxylic acids was used to generate the respective imprinted Au NP composites. Figure 5.13 shows the calibration curves observed upon the treatment of the imprinted Au NP composites (using (5), (6), or (7) as imprinting templates) with variable concentrations of EGDN. It was found that the three Au NP matrixes reveal comparable sensing capabilities for EGDN. Maleic acid as an imprinting template (7) yields the superior sensing matrix (highest reflectance changes along the EGDN concentration profile), while the second-best sensing matrix is formed upon the imprinting of (5) into the composite. The Au NP composite imprinted with (6) reveals the lowest sensing features. Thus, it was concluded that the imprinting of the dicarboxylic maleic acid in the Au NP composite is essential to yield an effective sensing matrix for EGDN.

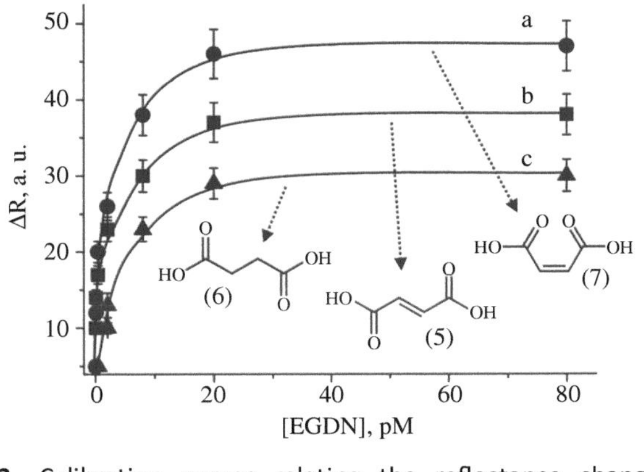

Figure 5.13 Calibration curves relating the reflectance changes to the concentrations of EGDN on (a) the maleic-acid-imprinted, (b) the fumaric-acid-imprinted, and (c) the succinic-acid-imprinted bis(aniline)-crosslinked Au NP matrixes. Error bars correspond to a set of $N = 5$ measurements. All measurements were performed in ethanol. Reprinted with permission from Ref. [71]. Copyright (2011) American Chemical Society.

Finally, the selectivity of the carboxylic-acids-imprinted matrixes was examined toward the analysis of the three explosives: PETN, NG, and EGDN. Figure 5.14A shows the calibration curves corresponding to the analysis of PETN (curve I), NG (curve II), and EGDN (curve III) by the citric-acid-imprinted Au NP composite. While PETN is detected with a sensitivity that corresponds to 200 fM, the NG and EGDN explosives yield detectable reflectance responses starting at concentrations above 100 pM and 400 nM, respectively. These results clearly indicate that the citric acid template that includes three carboxylic acid substituents and one OH functionality yields an imprinted contour that allows an effective binding for the four nitrate ester substituents associated with PETN. The EGDN that is sterically smaller and includes only two nitro substituents does not fit to the resulting imprinted sites, leading to low binding affinities and to poor performance of the matrix toward sensing EGDN.

It was also found that, compared to the other carboxylic acid imprint molecules, fumaric acid exhibited the highest selectivity toward EGDN among the different explosives. Figure 5.14B shows the calibration curves corresponding to the analysis of EGDN (curve a), NG (curve b), and PETN (curve c) by the fumaric-acid-imprinted Au NP composite. Evidently, EGDN shows the best sensing performance on the fumaric-acid-imprinted composite, whereas both NG and PETN are inefficiently sensed by this matrix, as reflected by the lower reflectance changes measured for these explosives at a similar concentration range. These results clearly indicate that the imprinting procedure is highly efficient and facilitates the analysis of picomolar, or subpicomolar, concentration range of the explosives and that, by the appropriate selection of the imprinting template, highly selective matrixes for each of the three explosives can be easily prepared.

This specific study has demonstrated the fabrication possibility of molecularly imprinted Au NP composites for the detection of PETN, NG, and EGDN explosives. The molecular imprinting process of site-specific matrixes was successfully achieved by applying di- or tricarboxylic acids as structural analogues for the nitrate ester substituents associated with the target explosives. The study has demonstrated that the structure of the carboxylic acids and their steric rigidification are of utmost importance to yield high-affinity sites for

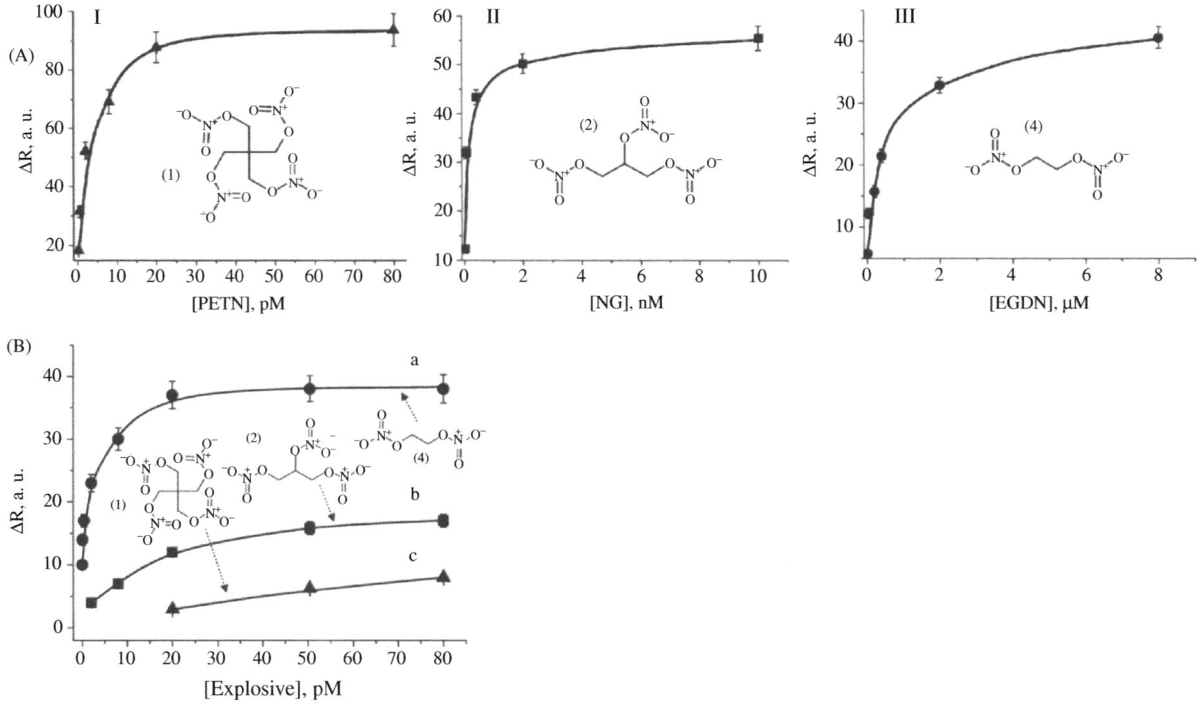

Figure 5.14 (A) Calibration curves relating the reflectance changes to the concentrations of (I) PETN, (II) NG, and (III) EGDN on the citric-acid-imprinted bis(aniline)-crosslinked Au NP matrix. (B) Calibration curves relating the reflectance changes to the concentrations of (a) EGDN, (b) NG, and (c) PETN on the fumaric-acid-imprinted bis(aniline)-crosslinked Au NP matrix. Error bars correspond to a set of N = 5 measurements. All measurements were performed in ethanol. Reprinted with permission from Ref. [71]. Copyright (2011) American Chemical Society.

the respective explosives. The electropolymerization method used to synthesize the imprinted Au NP complexes provides an effective approach to prepare predesigned specific recognition matrixes for the respective explosives. Also, by employing the oligocarboxylic acid templates, sensing platforms for other explosives may be produced and, particularly, the multiplexed analysis of explosives may be performed.

5.4 Conclusion

Two specific studies described in this chapter exemplify the versatility of SPR spectroscopy for building the highly sensitive and selective bio- and chemosensors using the molecular imprinting approach. The proposed techniques combine a number of unique advantages, including real-time small molecules detection, possibility of sensitive matrix adjustment to the analyte structure, analysis in different solvents, etc.

In the first study, specific imprinting of recognition sites for the NAD^+, $NADP^+$, NADH, and NADPH cofactors in an acrylamide–acrylamidophenylboronic acid copolymer film and utilization of SPR spectroscopy as a means to follow the association of the different cofactors to the imprinted polymer films were demonstrated. The study has shown that the SPR method combined with imprinted polymers can identify the selective binding of the structurally related cofactors to the respective imprinted polymers. The use of the NADH-imprinted film and SPR to follow the kinetics of the NAD^+-dependent lactate dehydrogenase biocatalyzed oxidation of lactate was also demonstrated. This opens new possibilities to characterize $NAD(P)^+$ and NAD(P)H biocatalyzed reactions by optical detection means.

In the second study, a novel method was developed to improve the low-molecular-weight analyte detection efficiency by an imprinted matrix. The imprinting process was successfully achieved by applying di- or tricarboxylic acids as structural analogues for the nitrate ester substituents associated with the target explosives. Additionally, this approach exploited the fabrication of imprinted composites with embedded bis(aniline)-crosslinked Au NPs on the

Au surface of SPR chip, which led to a substantial enhancement of the coupling between a localized surface plasmon of nanoparticles and a surface plasmon-polariton wave. The proximity of Au NPs to the specific binding sites increases the sensitivity of the method, resulting in even a minimal amount of an analyte close to the site capable of changing the equilibrium in the system, which leads to a change in the SPR angle. The study revealed the preparation of highly sensitive and selective imprinted composites, and the effectiveness of the proposed method was demonstrated on the example of the registration of ultralow (down to femtomolar) concentrations of PETN, NG, and EGDN explosives in ethanolic solutions.

References

1. Haupt, K. and Mosbach, K. (2000). Molecularly imprinted polymers and their use in biomimetic sensors, *Chem. Rev.*, **100**, pp. 2495–2504.
2. Zayats, M., Lahav, M., Kharitonov, A. B., and Willner, I. (2002). Imprinting of specific molecular recognition sites in inorganic and organic thin layer membranes associated with ion-sensitive field-effect transistors, *Tetrahedron*, **58**, pp. 815–824.
3. Lee, S. W., Ichinose, I., and Kunitake, T. (1998). Molecular imprinting of azobenzene carboxylic acid on a TiO_2 ultrathin film by the surface sol-gel process, *Langmuir*, **14**, pp. 2857–2863.
4. Whitcombe, M. J., Kirsch, N., and Nicholls, I. A. (2014). Molecular imprinting science and technology: A survey of the literature for the years 2004–2011, *J. Mol. Recognit.*, **27**, pp. 297–401.
5. Sellergren, B. and Shea, K. J. (1995). Origin of peak asymmetry and the effect of temperature on solute retention in enantiomer separations on imprinted chiral stationary phases, *J. Chromatogr. A*, **690**, pp. 29–39.
6. Wulff, G., Gross, T., and Schönfeld, R. (1997). Enzyme models based on molecularly imprinted polymers with strong esterase activity, *Angew. Chem. Int. Ed. Engl.*, **36**, pp. 1962–1964.
7. Hart, B. R. and Shea, K. J. (2001). Synthetic peptide receptors: Molecularly imprinted polymers for the recognition of peptides using peptide–metal interactions, *J. Am. Chem. Soc.*, **123**, pp. 2072–2073.
8. Schweitz, L., Andersson, L. I., and Nilsson, S. (1997). Capillary electrochromatography with predetermined selectivity obtained through molecular imprinting, *Anal. Chem.*, **69**, pp. 1179–1183.

9. Vallano, P. T. and Remcho, V. T. (2000). Highly selective separations by capillary electrochromatography: Molecular imprint polymer sorbents, *J. Chromatogr. A*, **887**, pp. 125–135.
10. Lahav, M., Kharitonov, A. B., and Willner, I. (2001). Imprinting of chiral molecular recognition sites in thin TiO_2 films associated with field-effect transistors: Novel functionalized devices for chiroselective and chirospecific analyses, *Chem. Eur. J.*, **7**, pp. 3992–3997.
11. Makote, R. and Collinson, M. M. (1998). Template recognition in inorganic–organic hybrid films prepared by the sol–gel process, *Chem. Mater.*, **10**, pp. 2440–2445.
12. Schweitz, L., Spégel, P., and Nilsson, S. (2000). Molecularly imprinted microparticles for capillary electrochromatographic enantiomer separation of propranolol, *Analyst*, **125**, pp. 1899–1901.
13. Brunkan, N. M. and Gagné, M. R. (2000). Effect of chiral cavities associated with molecularly imprinted platinum centers on the selectivity of ligand-exchange reactions at platinum, *J. Am. Chem. Soc.*, **122**, pp. 6217–6225.
14. Wang, P., Sun, X., Su, X., and Wang, T. (2016). Advancements of molecularly imprinted polymers in the food safety field, *Analyst*, **141**, pp. 3540–3553.
15. Yang, S., Wang, Y., Jiang, Y., Li, S., and Liu, W. (2016). Molecularly imprinted polymers for the identification and separation of chiral drugs and biomolecules, *Polymers*, **8**, 216.
16. Wang, W., Gao, S., and Wang, B. (1999). Building fluorescent sensors by template polymerization: The preparation of a fluorescent sensor for D-fructose, *Org. Lett.*, **1**, pp. 1209–1212.
17. Ichinose, I., Kikuchi, T., Lee, S. W., and Kunitake, T. (2002). Imprinting and selective binding of di- and tri-peptides in ultrathin TiO_2-gel films in aqueous solutions, *Chem. Lett.*, **31**, pp. 104–105.
18. Kugimiya, A. and Takeuchi, T. (2001). Surface plasmon resonance sensor using molecularly imprinted polymer for detection of sialic acid, *Biosens. Bioelectron.*, **16**, pp. 1059–1062.
19. Lai, E. P., Fafara, A., VanderNoot, V. A., Kono, M., and Polsky, B. (1998). Surface plasmon resonance sensors using molecularly imprinted polymers for sorbent assay of theophylline, caffeine, and xanthine, *Can. J. Chem.*, **76**, pp. 265–273.
20. Nishimura, S., Yoshidome, T., and Higo, M. (2002). Selective measurement of optical isomer by using molecular imprinting and surface plasmon resonance sensor, *Anal. Sci.*, **17**, pp. i1697–i1699.

21. Liao, S., Zhang, W., Long, W., Hou, D., Yang, X., and Tan, N. (2016). Adsorption characteristics, recognition properties, and preliminary application of nordihydroguaiaretic acid molecularly imprinted polymers prepared by sol–gel surface imprinting technology, *Appl. Surf. Sci.*, **364**, pp. 579–588.

22. Xing, R., Wang, S., Bie, Z., He, H., and Liu, Z. (2017). Preparation of molecularly imprinted polymers specific to glycoproteins, glycans and monosaccharides via boronate affinity controllable–oriented surface imprinting, *Nat. Protoc.*, **12**, pp. 964–987.

23. Wackerlig, J. and Schirhagl, R. (2015). Applications of molecularly imprinted polymer nanoparticles and their advances toward industrial use: A review, *Anal. Chem.*, **88**, pp. 250–261.

24. Cheng, W., Fan, F., Zhang, Y., Pei, Z., Wang, W., and Pei, Y. (2017). A facile approach for fabrication of core-shell magnetic molecularly imprinted nanospheres towards hypericin, *Polymers*, **9**, 135.

25. Wackerlig, J. and Lieberzeit, P. A. (2015). Molecularly imprinted polymer nanoparticles in chemical sensing: Synthesis, characterisation and application, *Sens. Actuators B Chem.*, **207**, pp. 144–157.

26. Kolla, P. (1997). The application of analytical methods to the detection of hidden explosives and explosive devices, *Angew. Chem. Int. Ed. Engl.*, **36**, pp. 800–811.

27. Laine, D. F., Roske, C. W., and Cheng, I. F. (2008). Electrochemical detection of triacetone triperoxide employing the electrocatalytic reaction of iron (II/III)-ethylenediaminetetraacetate and hydrogen peroxide, *Anal. Chim. Acta*, **608**, pp. 56–60.

28. Almog, J., Burda, G., Shloosh, Y., Abramovich-Bar, S., Wolf, E., and Tamiri, T. (2007). Recovery and detection of urea nitrate in traces, *J. Forensic Sci.*, **52**, pp. 1284–1290.

29. Tamiri, T., Rozin, R., Lemberger, N., and Almog, J. (2009). Urea nitrate, an exceptionally easy-to-make improvised explosive: Studies towards trace characterization, *Anal. Bioanal. Chem.*, **395**, pp. 421–428.

30. Krause, A. R., Van Neste, C., Senesac, L., Thundat, T., and Finot, E. (2008). Trace explosive detection using photothermal deflection spectroscopy, *J. Appl. Phys.*, **103**, 094906.

31. McQuade, D. T., Pullen, A. E., and Swager, T. M. (2000). Conjugated polymer-based chemical sensors, *Chem. Rev.*, **100**, pp. 2537–2574.

32. Andrew, T. L. and Swager, T. M. (2007). A fluorescence turn-on mechanism to detect high explosives RDX and PETN, *J. Am. Chem. Soc.*, **129**, pp. 7254–7255.

33. Goldman, E. R., Medintz, I. L., Whitley, J. L., Hayhurst, A., Clapp, A. R., Uyeda, H. T., Deschamps, J. R., Lassman, M. E., and Mattoussi, H. (2005). A hybrid quantum dot–antibody fragment fluorescence resonance energy transfer-based TNT sensor, *J. Am. Chem. Soc.*, **127**, pp. 6744–6751.

34. Zhang, H. X., Cao, A. M., Hu, J. S., Wan, L. J., and Lee, S. T. (2006). Electrochemical sensor for detecting ultratrace nitroaromatic compounds using mesoporous SiO_2-modified electrode, *Anal. Chem.*, **78**, pp. 1967–1971.

35. Wang, J., Bhada, R. K., Lu, J., and MacDonald, D. (1998). Remote electrochemical sensor for monitoring TNT in natural waters, *Anal. Chim. Acta*, **361**, pp. 85–91.

36. Wang, J., Lu, F., MacDonald, D., Lu, J., Ozsoz, M. E., and Rogers, K. R. (1998). Screen-printed voltammetric sensor for TNT, *Talanta*, **46**, pp. 1405–1412.

37. Yang, X., Du, X. X., Shi, J., and Swanson, B. (2001). Molecular recognition and self-assembled polymer films for vapor phase detection of explosives, *Talanta*, **54**, pp. 439–445.

38. Kannan, G. K., Nimal, A. T., Mittal, U., Yadava, R. D. S., and Kapoor, J. C. (2004). Adsorption studies of carbowax coated surface acoustic wave (SAW) sensor for 2,4-dinitro toluene (DNT) vapour detection, *Sens. Actuators B Chem.*, **101**, pp. 328–334.

39. McGill, R. A., Mlsna, T. E., Chung, R., Nguyen, V. K., and Stepnowski, J. (2000). The design of functionalized silicone polymers for chemical sensor detection of nitroaromatic compounds, *Sens. Actuators B Chem.*, **65**, pp. 5–9.

40. Larsson, A., Angbrant, J., Ekeroth, J., Månsson, P., and Liedberg, B. (2006). A novel biochip technology for detection of explosives–TNT: Synthesis, characterisation and application, *Sens. Actuators B Chem.*, **113**, pp. 730–748.

41. Riskin, M., Tel-Vered, R., Bourenko, T., Granot, E., and Willner, I. (2008). Imprinting of molecular recognition sites through electropolymerization of functionalized Au nanoparticles: Development of an electrochemical TNT sensor based on π-donor–acceptor interactions, *J. Am. Chem. Soc.*, **130**, pp. 9726–9733.

42. Riskin, M., Tel-Vered, R., Lioubashevski, O., and Willner, I. (2009). Ultrasensitive surface plasmon resonance detection of trinitrotoluene by a bis-aniline-cross-linked Au nanoparticles composite, *J. Am. Chem. Soc.*, **131**, pp. 7368–7378.

43. Riskin, M., Tel-Vered, R., and Willner, I. (2010). Imprinted Au-nanoparticle composites for the ultrasensitive surface plasmon resonance detection of hexahydro-1,3,5-trinitro-1,3,5-triazine (RDX), *Adv. Mater.*, **22**, pp. 1387–1391.

44. Wulff, G. (2002). Enzyme-like catalysis by molecularly imprinted polymers, *Chem. Rev.*, **102**, pp. 1–28.

45. Bossi, A., Bonini, F., Turner, A. P. F., and Piletsky, S. A. (2007). Molecularly imprinted polymers for the recognition of proteins: The state of the art, *Biosens. Bioelectron.*, **22**, pp. 1131–1137.

46. Haupt, K. (2001). Molecularly imprinted polymers in analytical chemistry, *Analyst*, **126**, pp. 747–756.

47. Bartlett, P. N., Tebbutt, P., and Whitaker, R. G. (1991) *Kinetic Aspects of the Use of Modified Electrodes and Mediators in Bioelectrochemistry. Series: Progress in Reaction Kinetics* **16**, (Pergamon Press, Oxford, New York).

48. Biade, A. E., Bourdillon, C., Laval, J. M., Mairesse, G., and Moiroux, J. (1992). Complete conversion of L-lactate into D-lactate. A generic approach involving enzymic catalysis, electrochemical oxidation of NADH and electrochemical reduction of pyruvate, *J. Am. Chem. Soc.*, **114**, pp. 893–897.

49. Burnett, J. N. and Underwood, A. L. (1965). Electrochemical reduction of diphosphopyridine nucleotide, *Biochemistry*, **4**, pp. 2060–2064.

50. Cunningham, A. J. and Underwood, A. L. (1967). Cyclic voltammetry of the pyridine nucleotides and a series of nicotinamide model compounds, *Biochemistry*, **6**, pp. 266–271.

51. Chardon-Noblat, S., Cosnier, S., Deronzier, A., and Vlachopoulos, N. (1993). Electrochemical properties of $[(C_5Me_5)\ Rh^{III}(L)Cl]^+$ complexes (L=2,2′-bipyridine or 1,10-phenanthroline derivatives) in solution in related polypyrrolic films. Application to electrocatalytic hydrogen generation, *J. Electroanal. Chem.*, **352**, pp. 213–228.

52. Bergel, A. and Comtat, M. (1992). Reduction of $NAD(P)^+$ by electrochemically driven $FADH_2$ and $FMNH_2$, *Bioelectrochem. Bioenerg.*, **27**, pp. 495–500.

53. Long, Y. T. and Chen, H. Y. (1997). Electrochemical regeneration of coenzyme NADH on a histidine modified silver electrode, *J. Electroanal. Chem.*, **440**, pp. 239–242.

54. Blaedel, W. J. and Jenkins, R. A. (1975). Electrochemical oxidation of reduced nicotinamide adenine dinucleotide, *Anal. Chem.*, **47**, pp. 1337–1343.

55. Katakis, I. and Domínguez, E. (1997). Catalytic electrooxidation of NADH for dehydrogenase amperometric biosensors, *Microchim. Acta*, **126**, pp. 11–32.

56. Miller, L. L. and Valentine, J. R. (1988). On the electron-proton-electron mechanism for 1-benzyl-1,4-dihydronicotinamide oxidations, *J. Am. Chem. Soc.*, **110**, pp. 3982–3989.

57. Gorton, L. (1986). Chemically modified electrodes for the electrocatalytic oxidation of nicotinamide coenzymes, *J. Chem. Soc., Faraday Trans. 1*, **82**, pp. 1245–1258.

58. Raitman, O. A., Chegel, V. I., Kharitonov, A. B., Zayats, M., Katz, E., and Willner, I. (2004). Analysis of NAD(P)$^+$ and NAD(P)H cofactors by means of imprinted polymers associated with Au surfaces: A surface plasmon resonance study, *Anal. Chim. Acta*, **504**, pp. 101–111.

59. James, T. D., Sandanayake, K. S., and Shinkai, S. (1995). Chiral discrimination of monosaccharides using a fluorescent molecular sensor, *Nature*, **374**, pp. 345–347.

60. Gabai, R., Sallacan, N., Chegel, V., Bourenko, T., Katz, E., and Willner, I. (2001). Characterization of the swelling of acrylamidophenylboronic acid–acrylamide hydrogels upon interaction with glucose by faradaic impedance spectroscopy, chronopotentiometry, quartz-crystal microbalance (QCM), and surface plasmon resonance (SPR) experiments, *J. Phys. Chem. B*, **105**, pp. 8196–8202.

61. Sallacan, N., Zayats, M., Bourenko, T., Kharitonov, A. B., and Willner, I. (2002). Imprinting of nucleotide and monosaccharide recognition sites in acrylamidephenylboronic acid–acrylamide copolymer membranes associated with electronic transducers, *Anal. Chem.*, **74**, pp. 702–712.

62. Wulff, G. and Gimpel, J. (1982). Über enzymanalog gebaute polymer, 16. Über den Einfluß der flexibilität der haftgruppen auf die racemattrennungsfähigkeit, *Makromol. Chem.*, **183**, pp. 2469–2477 (in German).

63. Wulff, G. and Haarer, J. (1991). Enzyme-analogue built polymers, 29. The preparation of defined chiral cavities for the racemic resolution of free sugars, *Makromol. Chem.*, **192**, pp. 1329–1338.

64. Jönsson, U., Ivarsson, B., Lundström, I., and Berghem, L. (1982). Adsorption behavior of fibronectin on well-characterized silica surfaces, *J. Colloid Interface Sci.*, **90**, pp. 148–163.

65. Malmsten, M. (1994). Ellipsometry studies of protein layers adsorbed at hydrophobic surfaces, *J. Colloid Interface Sci.*, **166**, pp. 333–342.

66. Kubitschko, S., Spinke, J., Brückner, T., Pohl, S., and Oranth, N. (1997). Sensitivity enhancement of optical immunosensors with nanoparticles, *Anal. Biochem.*, **253**, pp. 112–122.

67. Wink, T., van Zuilen, S. J., Bult, A., and Van Bennekom, W. P. (1998). Liposome-mediated enhancement of the sensitivity in immunoassays of proteins and peptides in surface plasmon resonance spectrometry, *Anal. Chem.*, **70**, pp. 827–832.

68. Zayats, M., Raitman, O. A., Chegel, V. I., Kharitonov, A. B., and Willner, I. (2002). Probing antigen–antibody binding processes by impedance measurements on ion-sensitive field-effect transistor devices and complementary surface plasmon resonance analyses: Development of cholera toxin sensors, *Anal. Chem.*, **74**, pp. 4763–4773.

69. Lyon, L. A., Musick, M. D., Smith, P. C., Reiss, B. D., Pena, D. J., and Natan, M. J. (1999). Surface plasmon resonance of colloidal Au-modified gold films, *Sens. Actuators B Chem.*, **54**, pp. 118–124.

70. Agarwal, G. S. and Gupta, S. D. (1985). Interaction between surface plasmons and localized plasmons, *Phys. Rev. B*, **32**, pp. 3607–3611.

71. Riskin, M., Ben-Amram, Y., Tel-Vered, R., Chegel, V., Almog, J., and Willner, I. (2011). Molecularly imprinted Au nanoparticles composites on Au surfaces for the surface plasmon resonance detection of pentaerythritol tetranitrate, nitroglycerin, and ethylene glycol dinitrate, *Anal. Chem.*, **83**, pp. 3082–3088.

Chapter 6

Electrochemical Surface Plasmon Resonance and Its Applications in Biosensing, Bioelectronics, and Material Science

6.1 Introduction

A large number of modern chemical and biological sensors are based on the surface plasmon resonance (SPR) phenomenon, and quite often research with SPR sensors is carried out under the conditions of external electric voltage applied to the metal–liquid sensor interface [1–4], which allows to expand the number of applications of SPR method. In particular, the method of electrochemical SPR (ESPR) spectroscopy was used to study electropolymerization [5, 6] and redox properties of polymers [7, 8], as well as to detect cofactors [2], glucose [1, 9], hydrogen peroxide [10], and other analytes [11]. An attractive idea for SPR biosensing is to use the applied electrical potential to manipulate adsorption and interaction properties of biomolecules carrying electric charges of different sign and magnitude. It should be borne in mind that the processes affecting the response of the SPR biosensor to biomolecular analytes upon the application of electric potential occur in the nanoscale area at the metal–liquid interface and require careful consideration in order to ensure the correctness of the SPR measurements. In particular, there is a problem in the correct interpretation of the results obtained

Molecular Plasmonics: Theory and Applications
Volodymyr I. Chegel and Andrii M. Lopatynskyi

ISBN 978-981-4800-65-5 (Hardcover), 978-0-429-29511-9 (eBook)
www.jennystanford.com

taking into account the effect of the electrical potential both on the object of the study and on the sensitive part of the sensor.

Studies on the influence of electrical potential applied to the metal–electrolyte interface on its optical properties via the excitation of surface plasma waves have been actively conducted over the past decades. In light electroreflection experiments, it was discovered that the optical properties of the metal electrode surface depend significantly on the potential applied to them [12, 13]. Later it was shown that the excitation energy of surface plasmon-polaritons in thin silver and gold films also depends on the magnitude of the applied potential [14]. This chapter gives a theoretical explanation for this effect, but only in the approximation of the change in the concentration of free electrons in the near-surface layer of metal when the electrical potential is applied. The studies in Refs. [15, 16] investigated the excitation of surface plasmon-polaritons on the silver–electrolyte interface, where the effect of the applied potential on the angular position of the SPR was observed. It was noted that at a constant light wavelength, the resonance was shifted toward higher values of the surface plasmon-polariton wave vector (larger angles of incidence) with an increase in the applied potential above the potential of zero charge of the electrode. In the above-mentioned conditions, an increase in the half-width of the SPR curves was also registered. In addition, a method for measuring the potential of zero charge of a metal electrode in an electrolyte solution by means of ESPR measurements was proposed in these works. An investigation of the gold–electrolyte interface using the ellipsometric method for detecting surface plasma waves was carried out in Ref. [17]. Using this method, the possibility of separating the influence of the applied voltage on the electrons in the metal and the ions in the electrolyte was shown. Additionally, the penetration depth of a static electric field into gold was demonstrated to be no more than 1 Å.

At the same time, the effect of the external electric potential on the response of SPR biosensor is insufficiently investigated, especially theoretically. In the studies in Refs. [18, 19], the SPR response was studied upon the cyclic change in the potential applied to a thin gold film in electrolytes of different composition with different pH levels, and redox reactions on the gold surface were identified. The paper in Ref. [19] presents a model for calculating the SPR response that considers the structure of the electric double layer at the metal–electrolyte boundary in the Gouy–Chapman approximation and the presence of an oxide layer on gold at significant positive potentials. It remains necessary to develop a more general theoretical model of the SPR response in the potential range corresponding to the double layer region, in which there are no electrochemical reactions present and which is most suitable for biosensor applications.

This chapter presents the results of specific studies on SPR sensor response peculiarities under the conditions of applied external electrical potential by means of ESPR technique and describes the applications of this method for the investigation of biosensor response manipulation, electropolymerization processes, redox properties, and switchable functions in various systems. The first study is focused on the experimental and theoretical analysis of ESPR sensor response and investigation of the biomolecular adsorption processes under the electrical potential applied to the gold–electrolyte interface. The second study deals with the ESPR transduction of the redox transformations in electropolymerized thin polyaniline films. The promising potential of the ESPR method is shown in the two last studies by the examples of cyclic control of hydrophilic/hydrophobic properties of Ag^+-thiolate monolayers and biocatalytic charging of gold nanoparticles associated with self-assembled monolayers on Au surface of the SPR chip.

6.2 Factor of Interfacial Electrical Potential for the SPR Sensor Response

6.2.1 General Theoretical Background

Electrochemical reactions may progress upon application of an external electrical voltage to the metal–electrolyte interface of the SPR sensor sensitive element [20]. If the metal electrode potential is sufficient for oxidation or reduction of the electrode, molecules, or ions on its interface, then an electrochemical reaction begins, and the current flows through the metal–electrolyte boundary. If the conditions on the interface are such that the electrochemical reactions do not take place, then the current does not flow across the interface and the electrical double layer (EDL) (Fig. 6.1) at the boundary behaves like a capacitor. The potential region where this behavior is observed is the well-known double layer region.

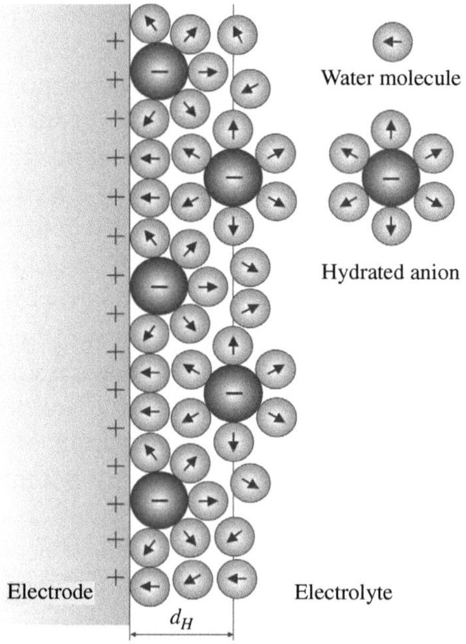

Figure 6.1 Model of electrical double layer on the positively charged electrode surface. Arrows indicate the dipole moments of water molecules. Reprinted from Ref. [20] under a Creative Commons Attribution-NoDerivatives 4.0 International License. Figure caption was adapted.

According to the double layer theory by Stern, the differential capacity of the EDL unit area C_{EDL} equals the capacity of two series-connected capacitors C_{H} and C_{GC} that correspond to the compact layer in the Helmholtz theory and the diffuse layer in the Gouy–Chapman theory [21]:

$$\frac{1}{C_{\mathrm{EDL}}}=\frac{1}{C_{\mathrm{H}}}+\frac{1}{C_{\mathrm{GC}}}$$

$$=\frac{d_{\mathrm{H}}}{\varepsilon_{\mathrm{el}}\varepsilon_0}+\frac{1}{(2\varepsilon_{\mathrm{el}}\varepsilon_0 z^2 e_0^2 N_{\mathrm{A}}\cdot 1000 c_{\mathrm{el}}/kT)^{1/2}\cosh(ze_0U/2kT)}, \quad (6.1)$$

where d_{H} is the compact layer thickness, $\varepsilon_{\mathrm{el}}$ is the relative static dielectric permittivity of the electrolyte, ε_0 is the vacuum dielectric permittivity, z is the ion charge, e_0 is the electron charge, N_{A} is the Avogadro constant, c_{el} is the electrolyte concentration in mol/l, k is the Boltzmann constant, T is the electrolyte temperature, and U is the potential at the distance d_H from the metal surface. Potential U is determined from Eq. (6.2) [22]:

$$U_0 = U + \sqrt{\frac{8kT \cdot 1000 N_A c_{el}}{\varepsilon_{el}\varepsilon_0}} \sinh\left(\frac{ze_0 U}{2kT}\right) \cdot d_H, \quad (6.2)$$

where U_0 is the total potential difference between the metal electrode and electrolyte. Then the surface charge density $\Delta\sigma$ that appears on the electrode surface is represented by Eq. (6.3) [21]:

$$\Delta\sigma = \int_{U_{pzc}}^{U_0} C_{EDL} dU, \quad (6.3)$$

where U_{pzc} is the potential of zero charge of the electrode in the present electrolyte.

Formation of a charge on the metal electrode surface during the charging of EDL leads to a change in the free electron concentration in the surface layer of metal, which is one of the effects taking place when applying the voltage to a metal–electrolyte system. As the static electric field penetrates the metal to the distance $d \sim 1$ Å, the change in the free electron concentration, which is defined by the equation $\Delta N = \Delta\sigma/e_0 d$, takes place in this thin layer only [14]. This results in the change in the free electron contribution to the relative dielectric permittivity of this layer, which is represented by Eq. (6.4) [14]:

$$\Delta\varepsilon_m = (1 - \varepsilon_m^f)\frac{\Delta N}{N}, \quad (6.4)$$

where ε_m^f is the free electron contribution to the relative dielectric permittivity of the metal and N is the free electron concentration in the metal. The free electron contribution to the relative dielectric permittivity of the metal is defined by Eq. (6.5) [23]:

$$\varepsilon_m^f = 1 - \frac{\omega_p^2}{\omega^2 - i\frac{\omega}{\tau}}, \quad (6.5)$$

where $\omega_p = \sqrt{Ne_0^2/\varepsilon_0 m^*}$ is the plasma frequency, m^* is the effective mass of an electron, ω is the light angular frequency, and τ is the free electron relaxation time.

Since the relative dielectric permittivity of the surface layer of a metal undergoes a change when the voltage is applied to the metal–electrolyte interface according to Eq. (6.4), the value of the SPR angle will also depend on the value of the applied voltage. Therefore, the theoretical calculation of the SPR angle will allow predicting the effect of applying external voltage on the optical properties of the metal–electrolyte interface and taking this into account when carrying out experimental studies.

However, thin films of noble metals, for example gold films, used in SPR sensors are made using the method of thermal vacuum evaporation and are characterized by a rough surface, which needs to be taken into account for theoretical calculation of optical properties. To this end, the metal–electrolyte separation boundary was proposed to be modeled as a statistical heterosystem consisting of three components: a rough metal surface, a surface layer of metal of 1 Å thickness, whose optical properties depend on the externally applied voltage, and the electrolyte (Fig. 6.2). Since there is no predominant component in this system that could play the role of the main matrix, the symmetrical Bruggeman approximation was used to calculate the effective complex refractive index of this heterosystem [24]:

$$\sum_i f_i \frac{N_i^2 - \tilde{N}^2}{N_i^2 + 2\tilde{N}^2} = 0, \tag{6.6}$$

where $i = 1, 2, 3$ is the sequence number of the component, f_i and N_i are the volume fractions and refractive indices for the corresponding components of the heterosystem.

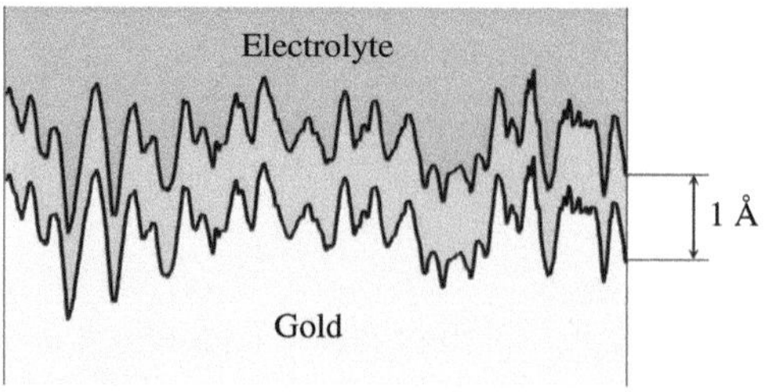

Figure 6.2 Structure of the near-surface metal layer in the SPR sensor sensitive element upon the application of external voltage.

6.2.2 Influence of Processes in Electrical Double Layer at the Surface of Sensitive Element on the SPR Sensor Response

Under the conditions of applying an external electrical potential to the gold–electrolyte interface, formation of SPR sensor optical response is a complex process. This is due to the simultaneous influence of both the detected analyte itself and the changes in the optical properties of the sensitive element of the SPR sensor induced by the applied potential. Therefore, it is necessary to investigate the influence of processes in the electrical double layer at the gold–electrolyte interface on the optical response of SPR biosensor when an external electrical potential is applied. This is promising due to the possibility of evaluating such influence in advance and making an appropriate correction of the results obtained under regular measurements, as well as identifying the necessary experimental parameters that may lead to the minimization of this effect. For this purpose, the ESPR measurements of the model electrolyte (0.1 M H_2SO_4 aqueous solution) were carried out and a mathematical model presented in Subsection 6.2.1 was used for calculating the response of the SPR sensor dependence on the value of the applied electrical potential.

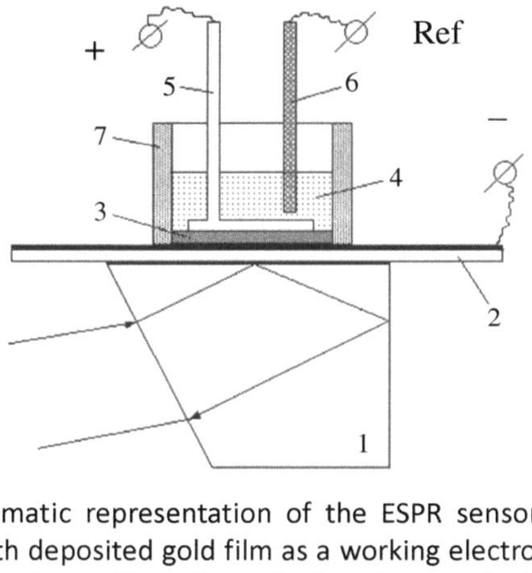

Figure 6.3 Schematic representation of the ESPR sensor. 1 – glass prism, 2 – glass plate with deposited gold film as a working electrode, 3 – polyaniline layer, 4 – electrolyte, 5 – auxiliary electrode, 6 – reference electrode, 7 – teflon cell.

Here, as well as in the majority of ESPR experiments with simultaneous SPR and electrochemical measurements, the classical three-electrode scheme was used with a metal film, carrier of plasmonic oscillations, as a working electrode, Pt auxiliary electrode and Ag-AgCl reference electrode (Fig. 6.3). As a result of ESPR measurements, cyclic voltammograms of the gold–electrolyte interface (Fig. 6.4) and temporal dependences of the angular position of SPR curves minima (Fig. 6.5) were obtained upon the potential sweep at the rate of 1 mV/s in the range of −0.6 to 1.4 V (here and later all potentials are reported versus the quasi-Ag/AgCl reference electrode).

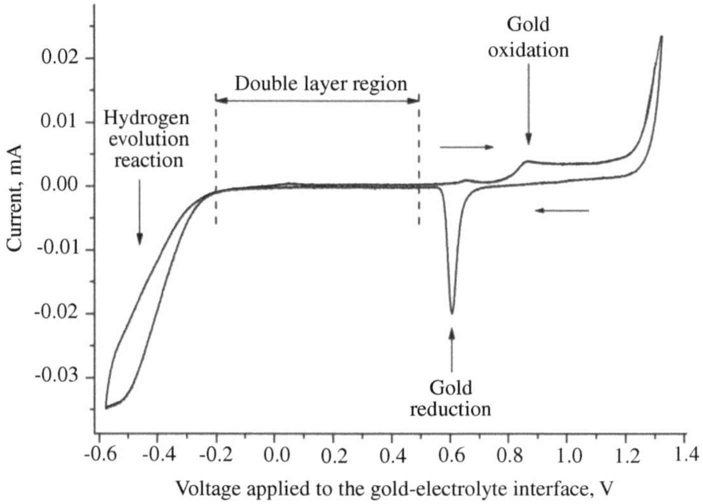

Figure 6.4 Measured cyclic voltammogram of the gold–sulfuric acid solution interface. Cyclic potential change direction is indicated by the arrows. Reprinted from Ref. [20] under a Creative Commons Attribution-NoDerivatives 4.0 International License. Figure caption was adapted.

Typical features of the electrochemical processes are clearly visible on the cyclic voltammogram shown in Fig. 6.4. During the change in the potential from −0.6 to 1.4 V at the potential difference greater than 0.7 V, the electrochemical reaction of gold surface oxidation takes place [25, 26]:

$$Au + 2H_2O \rightarrow AuOH{\cdot}H_2O + H^+ + e^-,$$

$$AuOH \rightarrow AuO + H^+ + e^-$$

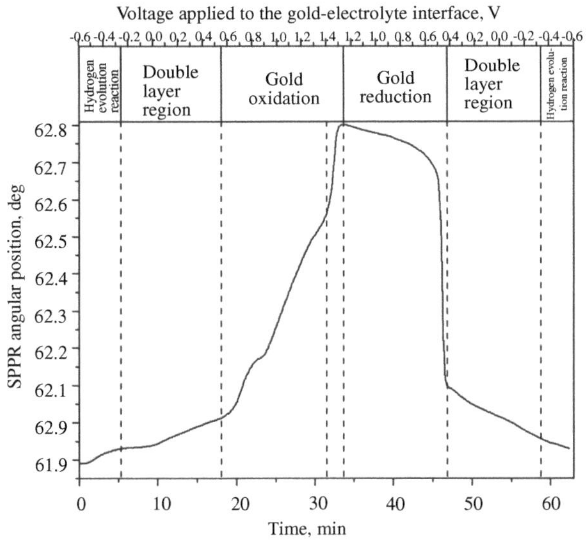

Figure 6.5 Measured temporal dependence of the SPR curves minimum angular position (sensogram). Dash line in the center indicates the moment when potential starts to change in the opposite direction. Reprinted from Ref. [20] under a Creative Commons Attribution-NoDerivatives 4.0 International License. Figure caption was adapted.

This leads to the formation of a thin gold oxide layer on the gold electrode surface, which results in an increase in the SPR angle (Fig. 6.5). During the change in the potential backward from 1.4 to −0.6 V at the potential difference about 0.7 V, the electrochemical reaction of the gold surface reduction begins, which is opposite to the above-mentioned process. This leads to the dissolution of the oxide film on the surface and to a decrease in the SPR angle (Fig. 6.5). At a potential difference lower than −0.2 V, the oxygen reduction and hydrogen evolution reactions begin on the gold electrode [27]. At a potential difference higher than 1.2 V, the oxygen evolution [28] reaction begins on the gold electrode.

The most important potential difference region for practical experiments is in the region where electrochemical reactions involving substances that the sensor is composed of are absent (double layer region) (Figs. 6.4 and 6.5). For the investigated system, this region ranges from about −0.2 to 0.5 V. Therefore, the individual measurements were performed in the potential difference region from −0.2 to 0.5 V with a different potential sweep rate, and their results are presented in Fig. 6.6.

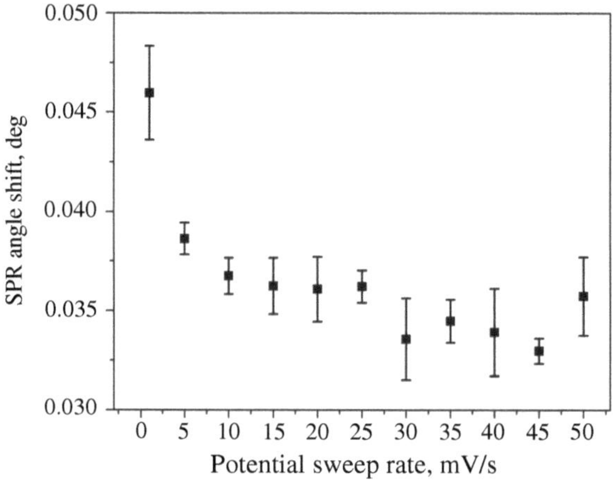

Figure 6.6 SPR angle shift with the change in the voltage applied to the gold–sulfuric acid solution interface from −0.2 V to 0.5 V for the potential sweep rate from 1 to 50 mV/s.

As it can be seen from Fig. 6.6, the shift in the SPR angular position with a change in the applied voltage from −0.2 V to 0.5 V increases with a decrease in the potential sweep rate. This may be due to the slow processes of diffusion of "heavy" SO_4^{2-} ions in an electrolyte solution under the action of an electric field between the electrodes and their subsequent accumulation on the surface of gold. In order to reduce the impact of these processes, the SPR measurements in the voltage range from −0.2 V to 0.5 V should be carried out at a potential sweep rate of not less than 10 mV/s. In particular, from the SPR sensogram presented in Fig. 6.7, it is evident that the processes that lead to a change in the SPR angular position at the application of an external voltage with a potential sweep rate of 25 mV/s are reversible, and there is no drift in the SPR angle. This means that when changing the voltage in the selected range, the effect of "heavy" ions is insignificant and there is no formation of an oxide film on the surface of gold due to electrochemical reactions, which is typically accompanied by a characteristic hysteresis during the repetition of the potential change cycles [29].

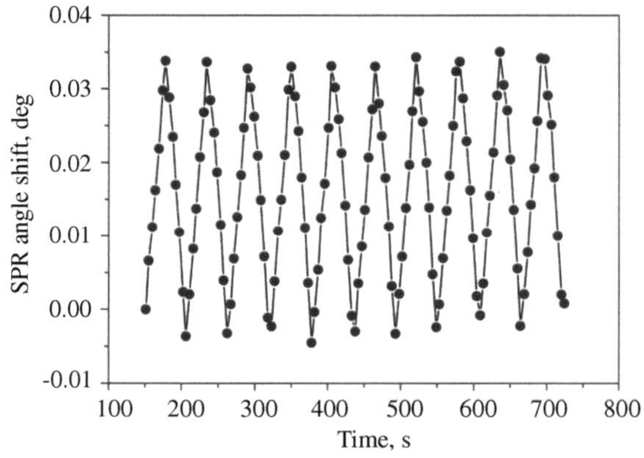

Figure 6.7 SPR sensogram obtained upon 10 cycles of change in the voltage applied to the gold–sulfuric acid solution interface from −0.2 V to 0.5 V for the potential sweep rate of 25 mV/s.

The study of the experimental SPR response in the double layer region allowed estimating the potential of zero charge value of the gold electrode in the electrolyte solution [15, 16]. Analysis of the dependences of SPR curve half-widths on the voltage between the gold electrode and electrolyte gave the potential of zero charge value, which equals $U_{pzc} = 0.17 \pm 0.04$ V versus the used reference electrode.

The theoretical modeling of the SPR curves for the potential difference in the double layer region was carried out using Eqs. (2.8–2.13, 6.1–6.6) for a multilayer system consisting of a semi-infinite glass layer (refractive index $n = 1.616$), a layer of chromium with a thickness of 5 nm ($n = 3.66 + 4.365i$ [30]), a layer of gold with a thickness of 45 nm ($n = 0.3 + 3.55i$ [31]), a surface layer of gold, whose refractive index is calculated using Eq. (6.6), and a semi-infinite electrolyte layer ($n = 1.3313$). The calculation of the parameters of the surface layer of gold was carried out using the results of atomic force microscopy (AFM) (Fig. 6.8) of the gold film surface prior to immersion in the electrolyte solution. In particular, an effective thickness of the surface layer of gold was obtained, which was assumed to be equal to the root mean square deviation in height for a rough surface of gold, which was 9.75 Å. Also, from the results of the AFM studies using the specially developed software for processing the AFM data of the relief elevation along the sample, the volumetric fractions of the rough surface of gold, the surface layer of gold in the thickness of 1 Å, and the metal-free space for the three-component surface layer of gold (Fig. 6.2) were evaluated, which were equal to 0.538, 0.103, and 0.359, respectively. It should be noted that the result of the simulation is more consistent with experimental data at a slightly lower volumetric fraction of the rough gold surface, which is equal to 0.5.

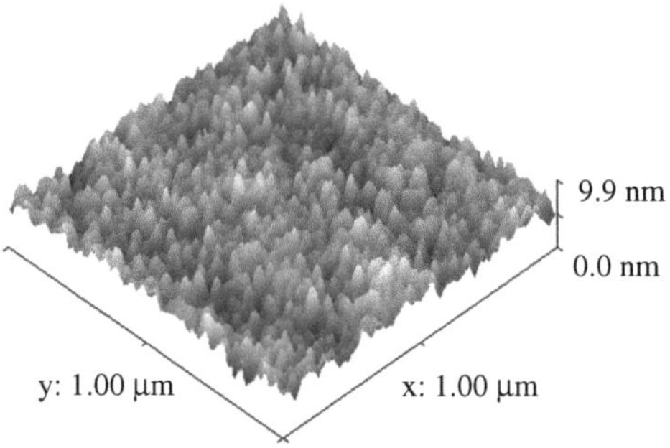

Figure 6.8 AFM image of the gold film rough surface, which was used as a sensitive element of SPR sensor. Reprinted from Ref. [20] under a Creative Commons Attribution-NoDerivatives 4.0 International License. Figure and figure caption were adapted.

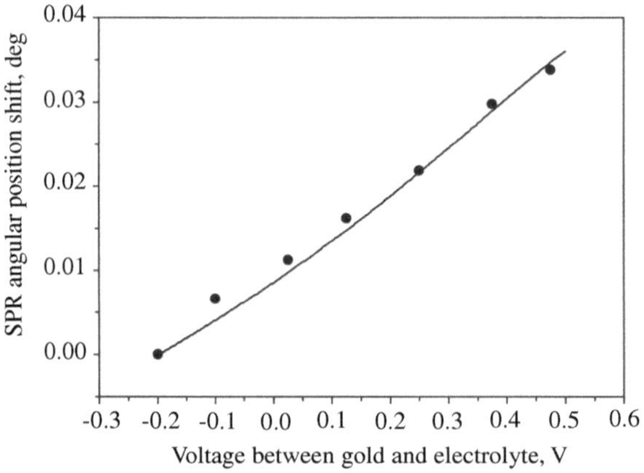

Figure 6.9 The dependence of the theoretically calculated angular position shift for the minimum of the reflected *p*-polarized light intensity (solid line) and the experimentally obtained dependence of the SPR angular position shift on the voltage applied to the interface between gold and sulfuric acid solution at a potential sweep rate of 25 mV/s (dots).

The results of calculation of SPR angle shift dependence on the voltage between the gold and the electrolyte that changes from −0.2 V to 0.5 V for the light wavelength of 650 nm and the corresponding experimentally obtained dependence at a potential sweep rate of 25 mV/s are depicted in Fig. 6.9. As it can be seen, the applied mathematical model allows for prediction of the SPR angle shift, which makes it possible to use it for the correction of the results of the SPR measurements under the conditions of the application of external voltage.

6.2.3 Influence of Applied Electrical Potential on the ESPR Biosensor Response upon the Registration of Biomolecular Processes

An SPR biosensor, as a very sensitive optical device, allows investigation of the influence of external factors that affect the sensitive sensor element and objects under study, such as temperature fluctuations, electric voltage applied, ambient pH changes, etc. In particular, the influence of the applied electrical potential on the course of biomolecular processes during the operation of SPR biosensor has been insufficiently studied. It is known that applying an electric voltage to the metal–electrolyte interface on the surface of a sensitive element of SPR biosensor changes the dispersion law of surface plasmon-polaritons that propagate along this interface [16]. This, in turn, leads to a shift and change in the shape of the SPR curves and influences the response of the SPR biosensor to the adsorption of biomolecules and the various biomolecular processes occurring on the surface of the sensitive element (thin gold film). Moreover, large biomolecules contain many constituents (for example, amino acid residues in the case of proteins) that carry various functional groups with different charge states, which give the biomolecules a specific electric charge interacting with external electric fields. Therefore, in this subsection, the results of the study of biomolecules adsorption processes on the example of proteins are presented with the application of the external electrical potential difference to the gold–electrolyte interface on the surface of the sensitive element of the ESPR biosensor, and the character of the optical response of the ESPR biosensor in such conditions was determined.

As objects for research, an enzyme trypsin (molecular weight about 23–24 kDa, isoelectric point pI = 10.1–10.8) and its inhibitor, obtained from soybeans (STI, molecular weight about 20–21 kDa, isoelectric point pI = 4.5) were selected. For the preparation of protein solutions, phosphate-buffered saline (PBS) solution, widely used in biochemical studies, was used. The measured pH value of the PBS buffer solution at 25°C was 7.25. Thus, dissolution of trypsin molecules in the used buffer solution leads to their positive total charge, since the isoelectric point of trypsin lies higher than the pH of PBS [32]. For STI molecules, the situation is opposite; when dissolved in PBS, they get a total negative charge. That is, the behavior of these biomolecules when located in the same electrical fields should be different.

Cyclic voltammetry measurements of gold–PBS system were performed to determine the range of voltages suitable for the study of biomolecular processes without overlapping of the electrochemical reactions influence with the SPR response (Fig. 6.10). The processes of electrochemical oxidation and reduction of gold were discovered, and it was found that these processes are absent when the applied voltage does not exceed 0.6 V. At negative potentials, the current flowing in the system began to increase from −0.2 V, which was associated with the reduction of oxygen dissolved in PBS [27]. Thus, only on the interval −0.2 to 0.6 V, there were no expressed electrochemical reactions, but only the charging of the double electric layer occurred [27], so this voltage range (the double layer region) was chosen for operation. It should be noted that for the gold–PBS system, the potential of zero charge is −275 mV [27]; thus, the near-surface layer of gold will always carry a positive charge within the selected voltage range.

An important advantage of the used ESPR biosensor was the possibility of simultaneous operation with two optical channels. This allowed using one of the channels as a reference during experiments to record the baseline, which registered the effects of external factors on the sensitive sensor element and the buffer solution surrounding the investigated biomolecules (temperature fluctuations, the effect of the applied voltage on the gold–electrolyte system, etc.). Comparative SPR measurements in both channels with staircase potential sweep throughout the investigated voltage range

(Fig. 6.11) have shown that the SPR response is almost reproducible. So all further results are presented taking into account the baseline, measured in the reference channel of the ESPR sensor.

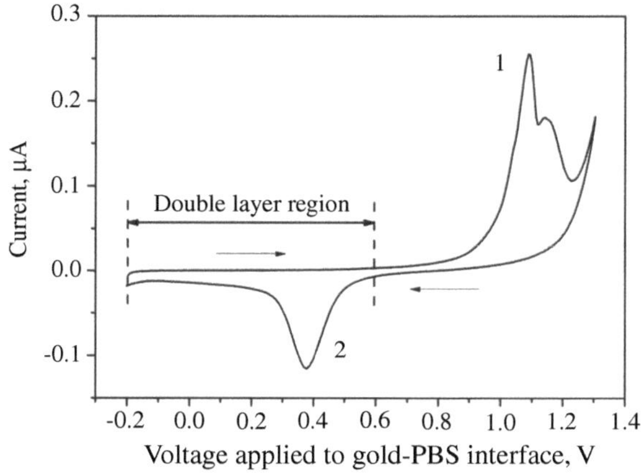

Figure 6.10 Measured cyclic voltammogram for the gold–PBS interface at a sweep rate of 50 mV/s: 1 – electrochemical oxidation of gold, 2 – electrochemical reduction of gold. The direction of the potential change is indicated by the arrows.

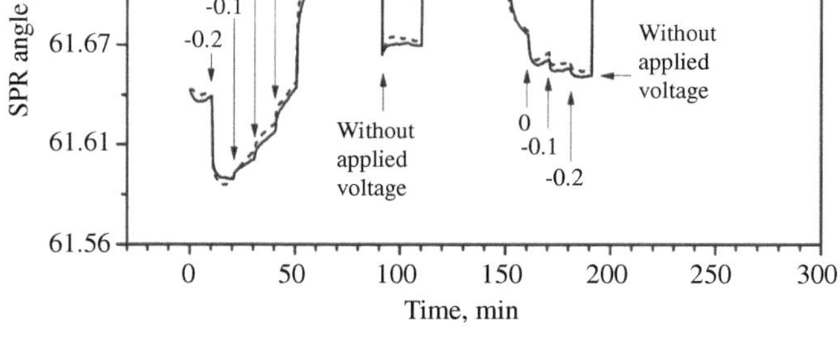

Figure 6.11 Measured SPR sensograms at a staircase potential sweep in the applied voltage range of –0.2 to 0.6 V. The moments of potential application and its value are indicated by arrows.

The measured sensograms of the ESPR biosensor response to the adsorption of biomolecules when the staircase potential sweep was applied to the "gold–solution of biomolecules" system are presented in Fig. 6.12. First, a gold electrode immersed in PBS buffer was maintained for a period of 15 min at a potential of 0.2 V, and then a solution of biomolecules was introduced into the working cell. The response of the ESPR biosensor to the adsorption of biomolecules was observed at the change in the applied potential in the range of 0.2–0.6 V. The corresponding dependences of the ESPR biosensor response upon the adsorption of biomolecules on the voltage applied to "gold–solution of biomolecules" system are shown in Fig. 6.13. A characteristic feature of the obtained dependencies is that, in the case of the addition of trypsin, the SPR response decreases with increasing value of the applied electrical voltage, while in the case of STI, the SPR response increases. This can be explained by a different charge sign these biomolecules bear, namely, trypsin has a positive charge, and STI has negative. Therefore, with the growth of the potential of the gold electrode, and hence with the increase in its positive charge, the nature of the adsorption of biomolecules varies: in the case of trypsin, the repulsive force increases between the surface of gold and biomolecules, and in the case of STI, the attraction force increases. It should be noted that the value of the SPR response to the adsorption of biomolecules correlates with their molecular weight.

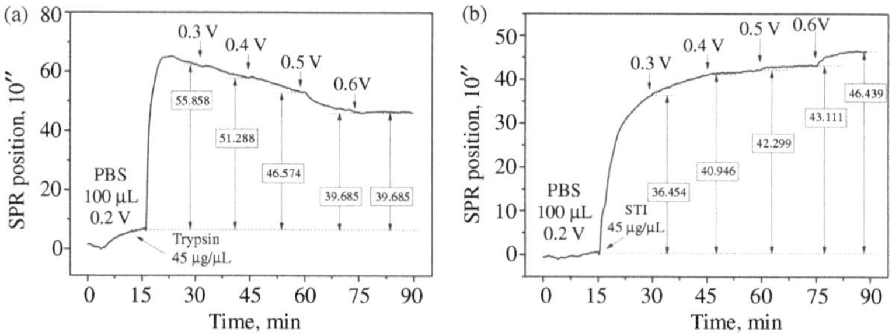

Figure 6.12 SPR sensograms obtained upon the staircase potential sweep in the range of applied voltage of 0.2–0.6 V when solution of (a) trypsin and (b) STI was added to the second channel of the biosensor. The moments of application of the potential and its value, as well as the time of introduction of biomolecules, are indicated by arrows. The SPR responses for each of the applied potentials are indicated by the numbers in the frames.

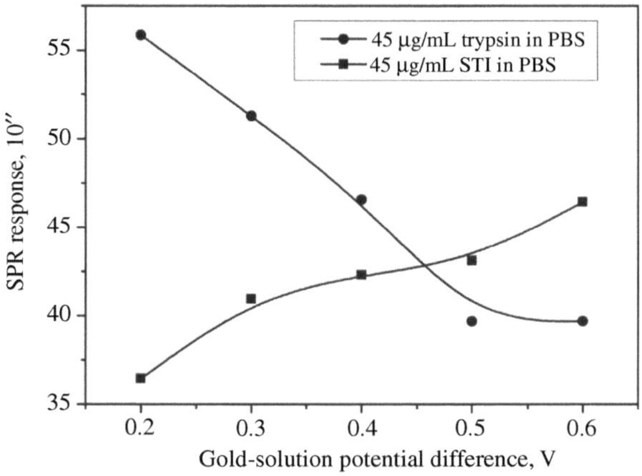

Figure 6.13 Obtained dependences of the ESPR biosensor response upon the adsorption of biomolecules on the voltage applied to the "gold–solution of biomolecules" system.

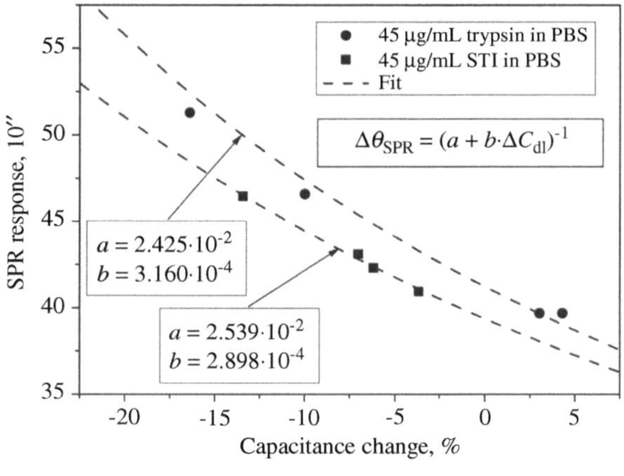

Figure 6.14 The dependence of the SPR response on the change in the normalized electrical double layer capacitance upon adsorption of trypsin and STI biomolecules (dots) and the results of fitting these dependences (dashed lines). The insets contain an expression for the function used for fitting and the values obtained for the corresponding parameters.

For the purpose of checking the obtained result, measurements of the capacity of the electric double layer, which depends on the presence of the adsorbed layer of biomolecules (decreases in the presence of the biomolecular layer [27]), were carried out. To accomplish this, an electrochemical impedance of a gold electrode was measured by means of electrochemical impedance spectroscopy in experimental conditions similar to those of the SPR experiment. The measured dependences of the SPR response and change in the capacitance of the electric double layer upon adsorbing biomolecules on the applied voltage between gold and solution of biomolecules correlate with each other. Namely, the SPR response $\Delta\theta_{SPR}$ and the change in the capacitance of the electric double layer ΔC_{DL} for the same value of the applied potential difference are related to a nonlinear dependence, which is well fitted to a rational type function $\Delta\theta_{SPR} = (a + b \cdot \Delta C_{DL})^{-1}$ (the adjusted coefficient of determination is 0.97 for trypsin and 0.99 for SIT, Fig. 6.14). The obtained results confirm the adsorption of protein molecules dependence on the applied electrical potential in the ESPR biosensor.

6.3 SPR Transduction of Redox Transformations in Thin Polymer Films

6.3.1 Investigation of Changes in Optical Properties of Gold Film-Redox Polymer System under the Influence of External Electrical Potential

SPR, in combination with electrochemical measurements, ESPR, attracted constant attention of researchers during the past decades [33]. The possibility of SPR technology for in situ evaluation of optical constants and thickness of thin films in real-time mode is very attractive for studies on various electrochemical processes, including the processes of electropolymerization [34–37].

This applies to both the need for the estimation of two main parameters, the refractive index n and the thickness of the layer d, which is usual for SPR measurements, and the influence of the external potential applied to the metal on the angular position of the resonance minimum and on the shape of the SPR spectrum. The possibility of deriving the values of n and d from the SPR measurements is studied fairly well and is solved under the condition of a correctly conducted experiment by fitting the experimental and theoretical SPR spectra [38]. Less studied is the effect of the influence of external potential on the angular position of the minimum and the shape of the SPR curve [13, 14, 16]. Application of the external potential to the plasmonic oscillations carrying metal film is associated with a change in the free electron concentration in the near-surface layer of the metal and the formation of an oxide film on it under the anodic decomposition of water and the effect of electrolyte ions [16, 17]. However, in the case of application of a potential up to 1 V to gold placed in an electrolyte, the influence of the external environment on the SPR can be clearly visible only if the electrolytic medium has expressed redox properties, that is, it can exist in two states, reduced and oxidized. Usually used in the practice of an electrochemical experiment, acidic or basic electrolytes do not possess such properties. In this case, the displacement of the angular position of the resonant minimum and the change in the shape of the SPR spectrum (SPR response) depends mainly on the formation of an oxide film on gold or its dissolution. The situation changes if the gold electrode is covered with a polymer film and, especially, if the film exhibits redox properties. When applying even small potentials under certain conditions, the SPR response becomes significant (up to several angular degrees) and depends primarily on the thickness of the film, its structure, and the nature of the redox properties. It is known that the optical parameters of redox substances may vary with the change in the thickness of the layer on the electrode during the electrochemical reaction [39]. This peculiarity additionally hinders the evaluation of properties of such materials. SPR provides additional possibilities in the investigation of similar structures, because this method is sensitive both to the concentration of redox molecules in the solution of the electrolyte and to the thickness of the adsorbed molecular layer. In its turn, cyclic voltammetry allows estimating the concentration gradient of redox particles near the surface of the working electrode. It is interesting to combine these two methods in the study of the properties of macromolecular structures with pronounced redox properties. In this subsection, the ESPR study of the gold–polyaniline (PA) structure is presented. This polymer is a redox substance and can be obtained in situ by electropolymerization under the conditions of the SPR experiment [40].

The kinetic dependences of SPR angular position were recorded simultaneously with cyclic voltammograms (CVs) [25]. The scan range for the CV was chosen so that it captures the region of the potential for the formation of oxide on the gold film (+1.2 V). This was important for explaining the results of ESPR experiments conducted using gold as a working electrode. Polyaniline films were deposited by in situ electropolymerization in the electrochemical cell of SPR sensor onto the precleaned 50 nm thick gold films thermally evaporated on glass slides. An external potential of 0.8 V was applied to a monomer solution (0.1 M solution of aniline in an electrolyte consisting of H_2SO_4, 0.1 M, and Na_2SO_4, 0.5 M) for 5 s. The resulting PA film was washed with an electrolyte solution to remove the monomer residues from the cell. Polyaniline film is a material that can exist in two states: positively charged oxidized $(An^{2+})_n$ and neutral reduced $(An)_n$ at potentials of 0.4 V and 0.0 V, respectively. It should be noted that the redox properties of the PA film immediately after electropolymerization and after several cycles of CV measurements differ due to some irreversibility of adsorption and diffusion processes during the studied electrochemical reaction. In the study considered, the results of parallel SPR-CV measurements for the PA film received immediately after the electropolymerization process are presented.

Figure 6.15 shows the kinetic dependences of the angular position of the SPR minima (sensograms) registered under the application of the external time-dependent potential in the range of −0.2 to 1.3 V to the gold electrode. The sensograms measured for bare gold and gold coated with a PA film having a thickness of about 150 nm were presented. The thickness of the PA layer was determined from the results of the fitting procedure for the experimental and theoretical SPR curves [38] for the curve obtained after electropolymerization, subsequent removal of the monomer residues by washing the cell with the buffer solution and 1 h long polymer structure stabilization. It is evident from Fig. 6.15 that application of a time-dependent potential induced the SPR response (curve 1, Figs. 6.15 and 6.16) for bare gold with an amplitude (ca. 25′), which is much smaller than that for gold coated with a PA film (ca. 170′, curve 2, Fig. 6.15). This fact was a direct confirmation of a noticeable change in the complex dielectric permittivity $\varepsilon_p(\omega)=\varepsilon_p'(\omega)+\varepsilon_p''(\omega)$ of the polymer induced by the applied potential. A small (up to 10′) gradual positive shift in the SPR minimum position in the sections a–c (−0.2 to 0.9 V) corresponds to the horizontal section of the CV curve (curve 2, Fig. 6.16) and can be explained mainly by the change in the near-surface free electron concentration of a gold film. Under the influence of the external applied potential, free electrons are shifted from the surface of the electrode into the metal, which can be interpreted as a change in the effective dielectric permittivity ε_m of the near-surface metal layer [17].

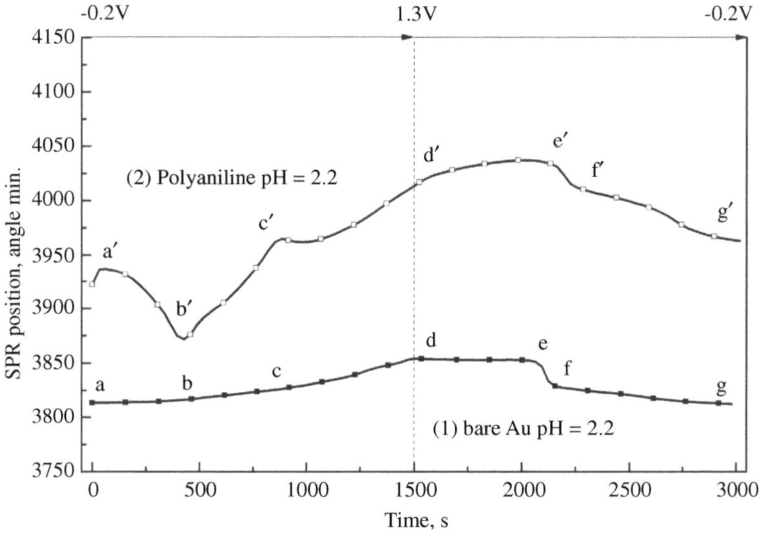

Figure 6.15 Kinetic dependences of the angular position of the SPR minima when external time-dependent potential was applied to the gold electrode (CV –0.2 to 1.3 V): (1) bare gold, pH = 2.2, (2) polyaniline film on gold, pH = 2.2.

Another contribution to the change of ε_m in the section a–c (Figs. 6.15 and 6.16) was due to the polarization processes on a gold electrode. In the section c–d (0.9–1.3 V), the determining factor of accelerated SPR minimum shift up to 20′ was the formation of a thin oxide layer on gold, which was confirmed by the peak observed near 1.2 V in the CV (curve 2, Fig. 6.16). On the reverse CV course, decrease in the potential applied to the gold electrode in the section d–e (1.3–0.6 V) did not lead to a noticeable change in the SPR minimum position, indicating the oxide layer screening effect and the resulting stability of the free electron concentration in the near-surface layer of gold. In the section e–f (0.7–0.6 V), oxide layer on gold dissolved and a resulting sharp negative shift of the SPR minimum position was observed. The corresponding peak on the CV curve (0.7 V) corresponded exactly to the signal decrease on the SPR sensogram. Increase in the free electron concentration and the electrode depolarization due to gradual decrease in the applied potential (section f–g, 0.7 to −0.2V) define further gradual reduction in the SPR minimum position.

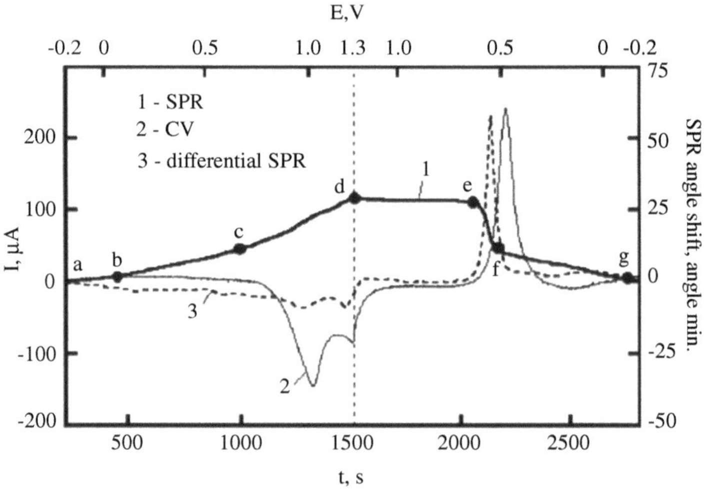

Figure 6.16 Kinetic dependence of the SPR angular position (curve 1), cyclic voltammogram (curve 2), and kinetic dependence of derivative $d\theta_{min}/dt$ (curve 3) when external time-dependent potential was applied to the bare gold electrode (−0.2 to 1.3 V, H_2SO_4, 0.1 M, and Na_2SO_4, 0.5 M, as electrolyte, pH = 2.2).

As it follows from the comparison of the two curves described above, cyclic voltammetry in the experiment under consideration was more sensitive to transient processes, while SPR spectrometry clearly registered the adsorption–desorption processes. An additional contribution to the evaluation of the kinetics of transitions on the gold–electrolyte interface can be easily obtained by differentiating the kinetic curve 1 (Figs. 6.15 and 6.16) describing the dependence of the SPR minimum angle on the applied potential. As it can be seen from Fig. 6.16, the resulting dependence $d\theta_{min}/dt$ (curve 3) correlates well with the magnitude of current in the CV dependence and, as a result, also tracks all transient processes recorded by cyclic voltammetry.

Formation of the PA film in the process of electropolymerization drastically changed the optical profile of the resulting layered structure. The ability of an oxidized form of polyaniline to absorb light, including the wavelength range of laser radiation used in the SPR device (λ = 670 nm), is known. The transition of part of the polyaniline molecules from the usual reduced state to oxidized and backward was achieved by applying the potential in the range of 0–0.4 V. The application of the alternating potential leads to sharp changes in the position of the SPR minimum (curve 2, Fig. 6.15) and to significant changes in the shape of the SPR angular spectrum (Figs. 6.17 and 6.18). In the initial section a′–b′ (−0.2 to 0.23 V), the SPR position sharply decreases by a significant amount (up to 65′) with a subsequent large increase in the section b′–c′. The further course of curve 2 in the segment c′–f′ basically repeats the direction of the course of curve 1 for bare gold. That is, the presence of an oxide film on gold imposes an imprint on the optical profile of the system and on the nature of redox transformations in polyaniline, although the major changes in the optical profile are still determined by the processes in the polymer. Characteristic hysteresis for the CV curve was also observed in the SPR sensogram on the reverse course of the curve (section d′–g′). However, unlike the CV, the SPR minimum did not return to the former angular position, which indicates an irreversible change in the optical constants of the layer during one cycle. This may be the result of both slow diffusion processes associated with the transport of heavy ions (sulfates) and the processes of long-term structural stabilization of the polymer. It was assumed that the slow diffusion of heavy ions during their movement to the gold electrode and back mainly determines the incomplete reversibility of the SPR sensogram.

As it can be seen from Fig. 6.15, the amplitude of the SPR minimum angular position shifts reached 170′, i.e., about 3°. Such a significant change in the optical constants of the layer is characteristic of redox substances and can be explained by the orientation polarization mechanism of the dipoles for polar solids and liquids. The connection between the average polarizability of the substance *a* and its optical constants is described by the Lorentz–Lorenz equation:

$$a = \frac{3}{4\pi N}\frac{n^2-1}{n^2+2}, \tag{6.7}$$

where *N* is the number of molecules per unit volume and *n* is the substance refractive index. The dipole moment of the molecule *p* is proportional to the polarizability of the substance and the effective electric field E [41]:

$$p = aE.$$

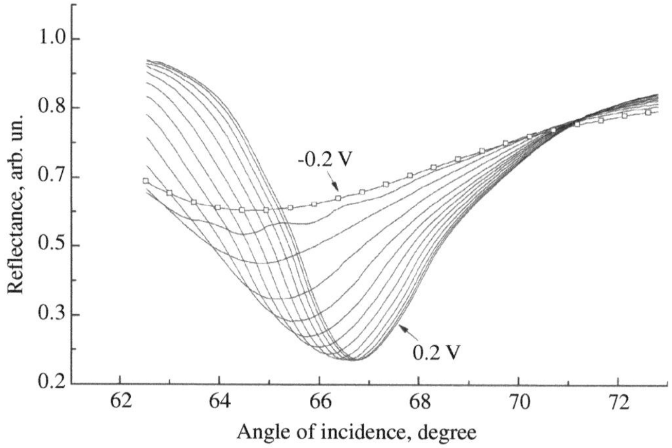

Figure 6.17 Shift in the minimum position and change in the shape of the SPR curve under the influence of the applied potential from −0.2 V to 0.2 V (section a–c, curve 2, Fig. 6.15).

The total polarization of the unit volume of an isotropic medium with N_d constant dipoles arising as a result of the reorientation of dipoles, each of which has the moment p, is equal to [42]:

$$P = N_d p^2 E / 3k_0 T, \tag{6.8}$$

where E is the applied electric field. Equations (6.7–6.8) suggest that when the polarization of a substance is induced by an electric field, the stronger the change in its optical properties, the greater the dipole moment of the molecule and the higher the concentration of dipoles.

It is known that polyaniline obtained as a result of electropolymerization has a porous gel-like structure [43]. The spatial orientation and delocalization of the electron shells of dipoles in such environment are significantly facilitated. In this case, the reversible property of the polyaniline molecule to be in two states, oxidized and reduced, depends on whether the proton addition reaction occurred, resulting in the formation of a hydrogen bond with a nitrogen atom exhibiting a strong electronegativity:

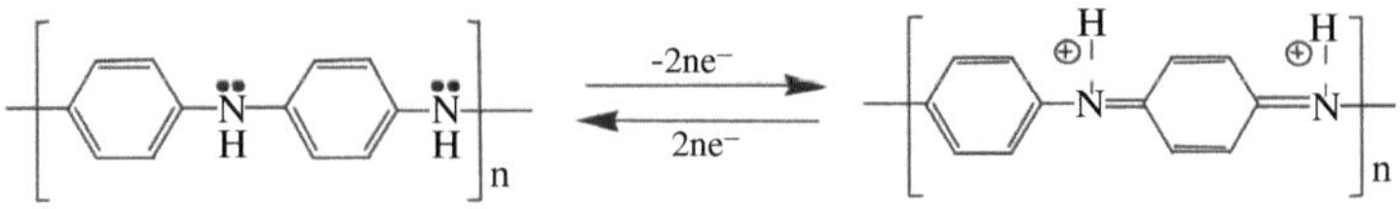

(6.9)

The result of this reaction is the change in the electronic structure of the polyaniline molecule, a sharp increase in its dipole moment due to polarization upon the oxidation with the addition of the proton and its decrease upon reduction. That is, the formation of dipoles induced by an electrochemical reaction was observed in this case. Simultaneously with polarization, the group orientation in the applied electric field of the formed dipoles resulted both in a change in the complex refractive index of the polymer in proportion to the number of molecules that have reacted and in the shrinking process through the formation of more densely packed structure.

According to the study in Ref. [44], the complex refractive index for the reduced polymer form is $n = 1.4 + 0.02i$. The comparison of the experimental and theoretical SPR curves for the oxidized polymer form showed a significant increase in the absorption coefficient ($n = 1.4 + 0.39i$). It should be noted that the value of n for redox materials may depend on the thickness of the film [39].

Obviously, with the application of a slowly varying potential, different stages of the polymer state have been dealt with, which differ in the concentration of oxidized and reduced polyaniline molecules. The electron and ion polarization processes characteristic of the particular investigated structure as a whole also contributed to the change in polarization and, respectively, the refractive index, but the main contribution, in this case, came from the mechanism of orientational polarizability of polyaniline dipoles.

6.3.2 SPR Registration of Structural Transformations in Polyaniline Films Initiated by the External Electric Potential

Experimental SPR curves for the polyaniline-modified gold surface upon the measurement of electrochemically induced processes in the polymer structure [7] are shown in Fig. 6.18 (see detailed description in the figure caption). The same figure (inset II) shows the kinetic dependence of the reflected laser light intensity (at a fixed angle of incidence of 65°) for the oxidation process of the PA film. The calculations show a bi-exponential dependence with kinetic constants equal to 9×10^{-3} s^{-1} and 5×10^{-4} s^{-1}. The fast-shrinking component was supposed to be related to the breakdown of hydrophobic polymer bonds that arise during the formation of a metastable structure upon the reduction of polyaniline (section a′–b′, curve 2, Fig. 6.15). Slow component was referred to the process of forming an organized, densely packed structure from which the water residues are removed.

Upon the PA reduction, a relaxation process was observed, which is accompanied by disordering of dipoles and restoration of the former volume (swelling) of the structure. Figure 6.18 (inset III) shows a reflectance sensogram (at a fixed angle of incidence of 65°) for the oxidation process of a polyaniline film (after a sudden application of the 0.4 V potential), which exhibits the relatively slow kinetics of the swelling process with a kinetic constant of 3×10^{-4} s^{-1}.

The consecutive application of the potential values of 0 V and 0.4 V to the polymer film, respectively, led to successive changes in its thickness, which implies the possibility to use polyaniline as a basis of artificial muscles or a controlled surface, taking into account the limitations in the speed of the full shrinking–swelling cycle. The fast component of this cycle is characterized by an instantaneous change in the refractive index and can be used for SPR transduction of the electronic signal, for example, in optoelectronic laser modulators [35] with low control potential. Figure 6.19A shows the measured chronoamperometric dependence of the investigated structure obtained as a result of its oxidation (0.4 V) and reduction (0.0 V). In this experiment, potential steps were applied to the PA film, which converted $(An)_n$ into $(An^{2+})_n$ and vice versa. The duration of the applied step did not allow the start of the swelling process, because the step was completed by applying the potential of the inverse transformation of the polymer to the state $(An)_n$, which, in turn, ended before the beginning of the shrinking process. Figure 6.19B shows the corresponding changes in light reflection intensity. As a result, by the cyclic application of the oxidative and reductive potential steps, the reversible SPR transduction of the polymer state was accomplished (Fig. 6.19B).

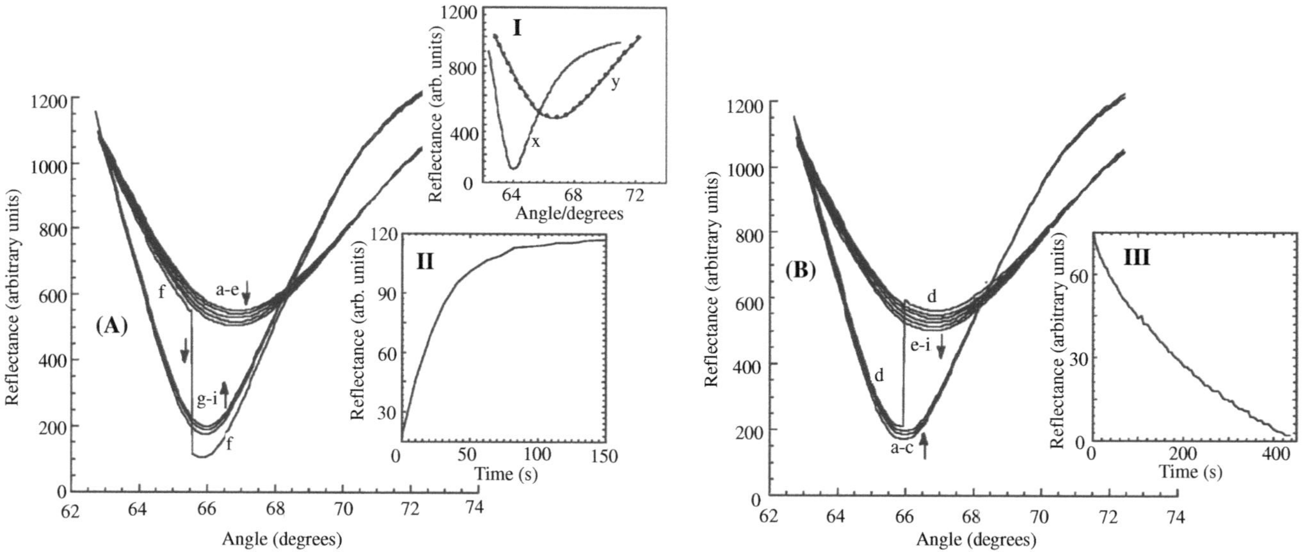

Figure 6.18 SPR spectra measured for the polyaniline-modified gold surface upon electrochemically induced shrinking and swelling processes. (A) The polymer was equilibrated at 0.0 V for 5 min prior to the measurements. (a–e) The spectra measured after the potential changed to 0.4 V. (f) The spectrum measured during the abrupt potential changes from 0.4 V to 0.0 V. (g–i) The spectra measured after the potential changed to 0.0 V. (B) The polymer was equilibrated at 0.4 V prior to the measurements. (a–c) The spectra measured after the potential was altered to 0.0 V. (d) The spectra measured during the abrupt potential change from 0.0 V to 0.4 V. (e–i) The spectra measured after the potential changed to 0.4 V. Inset I: (x) The SPR spectrum for the bare gold surface and (y) the SPR spectrum for the oxidized polymer film measured at 0.4 V (the dots show selected experimental points and the solid line is the best theoretical fit). Inset II: Time-dependent reflectance changes measured at a fixed angle of incidence ($\theta = 65°$) immediately after the applied potential was changed from 0.4 to 0.0 V. Inset III: Time-dependent reflectance changes measured with the fixed angle of incidence ($\theta = 65°$) immediately after the applied potential was changed from 0.0 to 0.4 V. Reproduced from Ref. [7] with permission from the Royal Society of Chemistry.

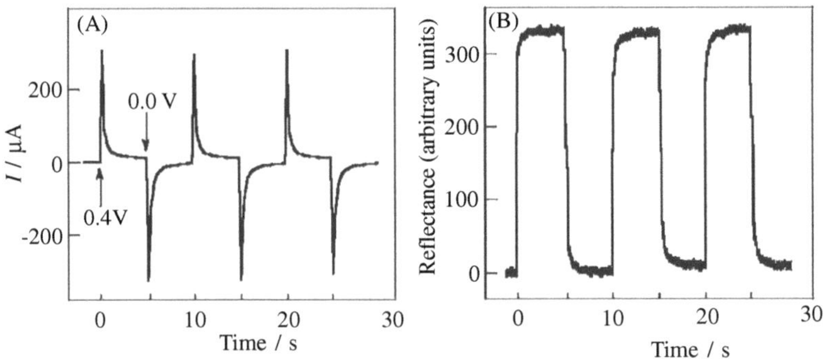

Figure 6.19 (A) Multi-potential step chronoamperometric measurements corresponding to the oxidative and reductive transformations of the polyaniline layer. (B) Time-dependent reflectance changes measured at a fixed angle of incidence (θ = 65°) upon the multi-potential step experiment. The arrows show the time of the application of the oxidative (0.4 V) and reductive (0.0 V) potentials. Reproduced from Ref. [7] with permission from the Royal Society of Chemistry.

6.4 Switchable Surface Properties through the Electrochemical or Biocatalytic Generation of Ag^0 Nanoclusters on Monolayer-Functionalized Electrodes

6.4.1 Control of Hydrophilic and Hydrophobic Properties of Surfaces

The control of the hydrophilic and hydrophobic properties of surfaces attracts substantial recent research activities [45]. Different applications of surface-regulated functions were suggested, including the development of self-cleaning surfaces [46], sensors [47], controlled delivery and transport [48], and more [49, 50]. The signal-triggered control of surface properties and, specifically, the reversible and cyclic regulation of the surface functions have attracted major interest. The electrochemical [51], photochemical [52], magnetic [53], chemical [54], and biocatalytic [55] activation of the surface properties was accomplished in recent years. For example, the redox activation of chemically modified electrodes between neutral and charged states was reported to reversibly switch the interface between hydrophobic and hydrophilic states, respectively [51]. Also, the electric-field-induced bending of charged units tethered to electrodes by long-chain alkane units was employed to control the hydrophilic/hydrophobic properties of surfaces [56]. Recently, the electrochemically driven mechanical shuttling of molecular components on surfaces was used to control the hydrophilic/hydrophobic properties of surfaces, and the translation of molecular machinery into macroscopic mechanical events (movement of water droplets) was demonstrated [57]. Several studies have addressed the wetting properties of nanoparticles associated with surfaces [58, 59]. Superhydrophobic behavior of the lotus leaf structure was demonstrated with nanoparticle-functionalized interfaces, thus revealing the effects of nanoparticles on the wettability properties of the modified surfaces.

The considered study deals with the formation of an Ag^+-coated, densely packed, monolayer of an aromatic dithiol on Au supports and the reversible control of the surface properties by the electrochemically induced generation and dissolution of Ag^0 nanoclusters [60].

6.4.2 Investigation of Cyclically Switchable Surface Properties

4,4′-Biphenyldithiol (BPDT) monolayer was electrochemically deposited on an Au electrode according to the reported procedure [61] and soaked in 10 mM Ag_2SO_4 solution in 0.1 M H_2SO_4 overnight with subsequent rinsing to yield an Ag^+ coating on the dithiol monolayer (Fig. 6.20).

Figure 6.21A shows the cyclic voltammograms of the Ag^+-functionalized surface at different potential scan rates. The cyclic voltammograms revealed several important features: (i) A cathodic wave corresponding to the reduction of Ag^+ ions associated with the interface to Ag^0 was detected, and the corresponding oxidation wave of the metal was observed at positive potentials. The coulometric assay of the reduction (or oxidation) wave indicated that the surface coverage of the Ag^+/Ag^0 redox species corresponds to 0.82 nmol/cm^2, suggesting that each BPDT unit was linked to a single Ag^+ ion and that the monolayer interface was completely saturated with the Ag^+ ions. (ii) The cyclic voltammograms revealed high stabilities upon cycling, implying that the reduction of Ag^+ or the oxidation of Ag^0 resulted in tightly bound surface species that are not dissolved or eliminated from the surface. (iii) The peak currents of the reduction or oxidation waves depended linearly on the scan rates implying that the redox-active species are surface confined (Fig. 6.21A, inset). (iv) The cyclic voltammograms of the Ag^+-functionalized surface were not kinetically reversible and were characterized by large peak-to-peak separation. The cathodic and anodic waves revealed kinetic nonsymmetricity, reflected by nonsymmetrical shifts in the cathodic and anodic waves. The reduction wave of Ag^+ to Ag^0 was significantly shifted upon the increase in the potential scan rate, while the oxidation wave was almost unaffected upon increasing the scan rate. These results suggest that the reduction of Ag^+ to Ag^0 revealed slow electron-transfer kinetics at the surface, whereas the oxidation of the Ag^0-modified surface was fast. Therefore, the middle point potentials, $E_{1/2}$, of the cyclic voltammograms were different at various potential scan rates, and they did not represent a thermodynamic potential of the redox system, E^0.

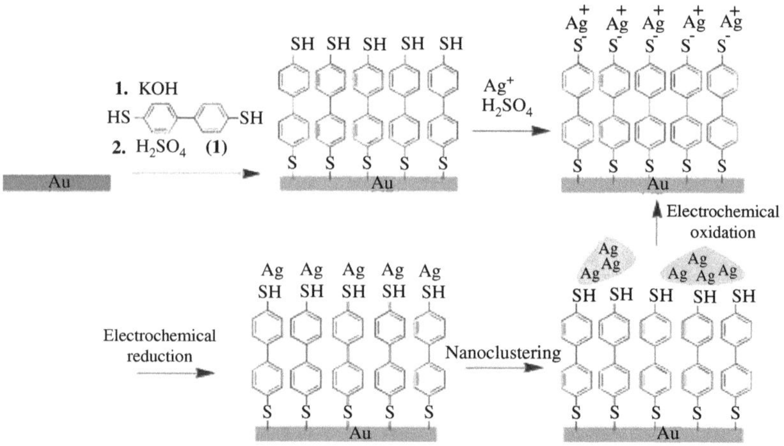

Figure 6.20 Assembly of the Ag^+-dithiol-monolayer-functionalized electrode, and the reversible electrochemical switching of the interface between the Ag^+ ionic state and the Ag^0 nanoclusters. Adapted with permission from Ref. [60]. Copyright (2006) American Chemical Society.

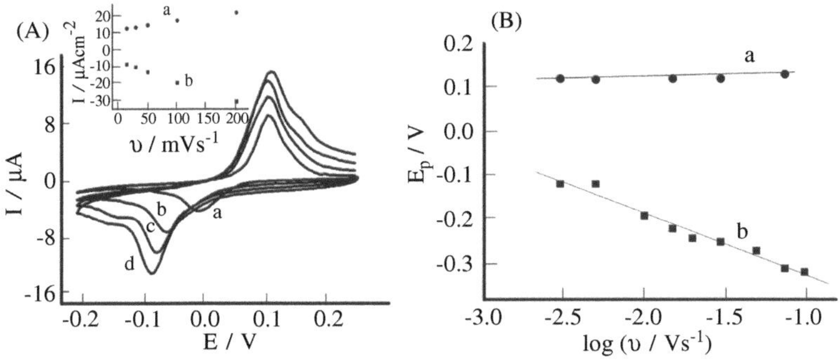

Figure 6.21 (A) Cyclic voltammograms of the Ag^+/BPDT/Au electrode recorded at different potential scan rates: (a) 5 mV/s, (b) 15 mV/s, (c) 30 mV/s, and (d) 50 mV/s. Inset: dependence of the anodic (a) and cathodic (b) peak currents of the cyclic voltammograms on the potential scan rate. (B) The dependence of anodic peak (a) and cathodic peak (b) potentials on the scan rate. Adapted with permission from Ref. [60]. Copyright (2006) American Chemical Society.

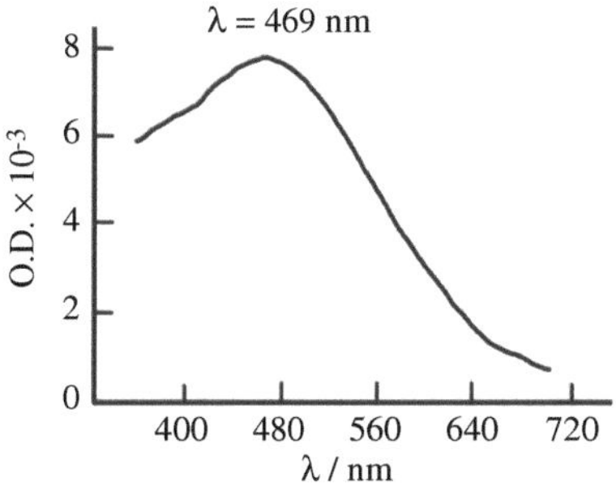

Figure 6.22 Differential absorption spectrum of electrochemically generated Ag^0 nanoclusters on BPDT/Au electrode obtained after subtraction of a spectrum of Ag^+ ions associated with the interface. Reprinted with permission from Ref. [60]. Copyright (2006) American Chemical Society.

Figure 6.21B shows the dependence of the cathodic and anodic peak potentials for the reduction and oxidation processes on the potential scan rate. Anodic and cathodic electron-transfer coefficients (α and β, respectively) were derived from the slopes of these dependences, and the ratio of α/β corresponded to 1/20. The crossing point of the two linear plots, extrapolated to slow potential scan rates, represents the thermodynamic potential for the redox process, E^0, and it corresponds to ca. 0.11 V. Further support that the electrochemical reduction of the Ag^+-BDPT monolayer led to the formation of Ag^0 nanoclusters on the BPDT monolayer was obtained by absorbance studies. The formation of the Ag^0-BDPT surface was accompanied with the appearance of an absorbance band at $\lambda = 469$ nm, consistent with the formation of Ag^0 nanoclusters [62] (Fig. 6.22).

6.4.3 SPR Study of Cyclically Switchable Surface Properties

To further understand the electrochemically induced Ag^+ ion to Ag^0 metal transitions, in situ ESPR measurements were applied. Figure 6.23A depicts the time-dependent SPR spectra upon applying a potential step from $E = 0.2$ to $E = -0.2$ V on the Ag^+-functionalized monolayer. At the final potential, the Ag^+ ions were reduced to Ag^0. The process was accompanied by a shift in the minimum reflectivity angles to higher values and the observation of a shallower SPR curve. Figure 6.23A, inset, shows the time-dependent changes in the reflectance at a constant angle of incidence of 63.0°. The major time-dependent changes in the SPR spectra proceeded for ca. 5 min. Interestingly, the time scale of chemical transformation occurring on the surface, and being probed by the SPR spectra, is by ca. 5 orders of magnitude slower than the primary electron transfer to Ag^+ that yields Ag^0, which proceeded on a time scale of ca. 15 ms, as was elucidated by the chronoamperometry experiments. This apparent discrepancy has been resolved by realizing that the specific chemical transformations occurring on the surface and probed by the different electrochemical/optical means are different. The reduction of the Ag^+ ions to Ag^0 atoms proceeded rapidly on the electrode, and this process was, indeed, associated with the electron transfer and followed electrochemically. The migration of the Ag^0 atoms on the surface and their clustering was, however, a substantially slower process (Fig. 6.20). Since the SPR spectra followed the phenomenon of coupling between the localized plasmon of the nanoclusters and the Au-surface plasmon, the kinetics of nanoclustering of the Ag^0 atoms was optically probed. Figure 6.23B shows the SPR spectra observed upon the application of a potential step from −0.2 to 0.2 V on the Ag^0-nanocluster-modified electrode. A substantially faster change in the SPR spectra was observed, and the original SPR spectrum of the Ag^+-modified monolayer was reobtained. Figure 6.23B, inset, shows the time-dependent changes in the reflectance at a constant angle of incidence of 63.0°. The major time-dependent changes in the SPR spectra proceeded within a time interval of ca. 20 s. Figure 6.23C shows the cyclic changes in the reflectance intensities (at 63.0°) upon the reversible switching of the monolayer configuration between the Ag^0 nanocluster and Ag^+ ion states, respectively. It should be noted that the alternate application of the potential values of −0.2 and 0.2 V on the BPDT-monolayer-functionalized Au electrode in the absence of the Ag^+/Ag^0 surface-confined redox system yielded only a minute change in the minimum reflectivity angle, as well as in the reflectance intensities. Thus, the observed changes in the SPR spectra originated from the reversible Ag^0 nanocluster formation and dissolution, respectively.

In the study of the effect of nanoparticles on the hydrophilicity/hydrophobicity of surfaces, it was proposed that the hydrophilic/hydrophobic properties of the surface will be controlled electrochemically during the cyclic transition of Ag^0 to Ag^+ [59]. Figure 6.24A shows the in situ electrochemically stimulated contact angle changes of a H_2SO_4, 0.1 M, droplet upon cyclic switching of the potential on the electrode between −0.2 and 0.2 V. At an applied potential of −0.2 V, where the Ag^0 nanoclusters were present on the monolayer surface, the contact angle revealed a low value in the region of 48–52°. On the other hand, after applying a constant potential of +0.2 V, where the Ag^+ ions were linked to the monolayer interface, a high contact angle corresponding to 64° ± 0.5° was observed. In a control experiment where the potential was cycled on the BPDT-modified electrode, which lacks the Ag^+ ions, between the values of −0.2 and +0.2 V, minute changes, $\leq\pm1°$, in the contact angle of the 0.1 M H_2SO_4 droplet were observed. This implies that the potential-induced contact angle changes in the Ag^+-functionalized monolayer originated from the redox-stimulated reduction of the Ag^+ ions to the Ag^0 nanoclusters and the reverse dissolution of the nanoparticles. The Ag^0-nanocluster-modified interface revealed hydrophilic properties, while the Ag^+-thiolate monolayer interface exhibited, based on the contact angle measurements, hydrophobic properties.

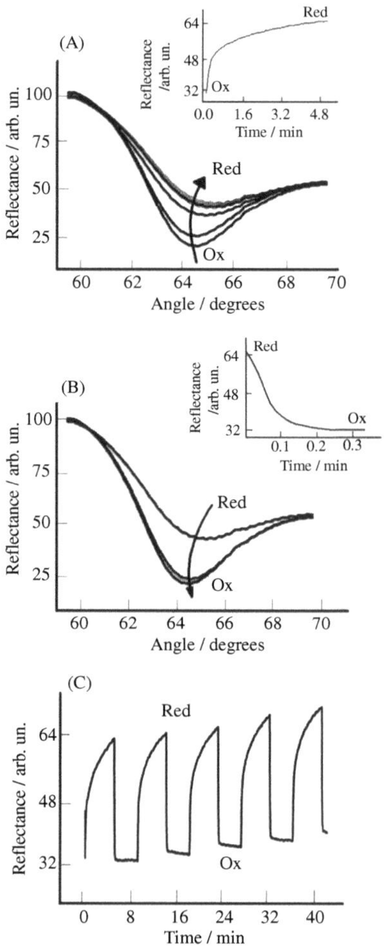

Figure 6.23 (A) SPR spectra of the $Ag^{+/0}$/BPDT/Au system at different time intervals after applying a potential step from 0.2 to –0.2 V. Inset: changes of the reflectance intensity of the modified electrode measured at a constant angle of 63.0° upon the potential step from 0.2 to –0.2 V. (B) SPR spectra of the $Ag^{+/0}$/BPDT/Au system at different time intervals after applying a potential step from –0.2 to 0.2 V. Inset: changes of the reflectance intensity of the modified electrode measured at a constant angle of 63.0° upon the potential step from –0.2 to 0.2 V. (C) Reversible changes of the reflectance intensity of the $Ag^{+/0}$/BPDT/Au system upon cyclic application of the potentials corresponding to –0.2 and 0.2 V on the modified electrode. "Red" represents the system in the reduced state containing Ag^0 nanoclusters; "Ox" represents the system in the oxidized state containing Ag^+ ions. Reprinted with permission from Ref. [60]. Copyright (2006) American Chemical Society.

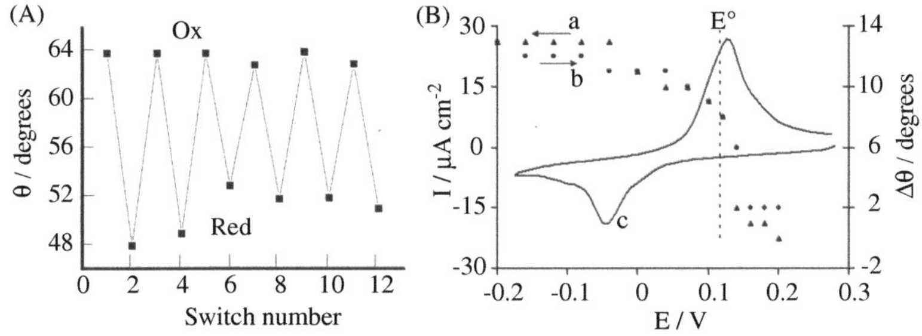

Figure 6.24 (A) Reversible changes in the contact angle of a 0.1 M H_2SO_4 aqueous droplet on the $Ag^{+/0}$/BPDT/Au system upon application of the potentials of (a) 0.2 V and (b) −0.2 V. At each point, the measurement was performed after holding the system under the corresponding potential for 5 min. "Red" represents the system in the reduced state consisting of Ag^0 nanoclusters; "Ox" represents the system in the oxidized state consisting of Ag^+ ions. (B) Superposition of the potential-controlled in situ contact angle values of a 0.1 M H_2SO_4 droplet on the $Ag^{+/0}$/BPDT/Au system upon: (a) potential steps of 20–25 mV applied from −0.2 to +0.2 V; (b) potential steps of 20–25 mV applied from 0.2 to −0.2 V; (c) the cyclic voltammogram of the system obtained at the scan rate of 20 mV/s in 0.1 M H_2SO_4 aqueous solution. Adapted with permission from Ref. [60]. Copyright (2006) American Chemical Society.

Figure 6.24B, curve a, shows the contact angles of the 0.1 M H_2SO_4 droplet on the Ag^+ monolayer upon the stepwise alteration of the potential applied on the electrode from 0.2 to −0.2 V, by augmentation of the potential step value by 20–25 mV. A sharp decrease in the contact angle value was observed at ca. 0.11 V, consistent with the E^0 value derived from the electrochemical measurements for the reduction of the Ag^+-thiolate layer to the hydrophilic Ag^0 nanoclusters (Fig. 6.24B, curve c). Figure 6.24B, curve b, shows the changes in the contact angle values of the 0.1 M H_2SO_4 aqueous droplet, positioned on the Ag^0-nanocluster interface, upon the stepwise change in the potential from −0.2 to 0.2 V. The contact angle characteristic to the Ag^0-nanoclustered electrode changed rapidly and increased in its value most steeply, at an applied potential of 0.11 V, consistent with the formation of a hydrophobic interface upon the oxidation of the metal nanoclusters to the Ag^+-thiolate monolayer state. The contact angle measurements performed upon the stepwise potential changes in the opposite directions did not show any hysteresis; thus, the droplet was equilibrated with the modified interface on the time scale of the experiment, and the derived contact angle behavior superimposes the thermodynamic equilibrium corresponding to the potential-controlled redox state of the modified surface.

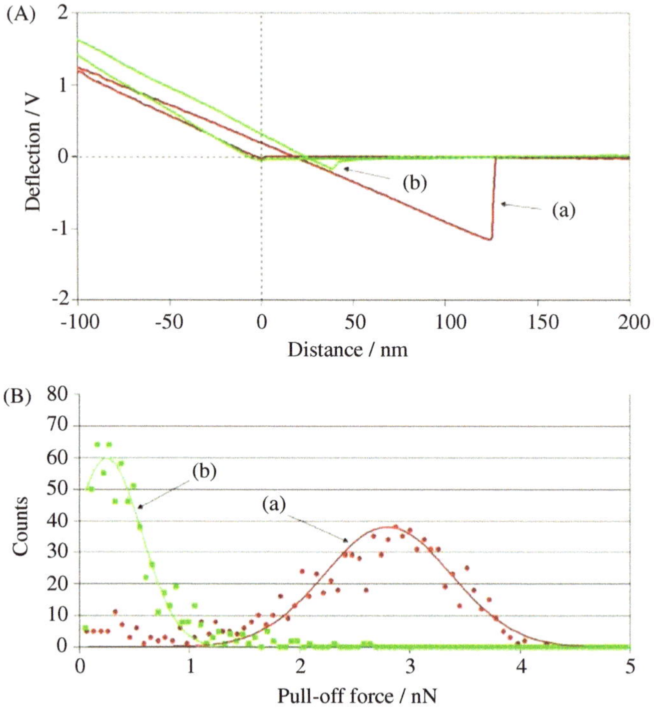

Figure 6.25 Force interaction between a hydrophobic AFM tip and the $Ag^{+/0}$ surface as a function of the oxidation state of Ag. (A) Typical deflection–distance curve on (a) Ag^+ and (b) Ag^0 surfaces in a 0.1 M Na_2SO_4 solution, pH = 3. (B) Distribution of the calculated pull-off force on (a) Ag^+ and (b) Ag^0 surfaces. The data points represent the frequency of occurrence of a specific pull-off force; the solid lines represent a Gaussian fit to these data. Adapted with permission from Ref. [60]. Copyright (2006) American Chemical Society.

Further support that the hydrophilic/hydrophobic properties of the interface were controlled by the redox transformation $Ag^+ + e^- \leftrightarrow Ag^0$ was obtained from AFM force measurements that follow the affinity interaction between the hydrophobic-modified tip and the electrochemically triggered Ag^+-thiolate monolayer surface. Toward this goal, the Si-AFM tip was modified with octadecyltrichlorosilane to yield a hydrophobic tip, and the force measurements were performed in 0.1 M Na_2SO_4, pH = 3. Figure 6.25A shows a typical force curve upon monitoring the Ag^+-thiolate-modified surface generated by the application of the potential of 0.2 V, curve a, and a representative force curve of the hydrophobic tip interacting with the Ag^0-nanocluster-modified monolayer after applying a potential of −0.2 V, curve b. The detachment of the hydrophobic tip from the surface required a force corresponding to 3.2 nN, whereas the disconnection of the hydrophobic tip from the Ag^0-nanocluster-modified interface, clearly, required a substantially lower force corresponding to 0.5 nN. Such force interactions were recorded on 3–4 different domains of the Ag^+- or Ag^0-nanocluster-modified surfaces with a total statistics of 675–900 measurements for each monolayer state. Figure 6.25B shows the derived histograms. The results indicated that the interaction of the hydrophobic tip with the Ag^+-thiolate surface, which exhibited a hydrophobic character, resulted in high force interactions, average force ca. 2.8 nN, while the Ag^0 surface revealed lower affinity toward the hydrophobic tip, force centered at 0.3 nN, consistent with weak interactions between the hydrophilic surface and the hydrophobic tip. The selective force interactions between the modified electrode and the surface might be reversibly switched by the cyclic transformation of the interface between the Ag^+ and Ag^0-nanocluster states.

The success to reversibly switch the hydrophobic/hydrophilic properties of the Ag^+-thiolate monolayer by electrochemical means suggested that one may couple biocatalysis and electrochemistry to reversibly control the surface properties of the Ag^+-thiolate-modified surface (Fig. 6.26). Alkaline phosphatase hydrolyses *p*-aminophenyl phosphate (**2**) to *p*-aminophenol (**3**). The latter product acts as reducing agent, and it reduces Ag^+ to Ag^0 nanoclusters, with the concomitant formation of the imino-quinone type product (**4**). Thus, one might transform by a biocatalytic process, driven by alkaline phosphatase, the conversion of the Ag^+-thiolated interface to the hydrophilic Ag^0-nanocluster interface, and upon the electrochemical oxidation of the surface, the hydrophobic Ag^+-thiolate layer would be regenerated (Fig. 6.26). Figure 6.27A shows the SPR spectrum of the Ag^+-thiolated monolayer prior to the interaction with alkaline phosphatase and **2**, curve a, and after treatment with the biocatalyst and **2**, curve b. The shift in the minimum reflectivity angle and the shallow pattern of the reflectivity were well consistent with the formation of the metal nanoclusters on the surface. Figure 6.27A, curve c, shows the SPR spectrum of the biocatalytically generated Ag^0-nanostructured interface after applying a potential step from −0.2 to 0.2 V. The SPR spectrum of the Ag^+-thiolate monolayer was regenerated, implying that the potential step oxidized the Ag^0 nanostructures to the Ag^+ state. Figure 6.27B shows the linear sweep voltammogram of the alkaline phosphatase generated Ag^0-nanostructured interface upon scanning the potential from −0.25 to 0.4 V. The anodic wave indicated that the electrochemical oxidation of the Ag^0 nanoclusters, indeed, occurred. The coulometric analysis of the anodic wave indicated that the full coverage of the thiolated monolayer by Ag^+ ions was accomplished, implying that the enzyme transformed all the Ag^+-thiolate units to the Ag^0 nanoclusters.

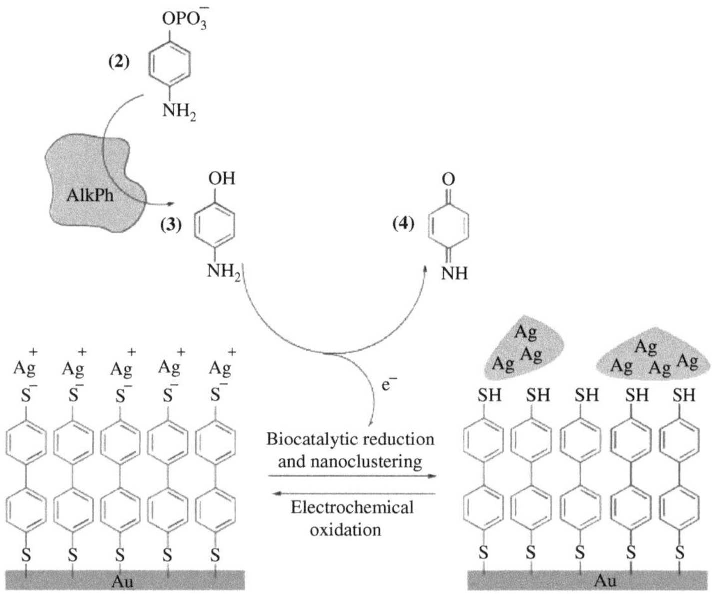

Figure 6.26 Biocatalytic generation of the Ag^0 nanoclusters on the functionalized surface, and electrochemical regeneration of Ag^+ ions associated with the interface. Reprinted with permission from Ref. [60]. Copyright (2006) American Chemical Society.

Contact angle measurements further confirmed the cyclic transition of the interface from the hydrophobic Ag^+-thiolate layer generated by electrochemical means to the Ag^0-nanoclustered surface produced by the biocatalytic process (Fig. 6.28). At point a, the monolayer existed in the hydrophobic Ag^+-thiolate state. The Ag^+-thiolate interface was then interacted with the biocatalytic alkaline phosphatase/**2** system, and the contact angle of the 0.1 M H_2SO_4 droplet was ex situ monitored on the resulting surface, point b. The lower contact angle value confirmed the formation of the Ag^0 nanoclusters, which exhibited enhanced hydrophilicity. The in situ electrochemical oxidation of the surface under the 0.1 M H_2SO_4 droplet restored the hydrophobic interface, point c. The points d, f, and h represent the contact angle of the Ag^0-nanocluster interface attained by repeated biocatalytic transformations with alkaline phosphatase/**2**. At points e, g, and i, the Ag^+-thiolate monolayer was generated in situ by the electrochemical oxidation of the clusters, to the Ag^+-thiolate state. At points j and k, the Ag^+-thiolate layer was treated either with alkaline phosphatase, or with the substrate **2** only, respectively. The interface was not transformed into a hydrophilic state, implying that only the combination of alkaline phosphatase/**2** yielded the hydrophilic surface. Indeed, treatment of the Ag^+-thiolate support with the alkaline phosphatase/**2** biocatalytic system regenerated the hydrophilic Ag^0-nanoclustered support, point l.

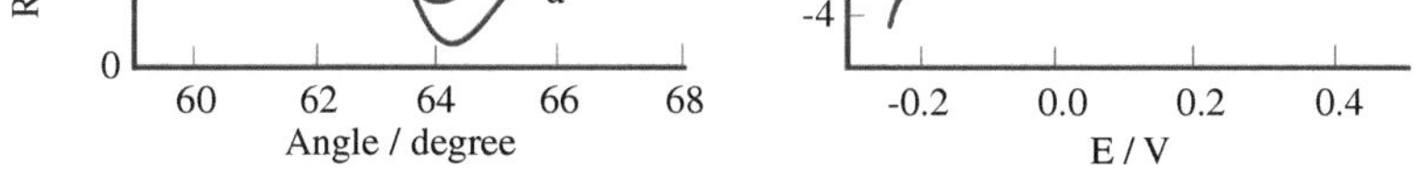

Figure 6.27 (A) SPR spectra corresponding to (a) the initial Ag^+-BPDT-monolayer-functionalized Au surface; (b) after the biocatalyzed reduction of the surface with alkaline phosphatase, 1 unit per mL, and **2**, 5 mM, for 1 h and the generation of Ag^0 nanoclusters on the surface; (c) after electrochemical oxidation of the Ag^0 nanoclusters and regeneration of the Ag^+-BPDT monolayer by applying a potential corresponding to 0.2 V. (B) Linear sweep voltammogram corresponding to the electrochemical oxidation of the biocatalytically generated Ag^0 nanoclusters. Potential scan rate, 20 mV/s. Adapted with permission from Ref. [60]. Copyright (2006) American Chemical Society.

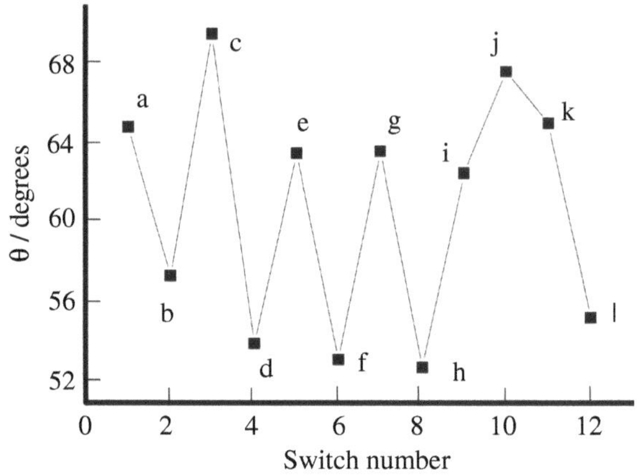

Figure 6.28 Reversible changes in the contact angle of a 0.1 M H_2SO_4 aqueous droplet on the $Ag^{+/0}$/BPDT/Au system upon biocatalytic reduction and electrochemical oxidation of the modified surface: (a) initial Ag^+-BPDT state; (b, d, f, h, and l) biocatalytically generated Ag^0-BPDT state in the presence of alkaline phosphatase, 1 unit per mL, and **2**, 5 mM; (c, e, g, and i) electrochemically oxidized Ag^+-BPDT state generated by applying a potential of 0.2 V; (j) control experiment in the presence of the Ag^+-BPDT state and alkaline phosphatase and absence of **2**; (k) control experiment in the presence of the Ag^+-BPDT state and **2**, 5 mM, and in the absence of alkaline phosphatase. Adapted with permission from Ref. [60]. Copyright (2006) American Chemical Society.

Other metal ions, e.g., Hg^{2+}, Cd^{2+}, or Pb^{2+}, as well as other functional monolayers, such as carboxylic acid-terminated monolayers, are anticipated to yield similar electroswitchable or biocatalytically switchable surface properties.

6.5 SPR Investigation of Au Nanoparticles Charging

6.5.1 Interplay between Electrical and Plasmonic Properties of Au Nanoparticle–Biomolecule Hybrids

The application of molecular components that transmit or store signals in electronic devices, i.e., molecular electronics, is a part of systems that can be considered in the range from information storage and processing to molecular photovoltaics and sensors [63–67]. Biomaterials such as proteins/enzymes, antigens/antibodies, receptors, and DNA, which demonstrate unique catalytic or recognition properties, could be important components of such molecular bioelectronic devices [68–71].

The electrochemical activation of enzymes [72, 73] is a common practice in bioelectrochemistry and in the development of biosensor devices [74] and biofuel cells [75]. The potentiometric analysis of substrates at enzyme-electrodes is usually accomplished by controlling the electrode potential by one of the reacting substrates [76], for example, the depletion of oxygen [77], or by chemical changes induced by the reaction products, for example, altering the pH of the system [78] as a result of the biocatalytic process.

The electrical contacting of redox enzymes and electrode supports was achieved by the surface reconstitution of apo-enzymes on relay-cofactor units associated with electrodes [79, 80]. For example, apo-glucose oxidase (apo-GOx) was reconstituted on a pyrroloquinoline quinone (PQQ)-flavin adenine dinucleotide (FAD) unit associated with an Au electrode [79]. In this system, the PQQ molecule operated as an electron relay providing efficient vectorial electron transfer from the FAD cofactor reconstituted into the glucose oxidase (GOx) enzyme to the electrode support. Recently, the electrical contacting of GOx was accomplished by the reconstitution of the apo-GOx on an FAD-functionalized Au nanoparticle (AuNP) (1.4 nm) that was linked to the Au electrode surface by a dithiol monolayer [81]. This study indicated that the electron transfer from the FAD cofactor to the electrode support occurred through the AuNP with its intermediate charging under the stable current generation conditions.

The quantum charging of AuNPs by electrochemical means is a subject of extensive research [82, 83]. The interactions of localized surface plasmon (LSP) of AuNPs and, specifically, of charged AuNPs with the surface plasmon (SP) excited on the surface of the bulk gold are of special interest and could be studied by the SPR technique.

The ability to probe physicochemical modifiers on surfaces by means of SPR turned the method into a useful analytical tool, and different biorecognition processes, such as protein binding or DNA hybridization, were followed by SPR measurements [84, 85]. In most of these studies, the chemical modification alters the refractive index and film thickness, and these changes affect the resulting SPR spectra.

In the presence of a metal nanoparticle layer, part of the SP energy is transferred outside through scattering within the nanoparticle layer. The interaction between the LSP associated with the noble metal nanoparticles and the SP on the metal–dielectric interface leads to a deformation of the dispersion curve of the SP and to a shift in the plasmon resonance angle [86]. Indeed, Au nanoparticles have been extensively used for the detection of bioaffinity-binding processes by virtue of the nanoparticle-enhanced SPR changes [87]. A significant enhancement of the SPR signal was achieved for immunosensing [88] and DNA sensing [89] when Au nanoparticles were used as labels. Also, the enhanced SPR read-out of redox reactions and bioelectrocatalytic processes was observed when the redox components that associated and dissociated to and from the electrode surface, respectively, were coupled with Au nanoparticles [90].

It is well established that charging the AuNP core with electrons, or the removal of electrons from the AuNP, results in significant shifts in the LSP band [91]. These spectral shifts in LSP band position were attributed to changes in the plasma frequency caused by the charge density increase resulting from the electrolytical charging of the metal nanoparticle [92]. The charging of the AuNP was anticipated to affect the coupling between the LSP and the SP of the bulk metal film, and thus changes in the SPR spectrum were expected upon the charging of the AuNP.

It is known that gold nanoparticles of size larger than 2 nm possess the plasmon band [93], and the shift in the plasmon peak position as a result of nanoparticles charging on an electrode surface or in solution was demonstrated in numerous studies [91, 92]. On the other hand, the plasmon peak is unidentifiable for AuNP of less than 2.0 nm effective diameter [93]. Therefore, changes in the electron density of these nanoparticles could not be traced by the shift in the peak position.

To address this issue, the considered study [94] reported on the assembly of the biocatalyst GOx on an array of AuNPs, mean diameter 1.4 nm, coupled to an SPR-active Au surface and the probing of the biocatalytic properties of the GOx/AuNP hybrid by means of SPR spectroscopy. The observed shifts in the plasmon angle (angle of minimum reflectance) were interpreted as originating from the enzyme-induced charging of the AuNP layer associated with the

Au support. It was demonstrated that the biocatalytic process in the presence of glucose charges the AuNPs, and that the charging process can be followed by SPR spectroscopy and by electrochemical means. To the best of the author's knowledge, this was the first example demonstrating the charging of metal nanoparticles with electrons by a biocatalytic process. Besides the fundamental interest of the system in the understanding of the "bio-pumping" of electrons into metallic nanoparticles, the results have important practical implications in developing new biosensor configurations. The method allowed tracing the AuNP charging event for such small Au nanoparticles that lack the characteristic absorbance peak.

6.5.2 Registration of Au Nanoparticles Charging as a Result of Interaction of Surface Plasmon and Localized Surface Plasmon

The reconstitution of apo-GOx on the FAD-functionalized AuNP (1.4 nm) and the assembly of the GOx/AuNP hybrid on an Au electrode were reported to electrically contact the enzyme and the electrode support [81], so similar approach was exploited in the considered study. The Au-coated SPR chips were modified with the dithiols 1,2-benzenedithiol or 1,9-nonanedithiol of different length by interacting the chips with ethanolic solutions of the respective dithiol, 10 mM, for 10 h, followed by rinsing of the chips with ethanol and water. The GOx-AuNP hybrid consisting of GOx reconstituted on the N^6-(2-Aminoethyl)-flavin adenine dinucleotide (amino-FAD)-functionalized AuNPs, ca. 1×10^{-6} M, was adsorbed onto the dithiol-modified Au electrode by interacting the modified electrode and GOx-AuNPs in a phosphate buffer overnight at 4°C.

The reconstitution of the enzyme on the AuNP aligned the FAD redox center with respect to the AuNP relay unit in a configuration that allows electrical communication. Figures 6.29A and 6.29B outline the method to prepare and assemble the GOx/AuNP hybrid on the Au surface. The AuNP (1.4 nm diameter) functionalized with a single *N*-hydroxysuccinimide ester group was reacted with the amino-FAD cofactor to yield the FAD-functionalized AuNP. The apo-GOx was reconstituted onto the modified AuNP to yield the GOx/AuNP hybrid. A monolayer of 1,4-benzenedithiol, a short dithiol, or of 1,9-nonanedithiol, a long dithiol, was assembled on the Au surface, and this acted as a bridging film that binds the GOx/AuNP system. Parallel microgravimetric quartz crystal microbalance experiments indicated that the surface coverages of the short thiol and of the long thiol monolayer were ca. 1.1×10^{-10} and 2.5×10^{-10} mol cm^{-2}, respectively. The surface coverage of the GOx/AuNP array in both systems was determined to be ca. 8.5×10^{-13} mol cm^{-2}, and this value translates to a surface coverage of 5.1×10^{11} hybrid species per cm^2. Taking into account the GOx average diameter of ca. 67 ± 5 Å, a densely packed monolayer of GOx exhibited a surface coverage of the enzyme corresponding to ca. 4.7×10^{-12} mol/cm^2. Thus, it was concluded that the deposition of GOx/AuNP yielded a random densely packed GOx monolayer (ca. 60% of the ordered packing) [95].

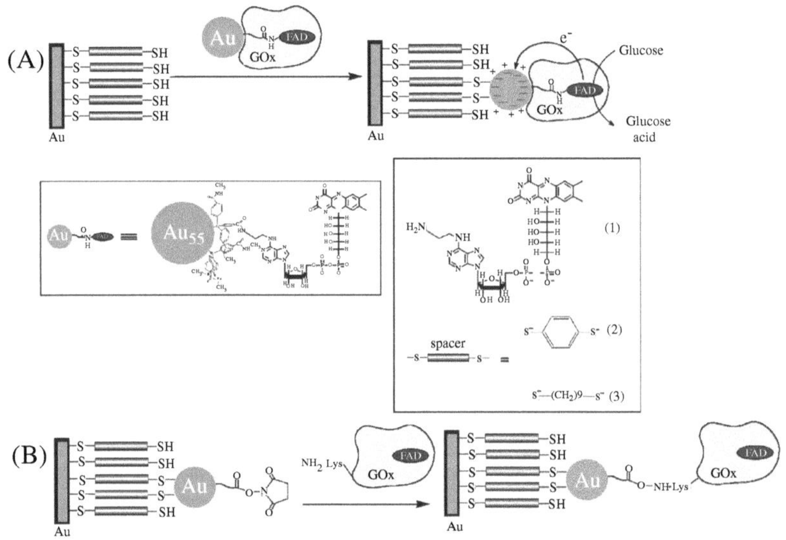

Figure 6.29 (A) Assembly of the electrically contacted reconstituted GOx/AuNP systems on Au electrodes by means of dithiol bridging units. (B) Assembly of the random covalently linked GOx/AuNP system on the Au electrode. Adapted with permission from Ref. [94]. Copyright (2004) American Chemical Society.

The GOx/AuNP hybrid system assembled on an Au surface was characterized by in situ ESPR spectroscopy. Figure 6.30A shows the SPR spectra of the GOx/AuNP system bridged to the Au support by the long dithiol monolayer before the addition of glucose (curve a),

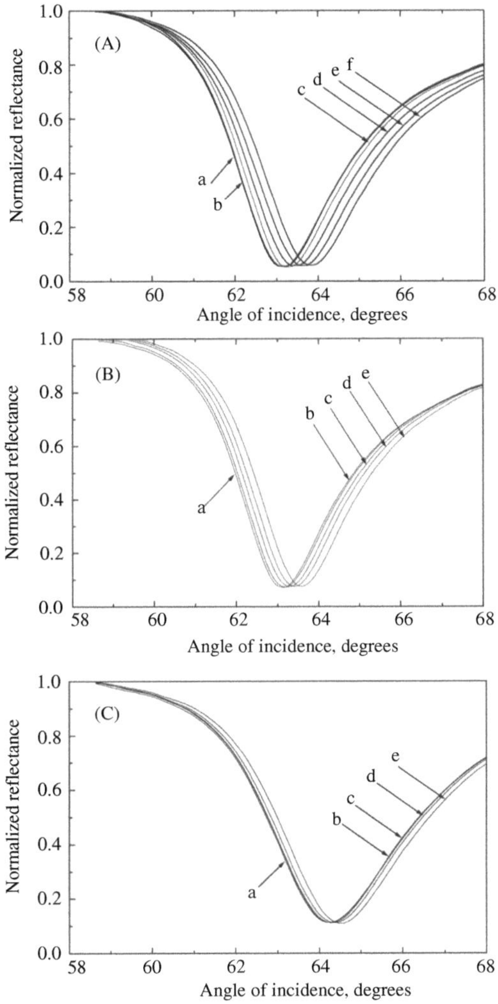

Figure 6.30 SPR spectra corresponding to the GOx-functionalized-AuNP arrays bridged to the Au electrodes with dithiol bridging units upon the addition of different concentrations of glucose: (a) 0 mM, (b) 0.3 mM, (c) 1.6 mM, (d) 8 mM, (e) 40 mM, (f) 100 mM. (A) Aligned GOx-reconstituted AuNPs bridged to the Au electrode with the long dithiol. (B) Aligned GOx-reconstituted AuNPs bridged to the Au electrode with the short dithiol. (C) Nonaligned GOx randomly covalently bound to the AuNPs bridged to the Au electrode with the long dithiol. Adapted with permission from Ref. [94]. Copyright (2004) American Chemical Society.

and after the addition of different concentrations of glucose in the range of 3×10^{-4} to 1×10^{-1} M (curves b–f). Clearly, the SPR angle shifts to higher values, and at a glucose concentration of 1×10^{-2} M, the SPR angle is shifted by $\Delta\theta_P = 39'$ (Fig. 6.31A, curve a). Figure 6.30B shows the SPR spectra of the analogous system, which was bridged with the short dithiol monolayer, before the addition of glucose (curve a), and after the addition of the same concentrations of glucose (curves b–e). As before, the addition of glucose caused the shift in the plasmon angle θ_P to higher values, but the changes in the plasmon angles were smaller than those in the previous system ($\Delta\theta_P = 27'$ in the presence of 1×10^{-2} M glucose) (Fig. 6.31A, curve b). Figure 6.30C shows the SPR spectra corresponding to a control system, where GOx was covalently linked through lysine residues to the active ester-functionalized AuNP, and the resulting GOx/AuNP hybrid system was associated to the long dithiol bridging monolayer (Fig. 6.29B). In this GOx/AuNP system, the enzyme was immobilized in a nonaligned configuration, and previous studies have shown that the FAD center lacks electrical communication with the AuNP in this randomly deposited enzyme [81]. From Fig. 6.30C, one can conclude that the addition of glucose leads to a shift in the plasmon angle θ_P, but these shifts were substantially smaller than those observed for the structurally aligned GOx/AuNP associated with the long dithiol-modified interface (Fig. 6.31A, curve c). For example, at a glucose concentration of 1×10^{-2} M, the nonaligned system yielded a change in the reflectivity angle of 14′ only, whereas the aligned GOx/AuNP system associated with the same long dithiol monolayer led to the change in $\Delta\theta_P = 39'$ under similar conditions.

To gain further understanding of the sources that led to the shifts in the SPR angles during the AuNP charging process, additional control experiments were performed comparing the changes in the values of the SPR angles θ_P upon treatment of the differently modified Au surfaces with various concentrations of glucose (Fig. 6.31A). Curves e and d correspond to the changes in the SPR angles $\Delta\theta_P$ of the Au surface modified with the long dithiol monolayer and to the long dithiol-modified surface functionalized with the amino-FAD-derivatized AuNPs upon the addition of different concentrations of glucose, respectively. Figure 6.31A, curve c, shows the changes in the values of the SPR angles upon treatment of the long dithiol-modified Au surface functionalized with the

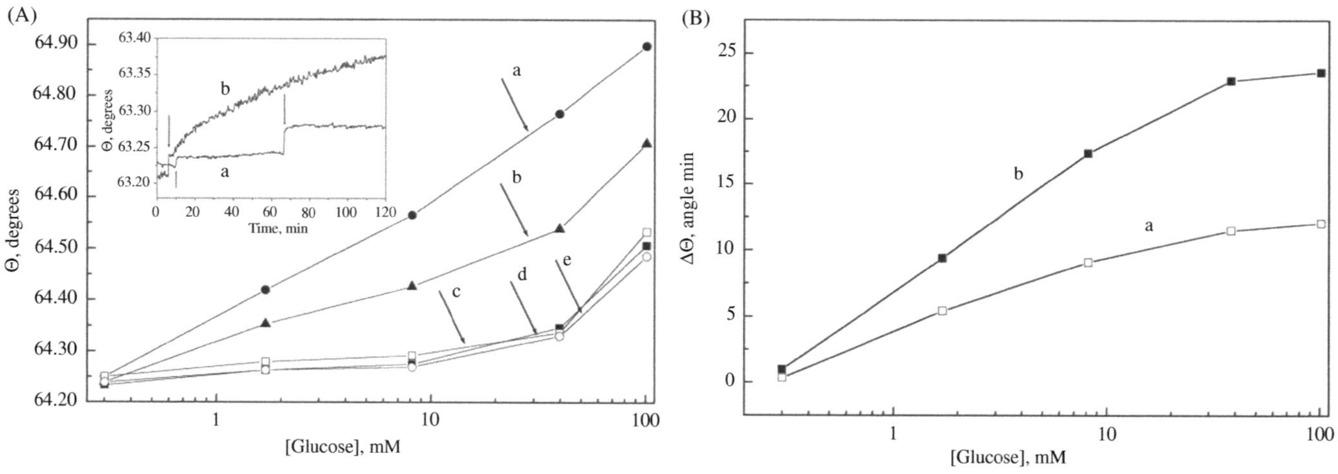

Figure 6.31 (A) Shifts in the SPR angles induced upon addition of different concentrations of glucose to the following: (a) The aligned GOx-reconstituted AuNPs bridged to the Au electrode with the long dithiol. (b) The aligned GOx-reconstituted AuNPs bridged to the Au electrode with the short dithiol. (c) The nonaligned GOx randomly covalently bound to the AuNPs bridged to the Au electrode with the long dithiol. (d) The NP-array bridged to the Au electrode with the long dithiol. (e) The Au electrode surface modified with the long dithiol monolayer. Inset: (a) Time-dependent changes in the SPR angle of the nonaligned GOx/amino-FAD-AuNP system bridged with the long dithiol monolayer upon the addition of 0.3 and 1.6 mM glucose (injection times are shown by the arrows). (b) Time-dependent changes in the SPR angle of the aligned GOx/AuNP system bridged with the long dithiol monolayer upon the addition of 1.6 mM glucose (injection time is shown by the arrow). (B) The relative changes in the SPR angle upon the addition of various glucose concentrations to the aligned GOx/AuNP system bridged to the Au electrode with: (a) the short dithiol linker, and (b) the long dithiol linker. Adapted with permission from Ref. [94]. Copyright (2004) American Chemical Society.

nonaligned GOx/AuNP system with different concentrations of glucose. Evidently, curve c overlaps with curves d and e, suggesting that the SPR spectrum changes in these systems originated only from the alteration of the refractive index of the solution containing different concentrations of glucose. On the other hand, the aligned GOx/AuNP structures linked to the Au surface through the short dithiol monolayer, curve b, or the long dithiol monolayer, curve a, revealed substantially higher changes in the values of the SPR angle shifts as compared to the changes originating from the alteration of the refractive index of the media upon addition of glucose. These enhanced changes in the SPR angles were attributed to the charging of the particles by the biocatalytic process (Fig. 6.29A). The dithiol monolayer between the AuNPs and the Au surface introduced a barrier for transporting the electrons from the AuNPs to the bulk Au surface. Thus, because of a slow electron-transfer rate between the AuNP and an Au electrode, electrons were effectively stored on the AuNP, leading to their steady-state charging. This changed the dielectric constant of the interface, leading to the SPR angle shifts. Figure 6.31A, inset, shows the different types of the time-dependent profiles of the SPR angle changes upon addition of glucose to the systems consisting of the aligned and nonaligned GOx/AuNP arrays bridged to the Au surface by the long dithiol. Curve a corresponds to the nonaligned GOx/AuNP system and shows the θ_P shift upon stepwise addition of glucose with concentrations of 3×10^{-4} and 1.6×10^{-3} M (the time at which glucose was injected is marked with arrows). The observed fast changes in the θ_P resulted from the immediate alteration of the refractive index of the bulk solution upon glucose addition. Curve b corresponds to the time-dependent changes in θ_P in the aligned GOx/AuNP system upon the addition of glucose. It shows the same fast change in the θ_P after glucose addition, 1.6×10^{-3} M (shown by the arrow), followed by a large gradual shift in θ_P that tends to level-off to a steady-state value. The observed gradual rise in θ_P, curve b, reflected a different mechanism that occurred only in the aligned GOx/AuNP system upon addition of glucose. While the instantaneous increase in the θ_P value was attributed to the change in the refractive index of the medium upon the addition of glucose, the subsequent slow changes in the θ_P values were attributed to the charging of the AuNPs by the biocatalytic process. This charging process was slow because it involved the charging of the particles by the biocatalytic process and the discharging of the AuNPs by tunneling the electrons to the bulk Au support, the process that led to a steady-state equilibrium charging of the AuNPs.

Figure 6.31B shows the net relative changes $\Delta\theta_P$ in the SPR angles originating from the charging of the AuNPs in the systems consisting of the aligned GOx/AuNP system associated with the short dithiol on the Au surface, curve a, and with the long dithiol on the Au surface, curve b, at different concentrations of glucose. These curves were obtained by the subtraction of the changes in the SPR angles originating from the alteration of the refractive index of the bulk solution at variable glucose concentration (Fig. 6.31A, curve c), from the total SPR shifts observed in the systems consisting of the aligned GOx/AuNPs linked to the Au surface by the spacers short or long dithiols (Fig. 6.31A, curves b and a, respectively). It is clear that the changes in the SPR angles increased upon elevating the glucose concentrations, and that the changes tended to level-off at a glucose concentration of ca. 1×10^{-2} M. Also, the changes in the SPR angles were more pronounced in the long dithiol system, as compared to the short dithiol system. For example, at a glucose concentration that corresponds to 1×10^{-2} M, the GOx/AuNP bridged to the short dithiol-functionalized Au surface revealed a value of $\Delta\theta_P = 12'$, whereas the GOx/AuNP system linked to the long dithiol-modified Au surface revealed a plasmon angle shift of value $\Delta\theta_P = 23.6'$.

These extra changes in the SPR angles upon addition of glucose in solution, beyond the changes induced by the bulk solution refractive index, were attributed to the biocatalytic charging of the AuNP (Fig. 6.29A). The aligned GOx/AuNP exhibited electrical communication between the enzyme and the AuNP. This enabled the catalyzed oxidation of glucose that resulted in the transfer of electrons to the AuNP. Current–voltage measurements of charge transport through molecules self-assembled onto gold electrodes revealed resistance values of the interfaces that corresponded to ca. 2.2×10^7 and ca. 5×10^8 Ω for 1,4-benzenedithiol [96] and 1,9-nonanedithiol [97] monolayers, respectively. Also, previous electrochemical studies have indicated that 1,4-benzenedithiol exhibits a relatively low tunneling barrier, whereas 1,9-nonanedithiol reveals a high tunneling barrier for the electron transfer from the AuNP to the Au electrode [81]. As a result, the steady-state charging degree of the AuNP was expected to be higher in the presence of the long dithiol monolayer. This was reflected by the enhanced shifts in the SPR spectra of the aligned GOx/AuNP system linked to the electrode by the long dithiol monolayer.

From Figs. 6.30 and 6.31, it is evident that the relative changes in reflectance, $\Delta R/R$, and half-width, Γ_w, of the SPR curves are much smaller than the changes in the SPR angle, θ_p. Thus, by analyzing the SPR spectra, it was concluded that the SPR angle, θ_p, could be selected as the parameter describing the main changes in the studied systems. Usually, the shift in the SPR angle, which is not accompanied by changes in reflectivity, is related to an increase in the real part of the corresponding dielectric constant, ε_1. Thus, for the theoretical explanation of the effect of AuNP charging on the SPR spectrum, an approach used for studying the effect of applying an electric potential to a gold film described in Subsection 6.2.1 can be used with certain modifications. However, in contrast to the "thin gold film–electrolyte" system, which was considered in Subsection 6.2.1, in this case it is necessary to consider the "nanoparticle–electrolyte" system, whose optical properties depend on the surplus charge received by the nanoparticle in accordance with the following considerations.

In the case of many metals, their dielectric properties are well described by the simple free electron Drude model [98]. According to this model, the response of a metal to an electromagnetic field of angular frequency ω is described by the dielectric constant $\varepsilon(\omega) = \varepsilon'(\omega) + i\varepsilon''(\omega)$, which includes the real and imaginary parts, respectively, and it may be written as Eq. (6.10):

$$\varepsilon(\omega) = 1 - \frac{\omega_\mathrm{p}^2}{\omega_\mathrm{d}^2 + \omega^2} + i\frac{\omega_\mathrm{p}^2\omega_\mathrm{d}}{\omega(\omega_\mathrm{d}^2 + \omega^2)}, \qquad (6.10)$$

where ω_p is the bulk plasma frequency and ω_d is the damping frequency, related to the mean free path of the conduction electrons in the bulk. Gold can be considered a free-electron-like metal for light of wavelength above 600 nm, and thus a simplified expression for the real part of $\varepsilon(\omega)$ can be formulated in the form of Eq. (6.11):

$$\varepsilon'(\omega) = \frac{\omega_\mathrm{p}^2}{\omega_\mathrm{d}^2 + \omega^2}, \qquad (6.11)$$

where the plasma frequency ω_p is proportional to the density of electrons, N, Eq. (6.12):

$$\omega_p^2 = Ne^2/m\varepsilon_0, \tag{6.12}$$

where e is the electron charge, m is the effective mass of the electron, and ε_0 is the permittivity of free space.

Importantly, the dielectric constant of metals depends on the plasma frequency ω_p (Eq. (6.10)) and, therefore, on the density of the conduction electrons, N. Therefore, when the biocatalyzed transfer of electrons into the particle occurs, the density of the free electrons, N, is increased, probably near the particle surface (Thomas–Fermi layer), and consequently the dielectric constant, ε', is altered according to Eq. (6.13):

$$\Delta\varepsilon'(\omega)/\varepsilon'(\omega) \sim \Delta N/N, \tag{6.13}$$

where ΔN is the change in the density of electrons in AuNP.

It is known that the SPR angular shift $\Delta\theta_p$ is a function of the immobilized layer thickness and of the optical (dielectrical) difference of the layer as compared to the surrounding medium, $\varepsilon_{\text{layer}} - \varepsilon_{\text{solution}}$ [99]. The changes in the dielectric constant of the AuNP layer will result in the changes in the dielectric constant of the interface and will lead to the shifts in the SPR angle position. Thus, the steady-state charging of the Au nanoparticles with the GOx-generated electrons and the change in the dielectric constant of the AuNP layer could lead to the enhanced SPR angle shifts observed in the considered study.

For further understanding of the AuNP charging mechanism, double layer differential capacitances, C_{DL}, of the interfaces composed of the aligned GOx/AuNP systems on the dithiol-functionalized Au surfaces were derived from non-Faradaic impedance spectra measurements [22]. The impedance spectra were fitted to the equivalent circuit composed of a constant phase element (CPE) (Eq. (6.14)):

$$\text{CPE} = A^{-1}\,(j\,\omega)^{-n}, \tag{6.14}$$

where the physical meaning of A depends on the value of the parameter n, and ω is the frequency of the alternative voltage. The constant phase element reflects nonhomogeneity of the layer, and the extent of its deviation from the capacitance is controlled by the parameter n in Eq. (6.14). The CPE has the meaning of capacitance, and the coefficient A becomes equal to the C_{DL} when $n = 1$ [22]. For the GOx/AuNP system on the dithiol-functionalized Au surfaces, the value for n was found to be ca. 0.93, and thus the CPE was considered mainly as capacitance. The value of the interfacial differential capacitance, C_{DL}, in the absence of glucose was found to be ca. 2.1 and 19.1 μF/cm^2 for the long dithiol and the short dithiol, respectively. This was consistent with the fact that the long dithiol forms a dense hydrophobic monolayer, whereas the short dithiol monolayer includes pinholes that allow the permeation of water molecules into the monolayer structure.

Charging of the AuNPs associated with the interface should result in changes in the double layer capacitance. For example, in the recent electrochemical studies on self-assembled monolayers functionalized with ionizable headgroups, the influence of the degree of ionization of the terminal headgroups on the interfacial capacitance was demonstrated [100]. The charge accumulation in the double layer upon ionization of acidic or basic groups associated with the monolayer resulted in the increase in the total interfacial capacitance, and this was confirmed by impedance spectra measurements [101]. The potentials of the GOx/AuNP-modified electrodes, prior to the addition of glucose, were close to the zero-charge potential (E_{zcp}) values and thus provided the minimum values of C_{DL}. Upon the addition of glucose, the biocatalyzed charging of the AuNPs resulted in a negative potential shift versus the initial E_{zcp}, and this was anticipated to yield an increase in C_{DL}. Indeed, the expected increase in the C_{DL} upon the biocatalyzed glucose oxidation was experimentally observed in the presence of the aligned GOx/AuNP system. The time-dependent changes in the C_{DL} upon addition of glucose to the system were followed by the impedance measurements at a fixed frequency $f = 1$ kHz ($\omega = 2\pi f$). Figure 6.32A, curve a, shows the time-dependent differential capacitance changes in the aligned GOx/AuNP system bridged to the Au electrode by the long dithiol upon the addition of 100 mM of glucose. The capacitance increased from ca. 2.1 to 3.6 μF/cm^2. The time profile of the differential capacitance changes was very similar to the time dependence of the SPR angle shifts upon addition of glucose (Fig. 6.31A, inset, curve b), suggesting that the parameters originated from the same phenomenon in the system (the biocatalyzed charging of the AuNPs). The extent of the capacitance increase was dependent on the glucose concentration. Figure 6.32A, inset, shows the capacitance increase, ΔC_{DL}, as a function of the concentration of added glucose to the system consisting of the aligned GOx/AuNPs bridged to the Au electrode by the long dithiol linker. For comparison, Fig. 6.32A, curve b, shows the capacitance of the nonaligned GOx/AuNP system linked by long dithiols to the Au surface. Clearly, no capacitance changes occurred on the interface of the nonaligned GOx/AuNP system, indicating that no biocatalyzed charging of the AuNP took place. These results were consistent with the SPR experiments.

Figure 6.32B shows the results of the same experiment upon following the capacitance changes in the aligned GOx/AuNP system linked to the Au surface by the short dithiol monolayer in the presence of added glucose, 100 mM, curve a. The capacitance increased from ca. 19 to 24.7 $\mu F/cm^2$. Similarly, the nonaligned GOx/AuNP system did not reveal any significant changes in the capacitance upon the addition of glucose (Fig. 6.32B, curve b). The results observed with the nonaligned GOx/AuNP systems imply that the added glucose had no effect on the capacitance of the system. Also, it was observed that the initial capacitance of the long dithiol-bridged system, 2.1 $\mu F/cm^2$, prior to the addition of glucose, was substantially lower than the capacitance of the short dithiol-bridged GOx/AuNP system, ca. 19 $\mu F/cm^2$. Moreover, upon addition of glucose, 100 mM, to the long dithiol-bridged aligned GOx/AuNP system, the capacitance increased by ca. 80% to ca. 3.6 $\mu F/cm^2$, whereas in the short dithiol-bridged aligned GOx/AuNP system, the capacitance increased only by ca. 22% to the value of 24.7 $\mu F/cm^2$. These results were consistent with the larger shift in θ_p observed in the SPR measurements in the presence of the long dithiol linker as compared to the changes observed with short dithiol linker. That is, the long dithiol linker yielded a densely packed monolayer that provided an effective tunneling barrier for the transport of electrons from the AuNPs to the bulk Au electrode. This resulted in the effective charging of the AuNPs, which was reflected by the significant changes in the interfacial capacitance. In contrast, the short dithiol linker allowed the leakage of electrons from the AuNPs to the bulk Au electrode, resulting in the diminished charging of the AuNPs. Accordingly, only moderate changes in the capacitance values of the interface were observed.

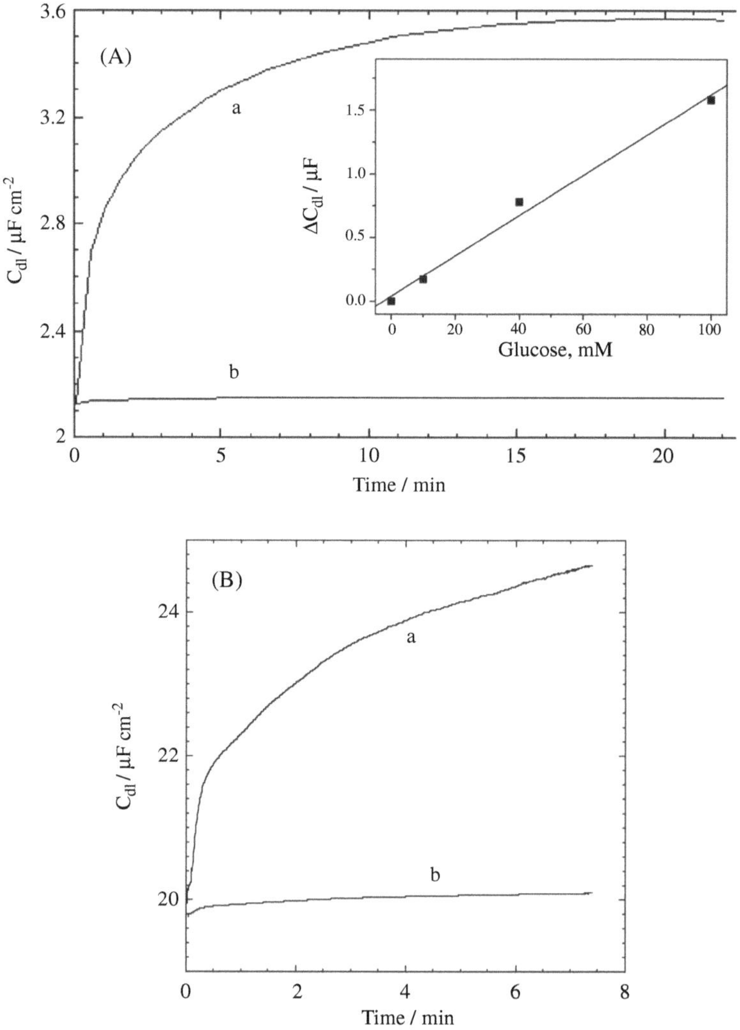

Figure 6.32 (A) Time-dependent capacitance changes in the GOx/AuNP system bridged by the long dithiol upon the addition of 100 mM glucose: (a) The aligned GOx/AuNP system. (b) The nonaligned GOx/AuNP system. Inset: Effect of various concentrations of glucose on the capacitance values of the aligned GOx/AuNP system bridged to the electrode with the long dithiol. (B) Time-dependent capacitance changes in the GOx/AuNP system bridged to the electrode by the short dithiol upon the addition of 100 mM glucose: (a) The aligned GOx/AuNP system. (b) The nonaligned GOx/AuNP system. The concentration of glucose was equal to 100 mM. Reprinted with permission from Ref. [94]. Copyright (2004) American Chemical Society.

6.5.3 Enzyme-Catalyzed Charging of Au Nanoparticles

Potentiometric measurements were employed to explore the nature of the charging processes of AuNPs. In these experiments, the potentials, E_G, generated on the bulk Au electrode functionalized with the aligned GOx/AuNPs and linked to the Au electrode by short or long dithiol spacer were measured in the presence of different concentrations of glucose. For example, upon addition of glucose, 8 and 100 mM, to the long dithiol-bridged aligned GOx/AuNP system, the measured electrode potentials, E_G, corresponded to −15 and −33 mV (versus saturated calomel electrode (SCE)), respectively.

Provided that the biocatalytic charging of the AuNP core leads to the observed shifts in the SPR angles, which were accompanied by the changes in C_{DL} and E_G, the reverse experiments were performed where a negative potential was applied on the same electrode that contained the GOx/AuNP array organized on either short or long dithiol spacer to charge the AuNP electrochemically. The charging of the AuNPs by the external potential was then monitored by the changes in the angular shifts in the SPR spectra. The potential of the modified Au electrode, E_A, was controlled by a potentiostat with the use of a counter electrode and a SCE reference electrode, and the SPR measurements were performed for each potential applied on the modified Au electrode. Figure 6.33 shows the changes in the SPR angles, θ_p, of the GOx/AuNP systems linked to the surface by short or long dithiols at different applied potentials. For example, the SPR angle shifts, $\Delta\theta_p$, similar to those observed in the presence of 100 mM of glucose, were generated upon the application of the potentials E_A of −230 and −360 mV (versus SCE) on the short and long dithiol-bridged GOx/AuNP systems, respectively. It should be noted that application of the potential of −360 mV (versus SCE) on the Au electrode coated with the long dithiol monolayer, which lacked the AuNPs, yielded only a minute shift of θ_p, ca. 2.6′. This confirmed that the observed changes in the SPR angle of the samples containing the AuNP array originated from the charging of the AuNPs as a result of the application of the negative potential to the electrode.

Under the externally applied potential on the bulk electrode, electron tunneling to the AuNP layer will occur, and the charging of the AuNPs will proceed until the potential equilibrium between the

Au electrode and the AuNP layer will be established. Charging of the AuNPs altered their θ_p; however, upon addition of 100 mM of glucose to the solution for the aligned GOx/AuNP array bridged via the long dithiol layer, the potential, E_G, generated on the bulk electrode was −33 mV. Meanwhile, a potential of −360 mV had to be applied on the bulk electrode to gain a similar change in the SPR angle of the system in the absence of glucose. Thus, it was concluded that a potential drop, ΔE, across the dithiol layer between the bulk Au electrode and the AuNPs resulted in this potential difference.

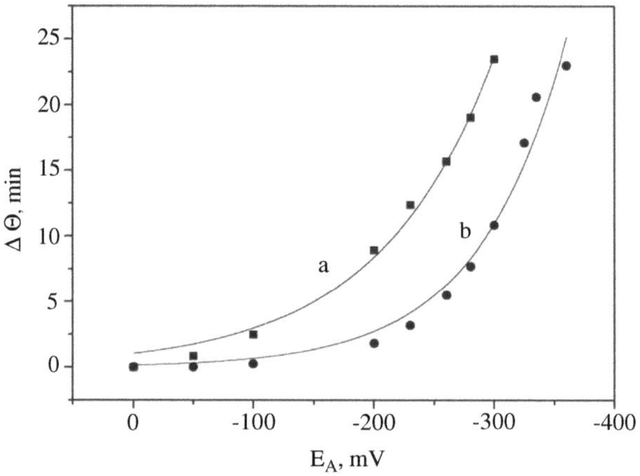

Figure 6.33 Changes in the SPR angle upon the application of variable potentials on the Au electrode modified with the GOx/AuNP array in the absence of glucose in solution: (a) The system bridged by the short dithiol monolayer. (b) The system bridged by the long dithiol monolayer. Reprinted with permission from Ref. [94]. Copyright (2004) American Chemical Society.

Thus, it was shown that the biocatalytic process and the applied external potential can independently charge the GOx/AuNP interface. The extent of charging was imaged by the changes in the SPR angle shifts. The charging process was, however, observed because of the existence of the molecular barrier between the GOx/AuNPs and the electrode that led to the potential drop across the interface. The results indicated that the potential drop across the long dithiol monolayer was higher than that across the short dithiol monolayer.

6.6 Conclusion

Specific studies described in this chapter demonstrate the versatility of the ESPR technique for various applications of molecular plasmonics in biosensing, bioelectronics, and material science. Proposed approaches exhibit a number of unique advantages provided by a combination of SPR and electrochemical methods, including biosensor response manipulation, optical characterization of redox events in thin-film structures, in situ studies of multiple functionalities of composite thin films induced by applied electrical potential, investigation of transient processes in bio-nanosystems, etc.

The mathematical model was proposed in one of the studies for calculating the SPR sensor response depending on the value of the applied external electric potential, which allowed predicting the experimentally measured SPR angle shift for the model electrolyte, which makes it possible to use it for correction of the results of the ESPR measurements. It was established that the SPR angle shift $\Delta\theta_{\mathrm{SPR}}$ and the change in the capacity of the double electric layer ΔC_{DL} on the gold surface during the adsorption of the model proteins under the conditions of the applied electrical potential are related by a nonlinear dependence of the type $\Delta\theta_{\mathrm{SPR}} = (a + b\cdot\Delta C_{\mathrm{DL}})^{-1}$. For the first time, the dependence of the degree of adsorption of biomolecules in the SPR sensor on the applied electrical potential was shown, which was confirmed by the electrochemical impedance spectroscopy measurements. This phenomenon can be used in biochemical analysis to control the interaction of biomolecules, amplification of response, blocking of unwanted adsorption, etc.

The influence of the external electrical potential application on the parameters of the SPR sensor was investigated, and an example of a parallel SPR-electrochemical study of a polyaniline film was given in another study, as well as a substantiation of physical processes observed upon the application of an external electrical potential to a "gold–polyaniline–electrolyte" system was provided. An SPR-based study of structural transformations in polyanaline films initiated by the external electric potential was carried out, which resulted in the estimation of kinetic constants for shrinking (bi-exponential dependence with kinetic constants of 9×10^{-3} s^{-1} and 5×10^{-4} s^{-1}) and swelling ($3 \times 10^{-4}\ s^{-1}$) of polyaniline film. The proposed modification of the SPR device enabling the ESPR investigations allowed the in situ electropolymerization of monomers directly on the sensitive surface of the sensor, which, in turn, opened up the possibility to evaluate in real time the kinetics of the polymerization process and to determine the optical constants and the geometry of the resulting polymer layers. Moreover, the developed ESPR technology allowed investigating the oxidation–reduction processes in redox materials.

Another study has demonstrated the cyclic control of the hydrophilic/hydrophobic properties of an Ag^+-thiolate monolayer by electrochemical means. In addition to the control of the macroscopic surface properties of the Ag^+/Ag^0-functionalized interface, interesting nanometric phenomena related to the switchable functions of the surfaces were elucidated. The kinetics of interfacial electron transfer that led to the reduction of the Ag^+-thiolate monolayer to the Ag^0 nanoclusters and to the oxidation of the Ag^0 nanoclusters to the Ag^+-thiolate interface were followed. Most interesting was the kinetic elucidation of the 2D lateral clustering of the Ag^0 atoms on the thiolate monolayer. Also, the demonstration that the electrochemical reduction and oxidation of the Ag^+ and Ag^0 nanoclusters on the surface proceeded laterally on a two-dimensional surface, without the diffusion of the component into the solution, was unique and noteworthy.

The last considered study has demonstrated a new phenomenon in bioelectronics, which involves the biocatalytic charging of AuNPs associated with self-assembled monolayers on surfaces and the read-out of the charging process by means of SPR spectroscopy and electrochemical means (potentiometric and capacitance measurements). Besides the basic understanding of the biocatalytic pumping of electrons into AuNPs, the phenomenon has important practical implications as it reveals new transduction methods. To accomplish the biocatalytic charging of the nanoparticles, it was shown that two prerequisites must be met: (i) The redox enzyme (in the present study GOx) should be electrically contacted with the AuNPs. This was accomplished by the reconstitution of apo-GOx on the FAD-functionalized AuNP, a process that orients and aligns the redox center of the enzyme with respect to the AuNP. (ii) The enzyme/AuNP hybrid system must be linked to a bulk Au surface by dielectric long-chain molecular bridging units. This molecular bridging element provides a tunneling barrier for the transport of electrons from the AuNPs to the bulk Au electrode. This tunneling barrier results in the steady-state equilibrium accumulation of electrons in the AuNPs. The biocatalytic charging of the AuNPs alters the plasma frequency and, as a result, changes the dielectric constant of the AuNPs. The changes in the dielectric constant of the AuNPs alter the dielectric properties of the whole interface, resulting in the pronounced shifts in the SPR minimum reflectivity. Thus, it was demonstrated that SPR spectroscopy provides a new optical means to observe both biocatalytic and electrochemical charging of AuNPs, even for the AuNPs, which do not exhibit the plasmon peak in the optical spectrum.

References

1. Gabai, R., Sallacan, N., Chegel, V., Bourenko, T., Katz, E., and Willner, I. (2001). Characterization of the swelling of acrylamidophenylboronic acid–acrylamide hydrogels upon interaction with glucose by faradaic impedance spectroscopy, chronopotentiometry, quartz-crystal microbalance (QCM), and surface plasmon resonance (SPR) experiments, *J. Phys. Chem. B*, **105**, pp. 8196–8202.
2. Raitman, O. A., Katz, E., Willner, I., Chegel, V. I., and Popova, G. V. (2001). Photonic transduction of a three-state electronic memory and of electrochemical sensing of NADH using surface plasmon resonance spectroscopy, *Angew. Chem. Int. Ed.*, **40**, pp. 3649–3652.
3. Yuan, Y., Guo, T., Qiu, X., Tang, J., Huang, Y., Zhuang, L., Zhou, S., Li, Z., Guan, B.-O., Zhang, X., and Albert, J. (2016). Electrochemical surface plasmon resonance fiber-optic sensor: In situ detection of electroactive biofilms, *Anal. Chem.*, **88**, pp. 7609–7616.
4. Guo, T. (2017). Fiber grating-assisted surface plasmon resonance for biochemical and electrochemical sensing, *J. Light. Technol.*, **35**, pp. 3323–3333.
5. Kang, X., Jin, Y., Cheng, G., and Dong, S. (2002). In situ analysis of electropolymerization of aniline by combined electrochemistry and surface plasmon resonance, *Langmuir*, **18**, pp. 1713–1718.
6. Sriwichai, S., Baba, A., Phanichphant, S., Shinbo, K., Kato, K., and Kaneko, F. (2015). In situ study of electropolymerized poly (3-aminobenzoic acid) thin film on BD-R and DVD-R grating substrates by electrochemical-

transmission surface plasmon resonance spectroscopy, *Int. J. Polym. Sci.*, **2015**, 650516.

7. Chegel, V., Raitman, O., Katz, E., Gabai, R., and Willner, I. (2001). Photonic transduction of electrochemically-triggered redox-functions of polyaniline films using surface plasmon resonance spectroscopy, *Chem. Commun.*, pp. 883–884.
8. Chen, D. and Hu, W. (2017). In situ investigation of electrochemically mediated surface-initiated atom transfer radical polymerization by electrochemical surface plasmon resonance, *Anal. Chem.*, **89**, pp. 4355–4358.
9. Tian, L., Qiu, J., Zhou, Y.-C., and Sun, S. G. (2010). Application of polypyrrole/GOx film to glucose biosensor based on electrochemical-surface plasmon resonance technique, *Microchim. Acta*, **169**, pp. 269–275.
10. Kang, X., Cheng, G., and Dong, S. (2001). A novel electrochemical SPR biosensor, *Electrochem. Commun.*, **3**, pp. 489–493.
11. Dahlin, A. B., Dielacher, B., Rajendran, P., Sugihara, K., Sannomiya, T., Zenobi-Wong, M., and Vörös, J. (2012). Electrochemical plasmonic sensors, *Anal. Bioanal. Chem.*, **402**, pp. 1773–1784.
12. McIntyre, J. D. E. (1973). Electrochemical modulation spectroscopy, *Surf. Sci.*, **37**, pp. 658–682.
13. McIntyre, J. D. E. and Peck, W. F. (1973). Optical spectroscopy of adsorbed intermediates in electrochemical reactions. Part 1. Surface field effects, *Faraday Discuss. Chem. Soc.*, **56**, pp. 122–137.
14. Kötz, R., Kolb, D. M., and Sass, J. K. (1977). Electron density effects in surface plasmon excitation on silver and gold electrodes, *Surf. Sci.*, **69**, pp. 359–364.
15. Gordon II, J. G. and Ernst, S. (1980). Surface plasmons as a probe of the electrochemical interface, *Surf. Sci.*, **101**, pp. 499–506.
16. Tadjeddine, A. (1989). Influence of the electrical polarization on the surface plasmon dispersion at metal-electrolyte interfaces, *Electrochim. Acta*, **34**, pp. 29–33.
17. Abeles, F., Lopez-Rios, T., and Tadjeddine, A. (1975). Investigation of the metal-electrolyte interface using surface plasma waves with ellipsometric detection, *Solid State Commun.*, **16**, pp. 843–847.
18. Abanulo, J. C., Harris, R. D., Sheridan, A. K., Wilkinson, J. S., and Bartlett, P. N. (2002). Waveguide surface plasmon resonance studies of surface reactions on gold electrodes, *Faraday Discuss.*, **121**, pp. 139–152.

19. Lioubimov, V., Kolomenskii, A., Mershin, A., Nanopoulos, D. V., and Schuessler, H. A. (2004). Effect of varying electric potential on surface-plasmon resonance sensing, *Appl. Opt.*, **43**, pp. 3426–3432.

20. Lopatynskyi, A. M., Lopatynska, O. G., Guiver, M. D., Poperenko, L. V., and Chegel, V. I. (2008). Factor of interfacial potential for the surface plasmon-polariton resonance sensor response, *Semicond. Phys. Quantum Electron. Optoelectron.*, **11**, pp. 329–336.

21. Brett, Ch. and Brett, A. (1994). *Electrochemistry: Principles, Methods, and Applications* (Oxford University Press, Oxford).

22. Bard, A. J. and Faulkner, L. R. (2001) *Electrochemical Methods: Fundamentals and Applications,* 2nd Ed. (John Wiley & Sons, Inc., New York).

23. Dressel, M. and Grüner, G. (2002) *Electrodynamics of Solids. Optical Properties of Electrons in Matter* (Cambridge University Press, Cambridge).

24. Dmitruk, N. L., Goncharenko, A. V., and Venger, E. F. (2009) *Optics of Small Particles and Composite Media* (Naukova Dumka, Kyiv).

25. Chegel, V. (2005). The electrochemically induced processes at the gold-polyaniline film interface: Surface plasmon resonance study, *Funct. Mater.*, **12**, pp. 355–361.

26. Garland, J. E., Assiongbon, K. A., Pettit, C. M., and Roy, D. (2003). Surface plasmon resonance transients at an electrochemical interface: Time resolved measurements using a bicell photodiode, *Anal. Chim. Acta*, **475**, pp. 47–58.

27. Moulton, S. E., Barisci, J. N., Bath, A., Stella, R., and Wallace, G. G. (2004). Studies of double layer capacitance and electron transfer at a gold electrode exposed to protein solutions, *Electrochim. Acta*, **49**, pp. 4223–4230.

28. Othman, S. H., El-Deab, M. S., and Ohsaka, T. (2011). Superior electrocatalytic activity of Au (110)-like gold nanoparticles towards the oxygen evolution reaction, *Int. J. Electrochem. Sci*, **6**, pp. 6209–6219.

29. Lioubimov, V., Kolomenskii, A., Mershin, A., Nanopoulos, D. V., and Schuessler, H. A. (2004). Effect of varying electric potential on surface-plasmon resonance sensing, *Appl. Opt.*, **43**, pp. 3426–3432.

30. Lide, R. D., Ed. (2004). *Handbook of Chemistry and Physics*, 84th Ed. (CRC Press, Boca Raton, FL).

31. Dmitruk, N. L., Fursenko, O. V., Kondratenko, O. S., and Romanyuk, V. R. (2003). Optical characterization of thin Au films by standard

and polaritonic ellipsometry, *Semicond. Phys. Quantum Electron. Optoelectron.*, **6**, pp. 349–353.

32. Stenesh, J. (1998) *Biochemistry* (Plenum Press, New York).
33. Iwasaki, Y., Horiuchi, T., Morita, M., and Niwa, O. (1999). Electrochemical reaction of $Fe(CN)_6^{3-/4-}$ on gold electrodes analyzed by surface plasmon resonance, *Surf. Sci.*, **427**, pp. 195–198.
34. Schlereth, D. D. (1999). Characterization of protein monolayers by surface plasmon resonance combined with cyclic voltammetry 'in situ', *J. Electroanal. Chem.*, **464**, pp. 198–207.
35. Jiang, Y., Cao, Z., Chen, G., Dou, X., and Chen, Y. (2001). Low voltage electro-optic polymer light modulator using attenuated total internal reflection, *Opt. Laser Technol.*, **33**, pp. 417–420.
36. Iwasaki, Y., Horiuchi, T., Morita, M., and Niwa, O. (1998). Analysis of electrochemical processes using surface plasmon resonance, *Sens. Actuators B Chem.*, **50**, pp. 145–148.
37. Koide, S., Iwasaki, Y., Horiuchi, T., Niwa, O., Tamiya, E., and Yokoyama, K. (2000). A novel biosensor using electrochemical surface plasmon resonance measurements, *Chem. Commun.*, pp. 741–742.
38. Beketov, G. V., Shirshov, Y. M., Shynkarenko, O. V., and Chegel, V. I. (1998). Surface plasmon resonance spectroscopy: Prospects of superstrate refractive index variation for separate extraction of molecular layer parameters, *Sens. Actuators B Chem.*, **48**, pp. 432–438.
39. Brennan, C. B., Sun, L., and Weber, S. G. (2001). Investigations of Prussian blue films using surface plasmon resonance, *Sens. Actuators B Chem.*, **72**, pp. 1–10.
40. Baba, A., Tian, S., Stefani, F., Xia, C., Wang, Z., Advincula, R. C., Johannsmann, D., and Knoll, W. (2004). Electropolymerization and doping/dedoping properties of polyaniline thin films as studied by electrochemical-surface plasmon spectroscopy and by the quartz crystal microbalance, *J. Electroanal. Chem.*, **562**, pp. 95–103.
41. Born, M. and Wolf, E. (1980) *Principles of Optics: Electromagnetic Theory of Propagation, Interference and Diffraction of Light*, 6th Ed. (Pergamon Press, Oxford).
42. Blakemore, J. S. (1985) *Solid State Physics*, 2nd Ed. (Cambridge University Press, Cambridge).
43. Volfkovich, Y. M., Sergeev, A. G., Zolotova, T. K., Afanasiev, S. D., Efimov, O. N., and Krinichnaya, E. P. (1999). Macrokinetics of polyaniline-based electrode: Effects of porous structure, microkinetics, diffusion, and electrical double layer, *Electrochim. Acta*, **44**, pp. 1543–1558.

44. Topart, P. and Hourquebie, P. (1999). Infrared switching electroemissive devices based on highly conducting polymers, *Thin Solid Films*, **352**, pp. 243–248.

45. Liu, Y., Mu, L., Liu, B., and Kong, J. (2005). Controlled switchable surface, *Chem.: Eur. J.*, **11**, pp. 2622–2631.

46. Feng, X., Feng, L., Jin, M., Zhai, J., Jiang, L., and Zhu, D. (2004). Reversible super-hydrophobicity to super-hydrophilicity transition of aligned ZnO nanorod films, *J. Am. Chem. Soc.*, **126**, pp. 62–63.

47. Liang, L., Feng, X., Liu, J., and Rieke, P. C. (1999). Preparation of composite-crosslinked poly (N-isopropylacrylamide) gel layer and characteristics of reverse hydrophilic–hydrophobic surface, *J. Appl. Polym. Sci.*, **72**, pp. 1–11.

48. Abbott, N. L. and Whitesides, G. M. (1994). Potential-dependent wetting of aqueous solutions on self-assembled monolayers formed from 15-(ferrocenylcarbonyl) pentadecanethiol on gold, *Langmuir*, **10**, pp. 1493–1497.

49. Liu, Y., Mu, L., Liu, B., Zhang, S., Yang, P., and Kong, J. (2004). Controlled protein assembly on a switchable surface, *Chem. Commun.*, pp. 1194–1195.

50. Sun, T., Wang, G., Feng, L., Liu, B., Ma, Y., Jiang, L., and Zhu, D. (2004). Reversible switching between superhydrophilicity and superhydrophobicity, *Angew. Chem. Int. Ed. Engl.*, **43**, pp. 357–360.

51. Wang, X., Katz, E., and Willner, I. (2003). Potential-induced switching of electrical contact by controlling droplet shapes at hydrophilic/hydrophobic interfaces, *Electrochem. Commun.*, **5**, pp. 814–818.

52. Wang, X., Zeevi, S., Kharitonov, A. B., Katz, E., and Willner, I. (2003). Probing photoinduced electron transfer reactions by in situ electrochemical contact angle measurements, *Phys. Chem. Chem. Phys.*, **5**, pp. 4236–4241.

53. Katz, E., Sheeney-Haj-Ichia, L., Basnar, B., Felner, I., and Willner, I. (2004). Magnetoswitchable controlled hydrophilicity/hydrophobicity of electrode surfaces using alkyl-chain-functionalized magnetic particles: Application for switchable electrochemistry, *Langmuir*, **20**, pp. 9714–9719.

54. Doron, A., Katz, E., Tao, G., and Willner, I. (1997). Photochemically-, chemically-, and pH-controlled electrochemistry at functionalized spiropyran monolayer electrodes, *Langmuir*, **13**, pp. 1783–1790.

55. Wang, X., Gershman, Z., Kharitonov, A. B., Katz, E., and Willner, I. (2003). Probing electrocatalytic and bioelectrocatalytic processes by contact angle measurements, *Langmuir*, **19**, pp. 5413–5420.

56. Lahann, J., Mitragotri, S., Tran, T. N., Kaido, H., Sundaram, J., Choi, I. S., Hoffer, S., Somorjai, G. A., and Langer, R. (2003). A reversibly switching surface, *Science*, **299**, pp. 371–374.

57. Katz, E., Lioubashevsky, O., and Willner, I. (2004). Electromechanics of a redox-active rotaxane in a monolayer assembly on an electrode, *J. Am. Chem. Soc.*, **126**, pp. 15520–15532.

58. Karuppuchamy, S. and Jeong, J. M. (2005). Super-hydrophilic amorphous titanium dioxide thin film deposited by cathodic electrodeposition, *Mater. Chem. Phys.*, **93**, pp. 251–254.

59. Zhai, L., Cebeci, F. C., Cohen, R. E., and Rubner, M. F. (2004). Stable superhydrophobic coatings from polyelectrolyte multilayers, *Nano Lett.*, **4**, pp. 1349–1353.

60. Riskin, M., Basnar, B., Chegel, V. I., Katz, E., Willner, I., Shi, F., and Zhang, X. (2006). Switchable surface properties through the electrochemical or biocatalytic generation of Ag^0 nanoclusters on monolayer-functionalized electrodes, *J. Am. Chem. Soc.*, **128**, pp. 1253–1260.

61. Rifai, S., Laferriere, M., Qu, D., Wayner, D. D. M., Wilde, C. P., and Morin, M. (2002). Electrodeposition of bilayers of dithiols, *J. Electroanal. Chem.*, **531**, pp. 111–118.

62. Patil, V. and Sastry, M. (2000). Formation of close-packed silver nanoparticle multilayers from electrostatically grown octadecylamine/colloid nanocomposite precursors, *Langmuir*, **16**, pp. 2207–2212.

63. Wei, Y. (1998). Molecular electronics: The future of bioelectronics, *Supramol. Sci.*, **5**, pp. 723–731.

64. Nitzan, A. (2001). Electron transmission through molecules and molecular interfaces, *Annu. Rev. Phys. Chem.*, **52**, pp. 681–750.

65. Stan, M. R., Franzon, P. D., Goldstein, S. C., Lach, J. C., and Ziegler, M. M. (2003). Molecular electronics: From devices and interconnect to circuits and architecture, *Proc. IEEE*, **91**, pp. 1940–1957.

66. Service, R. F. (2003). Molecular electronics. Next-generation technology hits an early midlife crisis, *Science*, **302**, pp. 556–559.

67. Park, J., Pasupathy, A. N., Goldsmith, J. I., Soldatov, A. V., Chang, C., Yaish, Y., Sethna, J. P., Abruna, H. D., Ralph, D. C., and McEuen, P. L. (2003). Wiring up single molecules, *Thin Solid Films*, **438**, pp. 457–461.

68. Davis, J. J. (2003). Molecular bioelectronics, *Philos. Trans. Royal Soc. A*, **361**, pp. 2807–2825.

69. Willner, I. (2001). Protein hinges for bioelectronics, *Nat. Biotechnol.*, **19**, pp. 1023–1024.

70. Willner, I. and Willner, B. (2001). Biomaterials integrated with electronic elements: En route to bioelectronics, *Trends Biotechnol.*, **19**, pp. 222–230.

71. Willner, I. (2002). Biomaterials for sensors, fuel cells, and circuitry, *Science*, **298**, pp. 2407–2408.

72. Armstrong, F. A. and Wilson, G. S. (2000). Recent developments in faradaic bioelectrochemistry, *Electrochim. Acta*, **45**, pp. 2623–2645.

73. Willner, I. and Katz, E. (2000). Integration of layered redox proteins and conductive supports for bioelectronic applications, *Angew. Chem. Int. Ed.*, **39**, pp. 1180–1218.

74. Wang, J. (1999). Amperometric biosensors for clinical and therapeutic drug monitoring: A review, *J. Pharm. Biomed. Anal.*, **19**, pp. 47–53.

75. Katz, E. and Willner, I. (2003). A biofuel cell with electrochemically switchable and tunable power output, *J. Am. Chem. Soc.*, **125**, pp. 6803–6813.

76. Zayats, M., Kharitonov, A. B., Katz, E., and Willner, I. (2001). An integrated relay/nitrate reductase field-effect transistor for the sensing of nitrate (NO_3^-), *Analyst*, **126**, pp. 652–657.

77. Rotariu, L., Bala, C., and Magearu, V. (2002). Yeast cells sucrose biosensor based on a potentiometric oxygen electrode, *Anal. Chim. Acta*, **458**, pp. 215–222.

78. Wang, J. Q., Chou, J. C., Sun, T. P., Hsiung, S. K., and Hsiung, G. B. (2003). pH-based potentiometrical flow injection biosensor for urea, *Sens. Actuators B Chem.*, **91**, pp. 5–10.

79. Katz, E., Riklin, A., Heleg-Shabtai, V., Willner, I., and Bückmann, A. F. (1999). Glucose oxidase electrodes via reconstitution of the apo-enzyme: Tailoring of novel glucose biosensors, *Anal. Chim. Acta*, **385**, pp. 45–58.

80. Raitman, O. A., Katz, E., Bückmann, A. F., and Willner, I. (2002). Integration of polyaniline/poly (acrylic acid) films and redox enzymes on electrode supports: An in situ electrochemical/surface plasmon resonance study of the bioelectrocatalyzed oxidation of glucose or lactate in the integrated bioelectrocatalytic systems, *J. Am. Chem. Soc.*, **124**, pp. 6487–6496.

81. Xiao, Y., Patolsky, F., Katz, E., Hainfeld, J. F., and Willner, I. (2003). "Plugging into enzymes": Nanowiring of redox enzymes by a gold nanoparticle, *Science*, **299**, pp. 1877–1881.

82. Hicks, J. F., Miles, D. T., and Murray, R. W. (2002). Quantized double-layer charging of highly monodisperse metal nanoparticles, *J. Am. Chem. Soc.*, **124**, pp. 13322–13328.

83. Hicks, J. F., Zamborini, F. P., and Murray, R. W. (2002). Dynamics of electron transfers between electrodes and monolayers of nanoparticles, *J. Phys. Chem. B*, **106**, pp. 7751–7757.

84. Berger, C. E. and Greve, J. (2000). Differential SPR immunosensing, *Sens. Actuators B Chem.*, **63**, pp. 103–108.

85. McDonnell, J. M. (2001). Surface plasmon resonance: Towards an understanding of the mechanisms of biological molecular recognition, *Curr. Opin. Chem. Biol.*, **5**, pp. 572–577.

86. Holland, W. R. and Hall, D. G. (1983). Surface-plasmon dispersion relation: Shifts induced by the interaction with localized plasma resonances, *Phys. Rev. B*, **27**, pp. 7765–7768.

87. Schultz, D. A. (2003). Plasmon resonant particles for biological detection, *Curr. Opin. Biotechnol.*, **14**, pp. 13–22.

88. Englebienne, P., Van Hoonacker, A., and Verhas, M. (2001). High-throughput screening using the surface plasmon resonance effect of colloidal gold nanoparticles, *Analyst*, **126**, pp. 1645–1651.

89. He, L., Musick, M. D., Nicewarner, S. R., Salinas, F. G., Benkovic, S. J., Natan, M. J., and Keating, C. D. (2000). Colloidal Au-enhanced surface plasmon resonance for ultrasensitive detection of DNA hybridization, *J. Am. Chem. Soc.*, **122**, pp. 9071–9077.

90. Zayats, M., Pogorelova, S. P., Kharitonov, A. B., Lioubashevski, O., Katz, E., and Willner, I. (2003). Au nanoparticle-enhanced surface plasmon resonance sensing of biocatalytic transformations, *Chem.: Eur. J.*, **9**, pp. 6108–6114.

91. Templeton, A. C., Pietron, J. J., Murray, R. W., and Mulvaney, P. (2000). Solvent refractive index and core charge influences on the surface plasmon absorbance of alkanethiolate monolayer-protected gold clusters, *J. Phys. Chem. B*, **104**, pp. 564–570.

92. Ung, T., Giersig, M., Dunstan, D., and Mulvaney, P. (1997). Spectroelectrochemistry of colloidal silver, *Langmuir*, **13**, pp. 1773–1782.

93. Alvarez, M. M., Khoury, J. T., Schaaff, T. G., Shafigullin, M. N., Vezmar, I., and Whetten, R. L. (1997). Optical absorption spectra of nanocrystal gold molecules, *J. Phys. Chem. B*, **101**, pp. 3706–3712.

94. Lioubashevski, O., Chegel, V. I., Patolsky, F., Katz, E., and Willner, I. (2004). Enzyme-catalyzed bio-pumping of electrons into Au-nanoparticles: A

surface plasmon resonance and electrochemical study, *J. Am. Chem. Soc.*, **126**, pp. 7133–7143.

95. Bourdillon, C., Demaille, C., Gueris, J., Moiroux, J., and Saveant, J. M. (1993). A fully active monolayer enzyme electrode derivatized by antigen–antibody attachment, *J. Am. Chem. Soc.*, **115**, pp. 12264–12269.

96. Reed, M. A., Zhou, C., Muller, C. J., Burgin, T. P., and Tour, J. M. (1997). Conductance of a molecular junction, *Science*, **278**, pp. 252–254.

97. Beebe, J. M., Engelkes, V. B., Miller, L. L., and Frisbie, C. D. (2002). Contact resistance in metal–molecule–metal junctions based on aliphatic SAMs: Effects of surface linker and metal work function, *J. Am. Chem. Soc.*, **124**, pp. 11268–11269.

98. Ashcroft, N. W. and Mermin, N. D. (1976) *Solid State Physics* (Holt, Rinehart and Wiston, New York).

99. Knoll, W. (1998). Interfaces and thin films as seen by bound electromagnetic waves, *Annu. Rev. Phys. Chem.*, **49**, pp. 569–638.

100. Andreu, R. and Fawcett, W. R. (1994). Discreteness-of-charge effects at molecular films containing acid/base groups, *J. Phys. Chem.*, **98**, pp. 12753–12758.

101. Schweiss, R., Werner, C., and Knoll, W. (2003). Impedance spectroscopy studies of interfacial acid–base reactions of self-assembled monolayers, *J. Electroanal. Chem.*, **540**, pp. 145–151.

Chapter 7

Studies of Conformational Changes in Molecular Systems Using Surface Plasmon Resonance

7.1 Introduction

Understanding conformational changes on molecular level is important because of their significant role in the technology of thin polymer films and crucial function in many biological functions and processes. For example, cooperative complexation between proteins and water-soluble polymers, which involves changes in the conformational state of the components, has been investigated because of its importance in colloid science as well as biological sciences [1, 2]. These conformational changes are often small and fast, making them difficult to measure and monitor. Surface plasmon resonance (SPR) is capable to detect rather small conformational changes in surface bound molecules [3]. However, in most cases, the results of SPR measurements alone are not enough to be sure whether conformational changes or some other processes have been detected, because there is always an uncertainty in which of the two parameters has changed, the thickness of investigated molecular layer or its refractive index. To solve this problem, additional research methods are necessary to obtain information on the above-mentioned parameters. Among the most suitable techniques to accomplish this task are microgravimetric methods, e.g., quartz crystal microbalance

Molecular Plasmonics: Theory and Applications
Volodymyr I. Chegel and Andrii M. Lopatynskyi

ISBN 978-981-4800-65-5 (Hardcover), 978-0-429-29511-9 (eBook)
www.jennystanford.com

(QCM) [4]; electrochemical methods, e.g., impedance spectroscopy [5]; or microscopic methods, e.g., atomic force microscopy (AFM) [6], scanning tunneling microscopy [7], or electron microscopy [8]. The molecular conformational processes such as swelling or shrinking for polymers and changing of secondary structure for biopolymers are usually exploited as parameters available for measurement by SPR and localized SPR (LSPR) methods [9–11].

This chapter presents the results of three specific practical studies exploiting the applications of SPR technology for the investigation of processes in (bio)molecular systems involving the conformational changes in their constituents. The first study is focused on the real-time SPR characterization of the cooperative interactions between aggregate-forming bovine serum albumin (BSA) and oppositely charged linear macromolecule poly(acrylic) acid, resulting in the estimation of the kinetic constants of conformational transformations for different pH values. The second study deals with the SPR transduction of the human olfactory receptor 17-40 stimulation with several odorants, revealing the possibility of developing a high-sensitive SPR-based bioelectronic nose. The third study reports on the characterization of conformational changes in acrylamidophenylboronic acid-acrylamide hydrogels upon interaction with glucose by SPR and complementary techniques, resulting in the development of a novel glucose-sensing approach. Application of molecular plasmonics methods for registration of conformational changes in polymers can also be found in other chapters, where they were used, for example, for selective sensing of small biomolecules (Chapter 5), following electrochemical transformations of redox materials (Chapter 6), or for development of metamaterials (Chapter 9).

7.2 Conformational Dynamics of Poly(Acrylic Acid)–BSA Polycomplexes at Different pH Conditions

7.2.1 Polyelectrolyte–Protein Polycomplexes in Colloid and Biological Sciences

The interaction between aggregate-forming proteins and oppositely charged linear macromolecules is known to result in the formation of polyelectrolyte–protein complexes, which exhibit a set of extremely interesting and practically important properties [12, 13]. The conformational transitions of water-soluble polymers were investigated in the studies in Refs. [14, 15] by using Raman spectroscopy, experimental titrations, fluorescence correlation spectroscopy [16], and other methods [17, 18]. The weak polyelectrolytes such as poly(acrylic acid) (PAA) are very sensitive to pH, and chains are almost uncharged at low pH but are fully charged at high pH. Authors of Ref. [19] reported that the conformation of PAA adsorbed on alumina estimated using a pyrene-labeled polymer changes appreciably with pH and affects the dispersion stability of polymer–protein complex solution. The pH-dependent complexation of PAA with phospholipid vesicles was characterized [20] using fluorescence polarization, differential scanning calorimetry, and surface pressure measurements of phospholipid monolayers. The complexation was pronounced below pH 4, when the polymer carboxyl groups are protonated, as shown by the binding of PAA to vesicles and the decrease in polymer mobility. Although it is very important to investigate colloidal stability by the adsorption of polymer–protein complexes in many applications, such as drug delivery, synthetic vaccines, very little work has been done to characterize the conformation of polymer–protein complexes adsorbed. Polyelectrolyte–protein polycomplexes (PCs) form as a result of the ion-exchange reaction between charged molecule of protein and ionized units of a macromolecule. There are two types of polymer–protein complexes: water-soluble nonstoichiometric and insoluble in water stoichiometric complexes, which may form depending on the binding conditions, such as pH, ionic strength, concentration, chemical structure of the components, and the reaction mixture composition [21, 22]. SPR is a surface-sensitive technique providing easy route to perturb a molecule and register its relaxation in real time [23]. This technique is ideally suited for observation of conformational changes in gel-formed polymer materials due to the possibility of easy detecting of the refractive index changes during real-time measurements of the studied object.

In this section, in situ study of the real-time conformational dynamics of PAA–BSA polycomplexes is considered, which was conducted under different pH of buffer solution using fluorescence spectroscopy and SPR methods [24].

7.2.2 Conformational Dynamics of BSA–PAA Complexes in SPR Study

To produce polymer–protein mixtures, various concentrations of the BSA solution in phosphate (PBS) or acetate (ABS) buffers were added to PAA. Experiments were carried out to analyze the effect of mixture order of the compositions by the titration of BSA solutions with different concentrations of PAA. In both cases, no differences were observed between adding the PAA solution onto the BSA solution or the BSA solution onto the PAA solution. The pH values were adjusted with 1 M NaOH. BSA/PAA ratios (n_{BSA}/n_{PAA}) were calculated using the equation of $n = CN_A/M$, where n is the number of the molecules in 1 mL; M is the molecular weight of components; N_A is the Avogadro number; and C is the concentration in g/100 mL. The SPR measurements were performed mainly with equal n_{BSA}/n_{PAA} ratio for most experiments.

The SPR technique monitors the changes in refractive index in close vicinity of the sensitive gold surface of the sensor. In studies of biomolecular interactions, changes in the molecular mass at the surface of the sensor associated with these interactions are usually the ones responsible for the changes in the refractive index and, therefore, the changes in the SPR signal. Hence, SPR detects the amount of the adsorbed molecules on the gold surface and molecules in the vicinity of the gold at the distance close to wavelength; in the present case, it is 650 nm. However, effective detection can be performed at about 200 nm from the gold surface. This point is important for the colloid liquids, where dispersed molecules are stable in the solution and the adsorption process depends on the parameters of colloid stability. When the colloidal solution is stable, SPR response depends mostly on the effective refractive index of colloid, because only separated molecules can adsorb on the surface of gold. However, when the stability of colloidal solution is disrupted, adsorption processes become the main parameters, influencing the SPR response.

The BSA/PAA complex exhibited a stable colloidal behavior at pH>5 even for different n_{BSA}/n_{PAA} values. For the pH 3.0–4.0, colloidal stability of complex strongly depended on the proportion of components of complex. In the 1.0–50 ratio range, the solubility of the polycomplex was changed and processes of aggregation and sedimentation were observed [23]. Two states of BSA/PAA PCs were studied with SPR, which exhibited insoluble (n_{BSA}/n_{PAA} = 1.0, Fig. 7.1A) and soluble (n_{BSA}/n_{PAA} = 20, Fig. 7.1B) properties. As depicted in the figures, the kinetics of SPR responses for PCs at the ratios of n_{BSA}/n_{PAA} = 1.0 and n_{BSA}/n_{PAA} = 20 exhibited the opposite behavior for purified (by centrifugation) and unpurified samples.

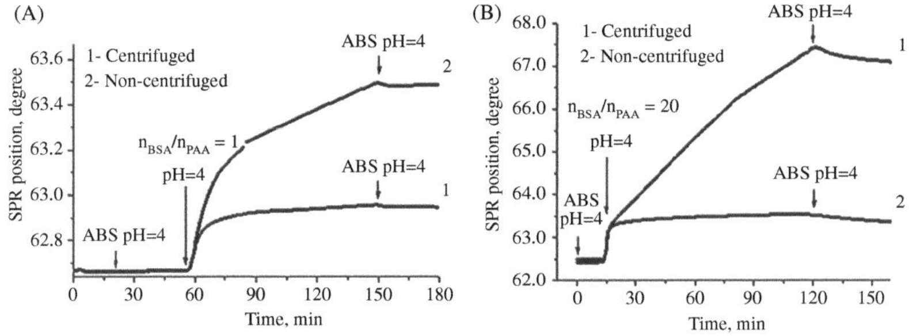

Figure 7.1 Temporal dependences of SPR angle position on the sedimentation of PCs, n_{BSA}/n_{PAA} = 1.0 (A) and n_{BSA}/n_{PAA} = 20 (B) at pH 4.0. Adapted with permission from Ref. [24]. Copyright 2008, WILEY-VCH Verlag GmbH & Co. KGaA, Weinheim.

This phenomenon was attributed to the process of aggregation of PCs with subsequent sedimentation and adsorption on the surface of gold. At low pH, the conformational processes in PCs with n_{BSA}/n_{PAA} = 1.0 were responsible for the existence of the insoluble aggregates with large size, which could not reach the sensitive surface of sensor due to steric hindrance (Fig. 7.1A, curve 2). After centrifugation, the aggregates with small size (supernatants) easily reached the surface (Fig. 7.1A, curve 1). As seen in Fig. 7.1B, the opposite processes were registered for PCs with the component ratio of n_{BSA}/n_{PAA} = 20. The considered conformational dynamics of PCs was observed only at low pH and disappeared at pH ≥ 5. As it can be seen also from Fig. 7.1A, the mass of the PAA–BSA complexes without centrifugation was about three times larger than that of the supernatants. Due to this, 70% of the aggregates remained in the pellet during centrifugation.

The sensitivity of SPR decays exponentially and increases at the surface of gold, so it is optimal to study the conformational dynamic of PCs placed directly on the surface of gold layer. In this case, one can register real-time changes in the structure of adsorbed BSA/PAA layer by simple changes in the pH of solution in flow mode.

The kinetics of changes in the structure of adsorbed BSA/PAA layer (n_{BSA}/n_{PAA} = 1.0) at the consequently changed pH is depicted in Fig. 7.2A. When the pH increased from 4.0 to 7.0, the carboxylic acid groups on the polymer backbone deprotonated and the electrostatic repulsion caused an increase in the thickness of the layer. As a result, the solvent molecules started reaching into the inner layers, while the effective refractive index of BSA/PAA layer decreased. It was assumed that the increasing pH caused the structure of PAA to open up (swelling). The dependence of plasmon resonance angular shift on the value of pH of buffer solution was presented in Fig. 7.2B for the case of acidic–basic transformation.

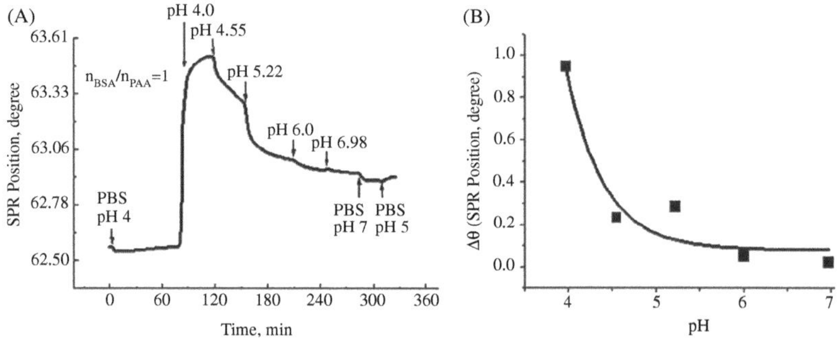

Figure 7.2 (A) Kinetic dependence of SPR angular position that reflects conformational changes in structures of the PAA–BSA mixture at the ratio of n_{BSA}/n_{PAA} = 1.0 when the overlayer pH was changed from acidic to basic (pH 4.0 → pH 6.98). The horizontal part of the graph represents gold surface without PAA/BSA layer. (B) Maximum of SPR responses depending on the wide range of pH. Adapted with permission from Ref. [24]. Copyright 2008, WILEY-VCH Verlag GmbH & Co. KGaA, Weinheim.

Figure 7.3 shows the transformation of PAA–BSA mixture from soluble to insoluble phase at the ratio of n_{BSA}/n_{PAA} = 1.0. As it follows from Fig. 7.3A, PAA was protonated and the overall thickness of the layer decreased while the pH is decreased; as a result, the refractive index after such transformation became higher. It was assumed that decreasing pH caused PAA to assume a coiled conformation due to the ensuing intramolecular hydrogen bond formation and the layer became more compact. Thus, the increasing SPR response showed the increasing mass of adsorbed molecules at the gold surface. The rate of structural transformation of PCs during pH variation

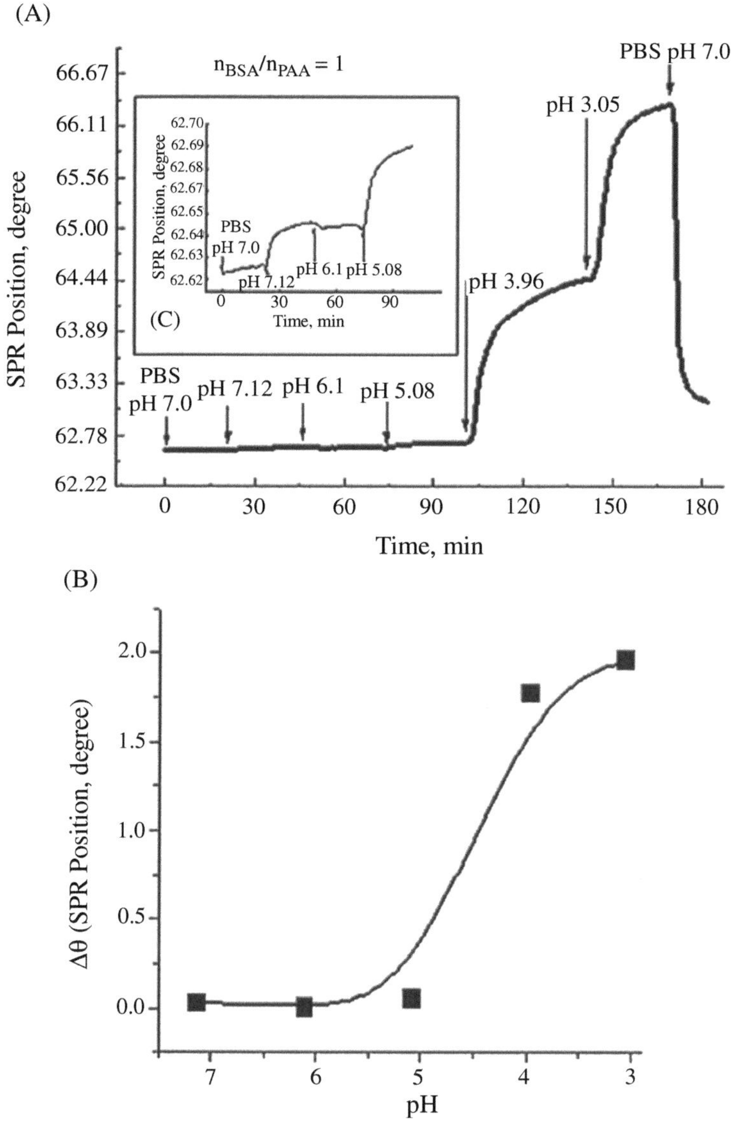

Figure 7.3 (A) Kinetic dependence of SPR angular position of PAA–BSA mixture at the ratio of $n_{BSA}/n_{PAA} = 1.0$ when the overlayer pH was changed from basic to acidic (pH 7.0 → pH 3.05) conditions. Here, BSA/PAA layer exists on the surface of gold before the start of measurements. (B) Respective maximum of SPR responses depending on the wide range of pH. (C) Inset: zoomed part of sensogram for basic pH. Reprinted with permission from Ref. [24]. Copyright 2008, WILEY-VCH Verlag GmbH & Co. KGaA, Weinheim.

changed dramatically between pH 4.0 and pH 5.0, which implied the existence of threshold charge state. As depicted in Figs. 7.2–7.3, the swelling of PAA–BSA complexes exhibited the lower rate constant K_d for the process of dissociation than for reverse associative process of shrinking of PCs layer. The higher rate constant K_a values in the interaction of PAA–BSA complexes without centrifugation compared with centrifuged complexes was explained in terms of an increase in binding capacity due to multivalent interaction. The rate constants of the observed kinetic transformations were calculated using standard fitting procedure and found to be changed in the range from 0.8×10^{-3} to 6.7×10^{-3} for different pH values and direction of change (Fig. 7.4).

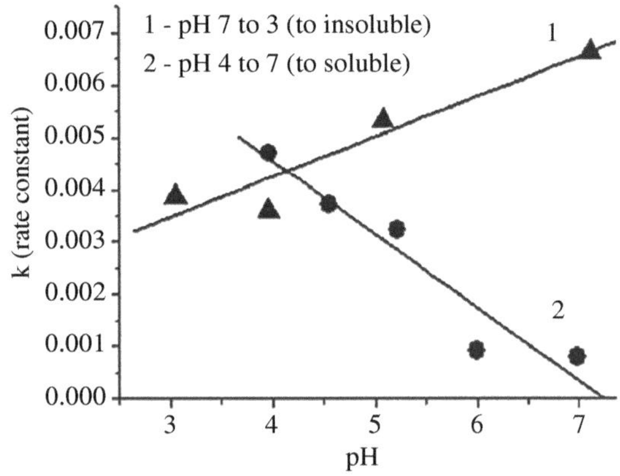

Figure 7.4 The calculated rate constants of structure transformation kinetics for BSA/PAA complex at the surface of gold due to change in pH of buffer solution from 3.0 to 7.0. Change of pH from 7.0 to 3.0 (1); from 4.0 to 7.0 (2). Adapted with permission from Ref. [24]. Copyright 2008, WILEY-VCH Verlag GmbH & Co. KGaA, Weinheim.

Fluorescence spectroscopy was also used to study the interactions and dynamics of protein–polymer complexes. Figure 7.5A shows the spectrofluoremetric titration of PAA–BSA mixture at the ratio of $n_{BSA}/n_{PAA} = 1.0$ and in a wide range of pH values (from pH 4.0 to pH 7.0) with 1M NaOH. Maximum fluorescence wavelength of this mixture depending on the pH is depicted in Fig. 7.5B. It is well-known that tryptophan (Trp) fluorescence of proteins varies with their conformational changes resulting in changes in fluorescence parameters, such as the emission maximum wavelength (λ_{max}), quantum yield, lifetime, and others [25, 26]. BSA (isoelectric point of BSA, pI_{BSA} = 4.9) contains two Trp [27]. One of them (spectral class 2 by Burstein [25] with λ_{max} = 340–342 nm) is located on the bottom of BSA hydrophobic cleft. The second Trp of class 3 (λ_{max} = 350–352 nm) with low quantum yield is superficial and completely accessible to aqueous solvent.

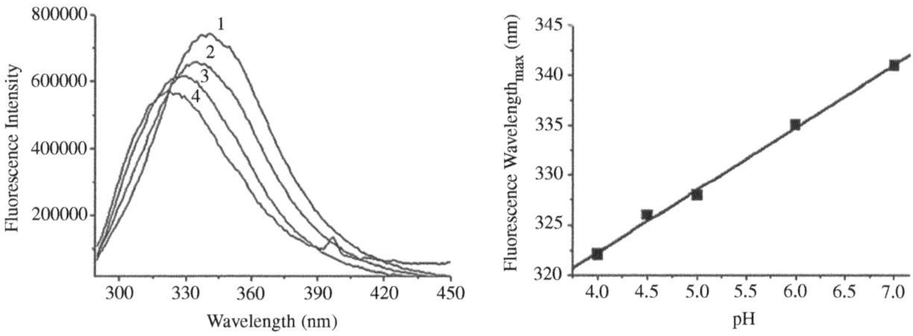

Figure 7.5 (A) Fluorescence spectra of PAA–BSA mixture at the ratio of n_{BSA}/n_{PAA} = 1.0 at 280 nm excitation wavelength: pH 7.0 (1); pH 6.0 (2); pH 5.0 (3); pH 4.0 (4). (B) Maximum fluorescence wavelength dependence on the pH. Adapted with permission from Ref. [24]. Copyright 2008, WILEY-VCH Verlag GmbH & Co. KGaA, Weinheim.

Fluorescence maximum shifted toward the red region (λ_{max} = 342 nm) at pH 7.0 as can be seen from Fig. 7.5A. The titration of n_{BSA}/n_{PAA} = 1.0 complex from pH 4.0 to pH 7.0 resulted in a 20 nm red shift (Fig. 7.5B). The emission maximum wavelength of BSA–PAA mixture at pH 7.0 was practically the same as that for pure BSA in solution (λ_{max} = 340 nm), which evidences the excess of free BSA in mixtures. The preexisting electrostatic repulsive forces between PAA and protein molecule prevented the formation of the stable polyelectrolyte complexes of BSA at pH 7.0. It was assumed that the large red shift obtained in the samples upon changing the pH of solution from 4.0 to 7.0 indicated that the BSA tryptophanyls became more accessible to aqueous solution, which might be due to the binding of the polymer with the protein.

7.3 Investigation of Human Olfactory Receptor 17-40 Interaction with Odorant Molecules by Means of Surface Plasmon Resonance

7.3.1 Registration of Odorant Molecules by Artificially Created Sensitive Structures (Bioelectronic Nose)

Recent advances in the pharmacology of olfactory receptors (ORs) result mainly from the development of drugs that can interact with ORs and initiate the physiological characteristics of the response of the receptor, triggering a chain of intracellular biochemical processes in the body [28]. There is a growing interest in the elaboration of biosensors based on ORs, employed in the membrane fraction or in the whole cell coupled to a solid transducer (bioelectronic nose) [29]. Such biosensor platform can be based on the direct monitoring either of developed drug binding to receptor or on an external monitoring of an organism's response caused by drug after binding to the receptor [30].

The pharmacological data available on mammalian ORs include various dose-response profiles. A typical adsorption curve of odorant octanal with a remarkably broad linear part (10^{-11}–10^{-3} M) was obtained by means of the QCM technique for rat OR I7 [31]. Signal profile of OR I7 from isolated olfactory neurons was sigmoid within the concentration range 10^{-7}–10^{-5} M of octanal [32]. Other data obtained from intracellular calcium and bioluminescence assays revealed the response pattern of OR I7 and OR 17-40 to be bell-shaped within the concentration range 10^{-14}–10^{-3} M of odorant helional [33].

This section covers the study of the registration of odorant molecules by artificially created sensitive structures using SPR and complementary methods [34]. The goal of this study was to investigate a pattern of G protein-coupled OR 17-40 response accompanying its biospecific interaction with odorant helional molecule using artificially created structures.

7.3.2 Investigation of the Interaction of Receptor OR 17-40 with Odorant Molecules Using SPR and Complementary Methods

Two different biofilm architectures were studied: Au + (16-mercaptohexadecanoic acid (MHDA) + 1,2-dipalmitoyl-*sn*-glycero-3-phosphoethanolamine-N-biotinyl sodium salt (biotinyl-PEA)) + neutravidin + biotinylated anti-cmyc monoclonal antibody (Ab) + OR 17-40 ("A1" biofilm), and Au + biotinylated Ab + OR 17-40 ("A2" biofilm).

To obtain self-organized heterogeneous layer onto gold in the case of A1 biofilm, 1 mM MHDA and 0.1 mM biotinyl-PEA were dissolved in ethanol and incubated with freshly cleaned SPR chip for 21 h at room temperature. MHDA was fixed onto Au via chemisorption, whereas biotinyl-PEA was inserted between long-chain thiols via hydrophobic interactions [35]. Such self-assembled monolayer (SAM) provided a good basis for the further anchoring of biomolecules to the surface. To elute unfixed molecules, the chip was rinsed with ethanol and dried under nitrogen flow. Neutravidin (0.5 μM in PBS) was bound to biotin through biospecific interaction; the rest of its specific sites served to attach biotinylated Ab also through biospecific interaction. In the process, due to the spatial structure of neutravidin, the Ab adopted a relative orientation toward the surface of gold, that is, the structure of the layer became more compact. In a comparative A2 biofilm, Abs were attached to the gold surface by random adsorption. Before the formation of any upper molecular layer, the previous one was rinsed with PBS for 5–15 min. In order to saturate all non-specific adsorption sites on modified surface, Ab layer was blocked by BSA (0.5 mg/ml in PBS). A non-specific to OR 17-40 odorant heptanal was used to control the selectivity of the biosensor structure.

Stock suspension of OR 17-40 in membrane fraction was diluted in PBS on ice down to the protein concentration 70 μg/ml, and 0.3 ml of this suspension was treated in the ultrasonic bath in ice-cold water for 20 min in order to obtain a homogeneous suspension of membrane vesicles called nanosomes due to their size [36]. Afterward, the suspension was immediately deposited onto the modified surface. One nanosome of 50 nm diameter could bear up to 10 ORs [37].

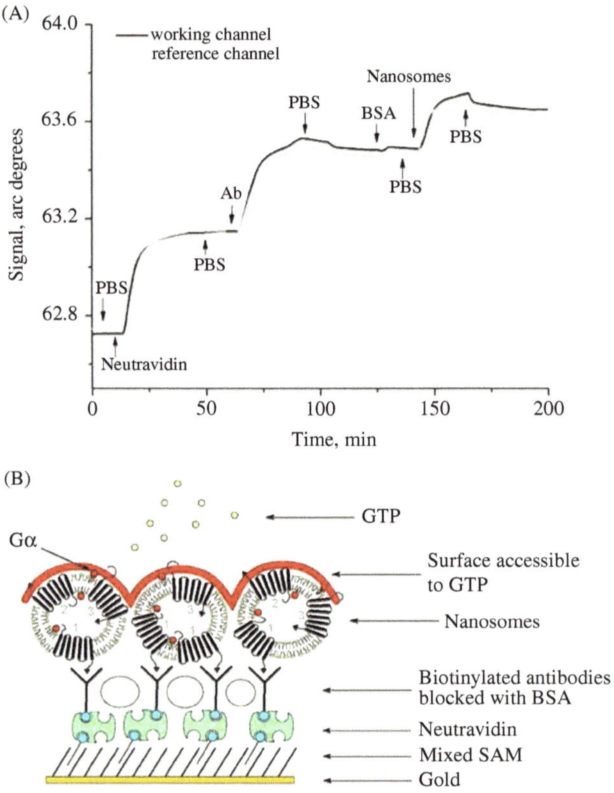

Figure 7.6 (a) Sensorgram of layer-by-layer assembly of A1 biofilm. (b) Scheme of A1 biofilm. Possible orientations (1, 2, 3) of $G\alpha_{olf}$ coupled receptors in a nanosome are described in Table 7.2. Adapted by permission from Springer Customer Service Centre GmbH: Springer *European Biophysics Journal,* Ref. [34], Copyright 2008.

Stock 0.1 M solutions of odorants were prepared freshly on the day of experiment in dimethyl sulfoxide; further dilutions (from 10^{-4} to 10^{-12} M) were obtained by successive 1:10 dilutions in PBS. The blank probes at the various dilutions were prepared replacing the odorant by PBS. Additionally, each odorant and blank probe contained 10 μM of guanosine 5′-O-[gamma-thio]triphosphate (GTP-γ-S) prepared on ice from the 1 mM solution. Measurement of the presence of odorant as an analyte was performed in the presence of GTP-γ-S, which acted as an activator of the OR 17-40 activity [36].

Receptors carried by nanosomes were immobilized via interactions of cmyc sequence with anti-cmyc monoclonal Ab attached to the gold in orientated or random way. As it was already mentioned, in the first case, Abs were uniformly attached to the neutravidin layer (Fig. 7.6). In the second case, random immobilization involved Abs' adsorption on the freshly cleaned gold (Fig. 7.7). SPR spectrum minimum shift for specific anchoring of biotinylated Ab to the neutravidin was about two times lower than a response to its direct adsorption on gold probably due to the limited quantity of biotinyl-PEA affinity sites at the surface. To estimate the thickness of each molecular layer, the experimental SPR spectra were fitted to the theoretical curves [38]. As a basis, an effective refractive index of n = 1.36 was used for protein layers [38] and of n = 1.46 for a membrane vesicle [39]. The calculated values of thickness are presented in Table 7.1.

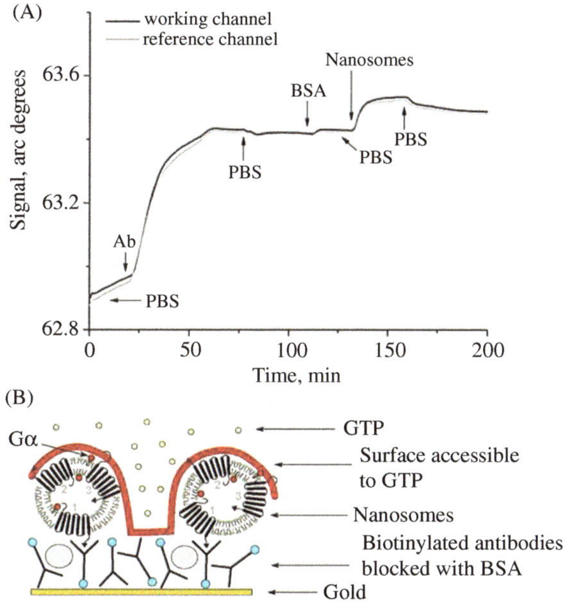

Figure 7.7 (a) Sensorgram of layer-by-layer assembly of A2 biofilm. (b) Scheme of A2 biofilm. Possible orientations (1, 2, 3) of $G\alpha_{olf}$ coupled receptors in a nanosome are described in Table 7.2. Adapted by permission from Springer Customer Service Centre GmbH: Springer *European Biophysics Journal,* Ref. [34], Copyright 2008.

Table 7.1 Calculated thickness of each molecular layer in A1 and A2 biofilms

Layer	A1, effective thickness, nm	A2, effective thickness, nm
MHDA	1.9 [40]	—
Neutravidin	15 ± 1	—
Biotinylated antibodies	12.5 ± 0.75	18.5 ± 1
BSA	0.5 ± 0.1	0.9 ± 0.2
OR 17-40–bearing nanosomes	11 ± 0.75 ($V = 1$)	11 ± 0.75 ($V = 0.25$)
Total thickness of multilayer, nm	40.9	30.4

V is a coefficient of surface coverage with lipidic biomaterial estimated from the AFM data shown in Fig. 7.9.

Source: Adapted by permission from Springer Customer Service Centre GmbH: Springer *European Biophysics Journal*, Ref. [34], Copyright 2008.

As it was mentioned in Section 4.3, the spatial orientation of immobilized Ab is crucial for analyte (here OR) capture since the latter is based on the highly specific interaction via cmyc tag. While the ratio Abs:nanosomes in the case of biofilm A1 could be close to 1 due to the oriented Ab layer, it should be much lower for the biofilm A2 suggesting that probably only a part of immobilized antibodies is properly oriented. Therefore, a random orientation of Ab layer resulted in a nanosome layer of comparatively low density (Table 7.1). At it was revealed by cyclic voltammetry (CV) measurements, A1 biofilm with an oriented architecture was highly insulating (Fig. 7.8). At the same time, electron transfer through the A2 biofilm was approximately 50% weaker in comparison with redox kinetics on bare Au (Fig. 7.8, inset).

In order to clarify this phenomenon, a biofilm similar to A2 consisting only of randomly adsorbed Ab was probed by means of CV under the same conditions. This layer of antibodies demonstrated an increased penetrability to redox couple after 12 h of contact with PBS (data not shown). Therefore, an increase in the insulating properties of A2 biofilm can be attributed to the nanosomes' fusion on the top of sensor surface. Since redox peaks did not completely disappear, one might conclude that the membrane vesicles did not merge into a continuous layer; therefore, the fusion of nanosomes on the electrode surface could be only partial, in agreement with reported AFM-based data [37].

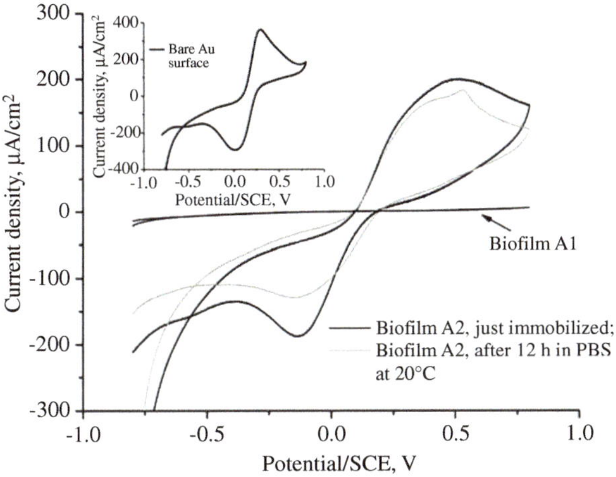

Figure 7.8 Cyclic voltammograms of biofilms A1 and A2. Inset: cyclic voltammogram of bare gold substrate. Adapted by permission from Springer Customer Service Centre GmbH: Springer *European Biophysics Journal,* Ref. [34], Copyright 2008.

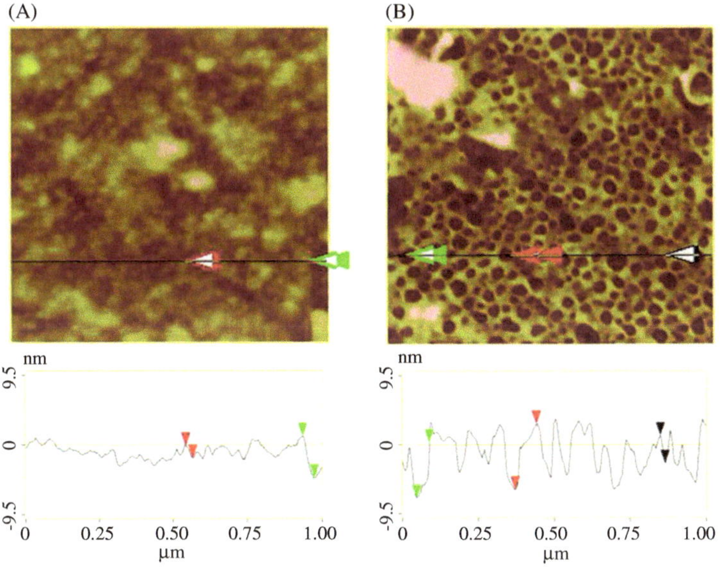

Figure 7.9 Atomic force microscopy images with cross-sectional profiles corresponding to working "spots" of A1 (a) and A2 (b) biofilms. Images were taken in tapping mode after the second day of using A1 and A2 biofilms for detection of odorants at room temperature. Reprinted by permission from Springer Customer Service Centre GmbH: Springer *European Biophysics Journal,* Ref. [34], Copyright 2008.

AFM images of biofilms were taken after the second day of odorant screening. Clear difference in relief porosity was observed between A1 (Fig. 7.9a) and A2 (Fig. 7.9b) structures. The surface profile of A1 biofilm was rather smooth due to the proper orientation of multilayer or to the collapsing of immobilized nanosomes, whether initial or after work with the surface. At the same time, the AFM image of biofilm A2 implies the shrinkage and clustering of the nanosome layer. In the calculations of thickness (Table 7.1), the coefficient of coverage of A1 biofilm with membrane biomaterial was taken as 1. The A2 biofilm surface coverage was estimated as 0.25 from the level of porosity observed.

Surface-grafted membrane fragments bearing $G\alpha_{olf}$ and ORs present a complex biorecognition unit where the above-described conformational changes of receptor and $G\alpha_{olf}$ are thought to occur upon OR stimulation with odorant. An olfactory signal is transmitted into sensory neurons via an interaction of OR with heterotrimeric G protein located on the cytoplasmic face of neuron ciliae membrane. Activated OR promotes the liberation of GTP-bound $G\alpha$ subunit from $G\beta\gamma$ dimer [41]. Possible orientations of $G\alpha_{olf}$-coupled OR in nanosome are shown in Figs. 7.6b and 7.7b, and a potential biorecognition efficiency of each configuration is schematized in Table 7.2.

Table 7.2 Biorecognition efficiency of various configurations (1, 2, 3) of G protein-coupled olfactory receptor in a nanosome (see Figs. 7.6b, 7.7b)

Configuration	**1**	**2**	**3**
Flexibility of N-ends in aromatic compounds, crucial for odorant binding	–	+	+
Accessibility of $G\alpha_{olf}$ to GTP-γ-S	–	–	+
Biorecognition efficiency	Low	Middle	High

Source: Adapted by permission from Springer Customer Service Centre GmbH: Springer *European Biophysics Journal* Ref. [34], Copyright 2008.

Two odorants, helional and heptanal, were tested on A1 and A2 biofilms in the concentration range from 10^{-12} to 10^{-5} M. Helional is documented as a cognate odorant for OR 17-40 [42]. SPR measurements were carried out in the differential mode, and

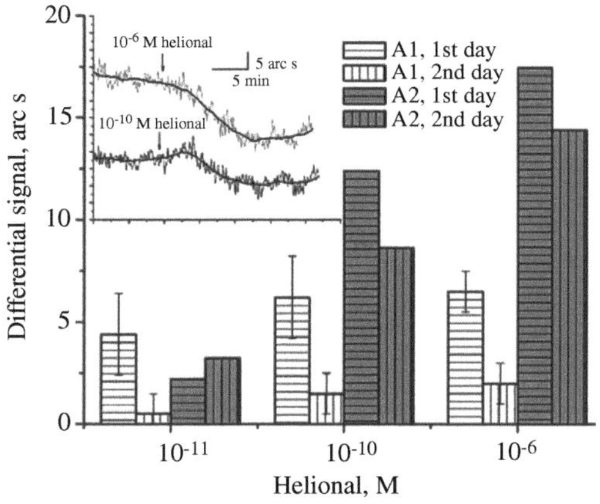

Figure 7.10 Dependences of SPR response for A1 and A2 biofilms on the helional concentration. The error values mentioned represent intersensor standard deviation (n = 2–3). Inset: typical kinetics of responses to helional obtained from the A2 biofilm in differential mode. Adapted by permission from Springer Customer Service Centre GmbH: Springer *European Biophysics Journal,* Ref. [34], Copyright 2008.

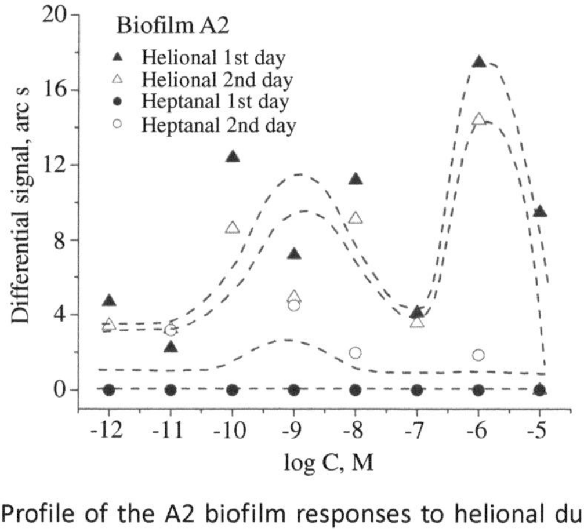

Figure 7.11 Profile of the A2 biofilm responses to helional during 2 days of work. Heptanal was used as a negative control. Data set was collected from the same sensor chip. Adapted by permission from Springer Customer Service Centre GmbH: Springer *European Biophysics Journal,* Ref. [34], Copyright 2008.

the registered kinetics of the SPR angular position change had a negative direction (toward the smaller angles of incidence of light) (Fig. 7.10, inset). Olfactory sensitivity of thicker A1 biofilm with comparatively dense nanosomes layer was weaker than that of the A2 (Fig. 7.10); therefore, further measurements of the response profile through the whole range of odorant concentrations were performed with A2 biofilm only. Sensitivity of receptors to helional after the overnight storage of SPR substrate coated with biofilm A1 decreased essentially, while A2 demonstrated only a slight relative decrease in response (Fig. 7.10). During 2 days of work with the same A2 biofilm, the same pattern of response to helional was observed (Fig. 7.11). Meanwhile, A2 sensitivity to the unrelated odorant heptanal remained insignificant somewhat increasing at 10^{-9} M after the overnight storage (Fig. 7.11).

7.3.3 Comparative Sensitivity Analysis for Different Types of Biofilm Architecture

As it can be seen from Table 7.1, the biofilm A2 based on the randomly oriented Abs was at least 27% thinner than the A1 biofilm. Low thickness of A2 biofilm and its high porosity (Fig. 7.9b) resulted in better accessibility of $G\alpha_{olf}$ protein to GTP-γ-S, which could explain a surprisingly better sensitivity of A2 to helional. The presence of pores in A2 initially originates from the random orientation of Abs and relates to their inability to bind a large amount of nanosomes. From this point of view, the thickness of both A1 and A2 nanosome layer is similar, whereas the filling ratio of the latter is smaller. The thickness of nanosome layer in both cases, close to twice a lipid bilayer thickness, demonstrates a flattening of the nanosomes down to partial collapsing.

The possible reasons of signal decrease after 12 h storage of SPR chip at room temperature are: (1) loss in receptor and/or $G\alpha_{olf}$ protein activity at room temperature, and (2) depletion of the available $G\alpha_{olf}$ protein pool. The α subunit of G protein has a molecular mass of ~40–50 kDa [43]; therefore, its desorption is reliably detected by the SPR technique. However, quite low amplitude of signals measured can be ascribed to the fewer number of receptors oriented in the direction allowing full access of $G\alpha_{olf}$ to GTP-γ-S. Indeed, the latter cannot penetrate the lipid bilayer to access $G\alpha_{olf}$ located inside the nanosome, whereas the hydrophobic odorants may penetrate the bilayer [44] to reach the receptor ligand binding pocket and activate the receptor. Another phenomenon that could contribute to the negative SPR angular shift was the intrinsic conformational change in activated OR. The latter is composed of a bundle of seven transmembrane α-helices connected by loops [45]. The odorant stimulation of OR seems to induce a rearrangement of helices leading to separation of transmembrane domains in the helix bundle [41]. Recently, it has been suggested that the signal changes observed by SPR can be also ascribed to the protein secondary structure changes. Thus, May and Russell [46] have correlated a decrease in SPR signal to the formation of β-sheets, turn or unordered protein secondary structures; compact helical structure was thought to possess higher refractive index and thus to increase an SPR signal. In this way, the OR conformational changes on SPR signals may also be taken into consideration.

Two maxima bell-shaped curves such as that found previously [36] can be superimposed to the experimental points (Fig. 7.11) to account for the functional response observed at the A2 biofilm stimulation by the odorant specific for OR 17-40. The exceptions are data for measurements at a concentration of 10^{-9} M with a rather low signal level, which require additional research. The presence of two maxima in the response profile can be explained by the presence of a significant number of specific binding sites with different receptor affinity in a receptor–odorant pair [36]. It should be noted that the detailed molecular mechanism of functional interactions between the odorant and the receptor is still the subject of research.

7.4 Characterization of Conformational Changes in Acrylamidophenylboronic Acid-Acrylamide Hydrogels upon Interaction with Glucose. Electrochemical Approach

7.4.1 Conformational Changes in Polymers as a Useful Signal for Sensor Development

Phase transitions of hydrogels and signal-triggered swelling processes of polymers are subjects of extensive research efforts [47,

48]. Volume changes of hydrogels are triggered by temperature [49], pH [50], solvent composition [51], gel composition [52], and light [53–55]. Several applications for reversible swellable polymers have been suggested such as their use as artificial muscles [56–58] or matrixes for the controlled release of substrates [59, 60].

Boronic acids, i.e., phenylboronic acid, act as a versatile ligand for the association of vicinal glycols, Eq. (7.1), and particularly sugars, e.g., glucose [61]. This property was used to develop various optical sensors [62, 63] or gel-sensitive sugar sensors [64]. Copolymers containing the boronic acid ligand were extensively examined, and the effects of added vicinal diols on the polymer properties were studied [65, 66]. The effect of glucose addition on the lower critical solution temperature of acrylamide boronic acid copolymers was reported [67]. The swollen copolymer consisting of poly(vinyl alcohol) and phenylboronic acid modified poly-*N*-vinyl-2-pyrrolidone formed upon addition of glucose was suggested as a sugar-triggered insulin-release membrane [68]. Phenylboronic acid copolymers were also employed as sensing matrixes for nucleotides [69] or sugars. The swelling of a phenylboronic acid hydrogel on a piezoelectric resonator was reported as a quantitative detection route for nucleotides. Similarly, glucose-responsive swelling of a boronic acid-functionalized poly(vinyl alcohol) membrane associated with an electrode was used to control the ion permeability through the membrane and the electrochemical transduction of glucose sensing [70].

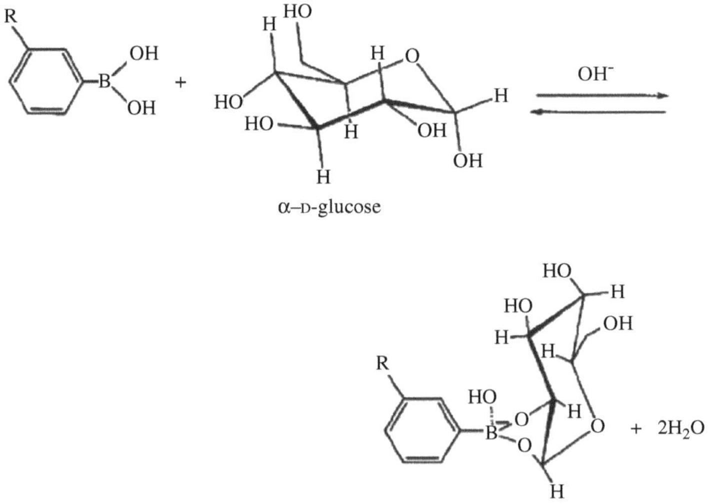

(7.1)

In this section, study on the electrochemical immobilization of a phenylboronic acid-functionalized acrylamide copolymer is considered. The glucose-responsive swelling of the polymer was characterized by electrochemistry (Faradaic impedance spectroscopy and chronopotentiometry), microgravimetric QCM measurements, and surface plasmon resonance analytes [71]. Besides the fundamental characterization of the polymer, the possibility of using the polymer as a sensing matrix for glucose was addressed, and the application of the various physical methods for analysis of the swelling processes of the polymer in the presence of glucose was described.

7.4.2 Electrochemical Formation of Acrylamidophenylboronic Acid-Acrylamide Hydrogel and Its SPR Characterization upon Interaction with Glucose

Acrylamide, *m*-acrylamidophenylboronic acid (**1**), and bisacrylamide as a crosslinker at a molar ratio of 45:5:1 were electropolymerized [72] on Au electrodes (Au wire, 45-nm-thick Au covered glass slide (SPR chip), and Au-quartz crystal in the electrochemical, SPR, and QCM experiments, respectively) in the presence of $ZnCl_2$, 0.2 M, as catalyst to yield the copolymer **2**. The electropolymerization was performed by potential cycling between 0.1 V and −1.4 V with the potential rest at −1.4 V for 20 s in each cycle (trapezoid sweep). The electrochemically induced polymerization of the acrylic monomers proceeds upon their reduction at −1.4 V, producing simultaneously metallic zinc at the electrode surface. When the potential returns to 0.1 V, the metallic zinc accompanying the polymer is electrochemically dissolved and the polymer's structure becomes more homogeneous than that of the polymer generated upon the electrolysis at a constant negative potential [72]. After the electropolymerization, the electrode was treated with 0.1 M HCl to ensure the complete dissolution of the metallic zinc. The elementary analysis of the copolymer **2** reveals the ratio 9:2 between acrylamide and *m*-acrylamidophenylboronic acid units. Thus, the copolymer is richer with the *m*-acrylamidophenylboronic acid units than the initial mixture of monomers by a factor of 2. The thickness of the polymer film was controlled by the number of the applied cycles.

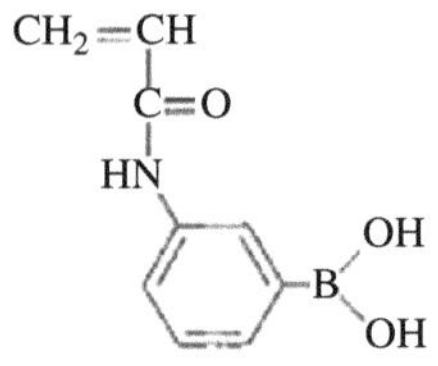

(1)

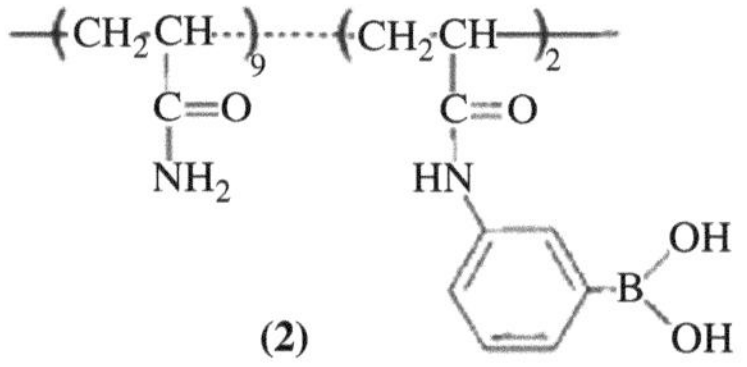

(2)

The SPR technique is extremely sensitive to the refractive index of layers present in the interfacial region at the gold–ambient boundary [73]. The technique has been employed widely to study processes in polymer and multilayer films [74–76], particularly to determine thin film thicknesses [77, 78]. Thus, SPR spectroscopy was applied in this study to characterize the copolymer **2** produced electrochemically.

Figure 7.12 shows the SPR spectra of the bare Au electrode in 0.05 M HEPES buffer, pH 7.4, curve a, after addition of the mixture of monomers containing $ZnCl_2$, curve b, and after the electropolymerization of the polymer film (five cycles) and removal of the Zn metal, in a 0.05 M HEPES buffer solution, pH 7.4, curve c. After the formation of the polymer film, the SPR curve minimum is shifted by ca. 9.7° in comparison to the bare Au surface. Theoretical fitting of the experimental data, curve c, indicates that the thickness of the polymer film corresponds to ca. 210 nm. Figure 7.12, curve d, shows the SPR spectrum of the Au surface covered with the copolymer **2** functionalized with the *m*-acrylamidophenylboronic acid units after addition of 25 mM glucose to the cell. The SPR spectrum, curve d, shows a distinct change in the shape and also a shift in the reflectance minimum, $\Delta\theta$ = 0.6°, consistent with the swelling process of the polymer. Theoretical fit of the experimental data, curve d, suggests that the thickness of the swollen polymer is ca. 320 nm. Thus, the addition of glucose, 25 mM, results in conformational changes in the copolymer molecule and swelling of the polymer film by 110 nm. The swelling of the polymer is fully reversible, and after the glucose solution is washed out and substituted with the background solution of 0.05 M HEPES buffer, pH 7.4, the polymer film shrinks to the original state, curve c.

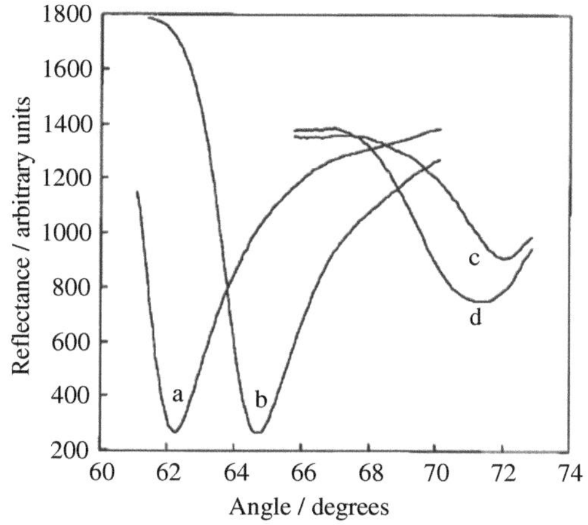

Figure 7.12 SPR spectra corresponding to the modification of the SPR chip Au surface with the functional copolymer **2** and to the effect of glucose-induced swelling of the copolymer: (a) a bare Au-coated glass slide; (b) a bare Au-coated glass slide in a solution composed of *m*-acrylamidophenylboronic acid, 0.2 M, acrylamide, 1.8 M, bisacrylamide, 0.04 M, and $ZnCl_2$, 0.2 M, prior to the electropolymerization; (c) the copolymer **2** produced on the SPR chip by five cycles of electropolymerization prior to the addition of glucose; (d) the copolymer **2** on the SPR chip after addition of 25 mM of glucose. Spectra a, c, and d were measured in 0.05 M HEPES buffer, pH 7.4. Adapted with permission from Ref. [71]. Copyright (2001) American Chemical Society.

Figure 7.13A shows the time-dependent changes in the reflectance minimum (sensogram) upon addition of the glucose solution, 25 mM, and Fig. 7.13B shows the respective sensogram recorded after substitution of the glucose solution with the background solution free of glucose. These two sensograms were measured in the open SPR cell upon equilibration of the polymer with the glucose solution and glucose-free buffer solution, respectively, and they represent the kinetic changes in the SPR spectra upon reversible swelling and shrinking of the polymer film, respectively.

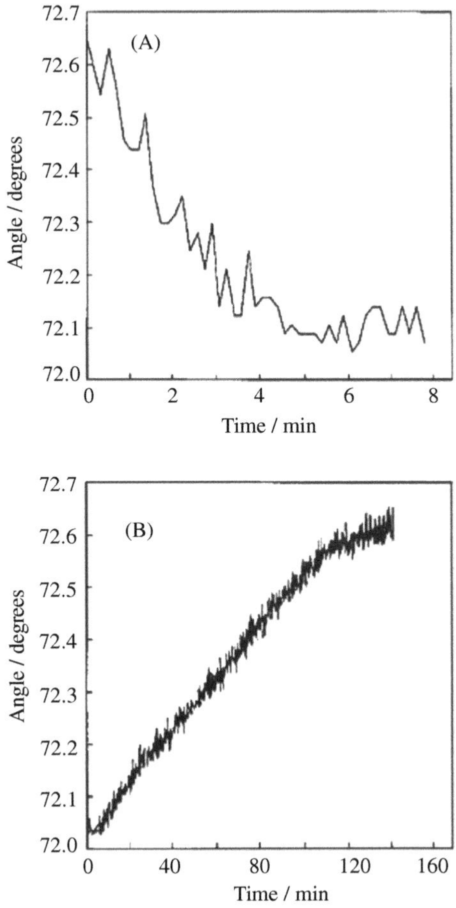

Figure 7.13 The time-dependent changes in the minimum reflectance angles (sensograms) of the polymer-functionalized SPR chips upon: (A) addition of the glucose solution, 25 mM; (B) substitution of the glucose solution with the background solution free of glucose. The SPR measurements were performed in the presence of 0.05 M HEPES buffer, pH 7.4. The copolymer **2** layer was assembled on the SPR chip by five cycles of electropolymerization. Adapted with permission from Ref. [71]. Copyright (2001) American Chemical Society.

Both these time-dependent SPR curves were fitted as the first-order kinetic processes, and the derived rate constants for the swelling, $k_{sw} = 1.7 \times 10^{-4}\ s^{-1}$, and shrinking, $k_{sh} = 2.3 \times 10^{-5}\ s^{-1}$, were calculated. It is evident that the swelling of the polymer layer is rapid, whereas its shrinking upon removal of the sugar is a slow process. Figure 7.14 shows the cyclic swelling of the polymer in the presence of glucose, 25 mM, and the shrinking of the polymer upon depletion of glucose with the buffer solution, respectively. It should be noted that the experiment described in Fig. 7.14 was performed under flow conditions in the SPR flow cell, and thus, the time ranges for the reversible shrinking–swelling processes are shorter than the time range depicted in Figs. 7.13A,B due to shear forces in the latter experiment.

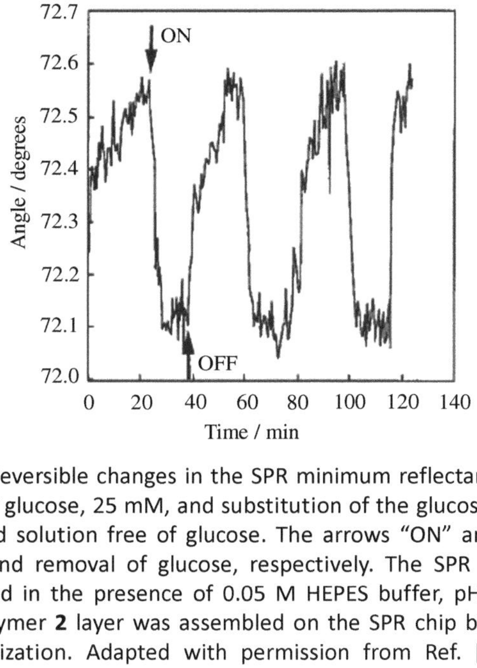

Figure 7.14 Reversible changes in the SPR minimum reflectance angle upon the addition of glucose, 25 mM, and substitution of the glucose solution with the background solution free of glucose. The arrows "ON" and "OFF" show the addition and removal of glucose, respectively. The SPR measurements were performed in the presence of 0.05 M HEPES buffer, pH 7.4, in a flow cell. The copolymer **2** layer was assembled on the SPR chip by five cycles of electropolymerization. Adapted with permission from Ref. [71]. Copyright (2001) American Chemical Society.

7.4.3 Electrochemical and QCM Characterization of Acrylamidophenylboronic Acid-Acrylamide Hydrogel upon Interaction with Glucose

The swelling of the polymer film upon addition of glucose and its shrinking as a result of the depletion of glucose are further supported by in situ constant-current chronopotentiometry/SPR measurements performed on the system. In chronopotentiometry experiments [79], a constant current is driven through the cell in the presence of a soluble redox probe, $K_3[Fe(CN)_6]/K_4[Fe(CN)_6]$ (1:1)-mixture, 10 mM, and the potential on the electrode alters as the resistance of the system changes during the electrochemical process. An overpotential η is thus required to retain the current I when the electrode resistance corresponds to R', Eq. (7.2) [80]:

$$\eta = IR' \tag{7.2}$$

Figure 7.15A shows the E-t transients observed at a copolymer-modified SPR chip in the presence of $K_3[Fe(CN)_6]/K_4[Fe(CN)_6]$ (1:1) redox mixture, 10 mM, upon the addition of glucose, 25 mM, to the SPR/electrochemical cell. While the copolymer-modified electrode reveals an overpotential of 86 mV as compared to a bare Au surface in the absence of glucose, curve a, the addition of glucose results in the time-dependent decrease in the overpotential required to reduce the redox mixture in solution, curves b and c, recorded after 5 and 8 min from the glucose addition, respectively. The overpotentials of 60 mV and 50 mV revealed after 5 and 8 min, respectively, can be translated to the electrode resistances, R', of 6.0 and 5.0 kΩ, respectively (Eq. (7.2)). Figure 7.15B shows the corresponding SPR curves recorded in situ at the same time intervals after the addition of glucose to the copolymer-functionalized surface layer. The film thickness increases by 60 and 100 nm upon addition of glucose after 5 and 8 min of interaction with glucose, curves b and c, respectively. The decrease in the electrode resistance as a result of the swelling of the hydrogel is attributed to the enhanced permeability of the polymer matrix to the redox label. Figure 7.15A, curves d and e, shows the time-dependent E-t chronopotentiometry transients recorded after the glucose solution was washed off and replaced with the redox probe solution free of glucose after 60 and 120 min, respectively. Figure 7.15B, curves d and e, shows the respective SPR spectra, recorded simultaneously with the chronopotentiometry transients after the glucose was washed off. Depletion of glucose from the polymer increases the overpotential for the electron transfer to the redox probe in the electrolyte and increases the electrode resistance (η = 70 and 78 mV after 60 and 120 min, respectively, translated to R' = 7.0 and 7.8 kΩ, respectively, according to Eq. (7.2)). The SPR spectra, Fig. 7.15B, curves d and e, clearly indicate that washing off of the glucose leads to the shrinkage of the polymer. By fitting the experimental SPR spectra, it was estimated that the glucose-swollen

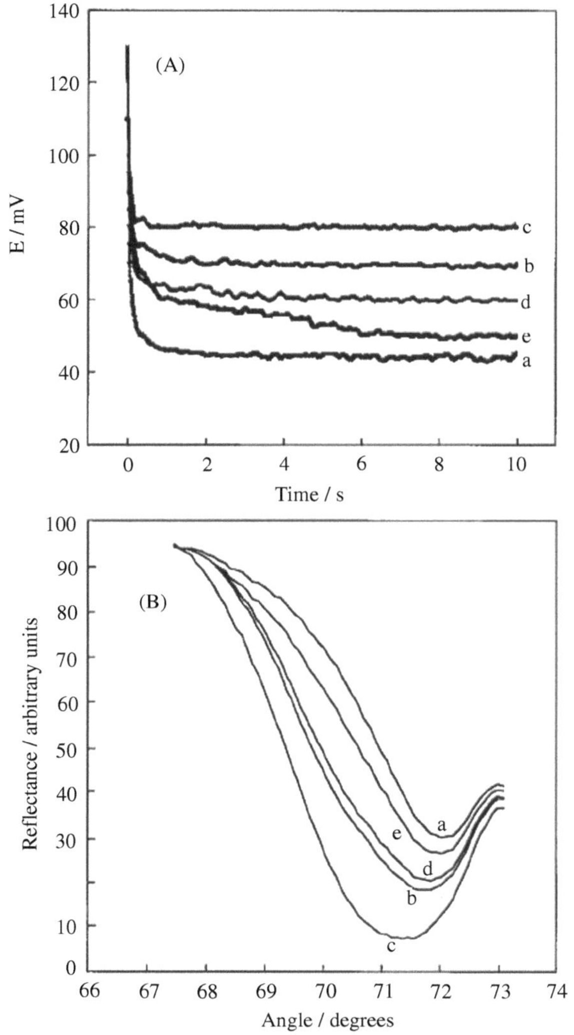

Figure 7.15 Swelling and shrinking processes induced in the copolymer **2** upon addition and depletion of glucose to and from the polymer film followed by chronopotentiometry and SPR spectroscopy: (A) Chronopotentiometric transients and (B) SPR spectra corresponding to (a) the copolymer-functionalized Au electrode prior to the addition of glucose, the functionalized electrode after interaction with 25 mM glucose for (b) 5 min and (c) 8 min, and after removal of glucose for (d) 60 min and (e) 120 min. The copolymer **2** film was generated by 60 cycles of electropolymerization. A constant current of 1 μA was applied during the chronopotentiometric measurements. Adapted with permission from Ref. [71]. Copyright (2001) American Chemical Society.

polymer is shrunken by ca. 60 nm and ca. 90 nm upon depletion of glucose, respectively. Thus, the increase in the overpotential associated with the redox process of $Fe(CN)_6^{3-/4-}$ is attributed to the shrinkage of the polymer, a process that decreases the polymer film porosity and thus perturbs the diffusion of the redox label through the polymer.

The influence of the swelling and shrinking processes of the copolymer **2** containing *m*-acrylamidophenylboronic acid units upon addition of glucose to the copolymer, and the depletion of sugar from the copolymer, respectively, is also characterized by Faradaic impedance spectroscopy [81, 82]. The impedance features of the electrodes Z_{re} and Z_{im} are recorded as a function of the voltage frequency. The plot of Z_{im} versus Z_{re}, observed at different frequencies (Nyquist plot), leads to the Faradaic impedance spectrum that for a modified electrode usually includes a semicircle part (for high frequencies) followed by a linear part characteristic of lower frequencies, and corresponding to diffusion-controlled electron transfer [81, 82]. The diameter of the semicircle domain corresponds to the electron-transfer resistance, R_{et}, for the reduction/oxidation of the redox label at the electrode interface. It should be noted that the electron-transfer resistance, R_{et}, derived from the Faradaic impedance spectroscopy and the electrode resistance, R', derived from chronopotentiometry are quite similar, provided the measurements are performed under carefully selected conditions [80].

An Au-wire electrode was modified with the copolymer **2** under the conditions similar to those used in the SPR and chronopotentiometry (60 cycles of electropolymerization). Figure 7.16A shows the Faradaic impedance spectra of the copolymer-**2**-functionalized electrode measured in the presence of a redox probe, $[Fe(CN)_6]^{3-}/[Fe(CN)_6]^{4-}$ (1:1) 10 mM in 0.05 M HEPES buffer, pH 7.4, prior to the addition of glucose, curve a, and after the addition of glucose, 25 mM, curve b. The polymer-modified Au-wire electrode reveals an interfacial electron-transfer resistance of R_{et} = 138 kΩ prior to glucose addition, and the addition of glucose, 25 mM, decreases the interfacial electron-transfer resistance to R_{et} = 93 kΩ. The decrease in the interfacial electron-transfer resistance is attributed to the swelling of the polymer, a process that facilitates the electron transfer between the redox label and the conductive support. Figure 7.16A, curve c, shows the Faradaic impedance spectrum of the electrode after the depletion of glucose from the polymer. The electron-transfer resistance at the electrode increases to the original value prior to the addition of glucose, implying that the polymer underwent a shrinking process that perturbs the interfacial electron-transfer process. The extent of swelling of the polymer upon addition of glucose and the secondary shrinking of the polymer upon washing off of the boronate-linked glucose is controlled by two factors: (i) the initial thickness of the polymer matrix and (ii) the content of glucose added to the polymer. The thickness of the polymer is dominated by the time of electrolysis.

Figure 7.16B shows the SPR spectra of two copolymers **2** generated upon five cycles (curve a) and three cycles (curve d). These spectra were measured prior to glucose addition, and the calculated thicknesses of these polymers corresponded to 210 and 130 nm, respectively. The addition of glucose, 25 mM, to the polymer-functionalized electrodes results in an increase in the polymer thicknesses to 320 and 200 nm, curves b and e, respectively. The subsequent depletion of glucose from the polymers restores the original shrunken polymers, curves c and f, respectively. Figure 7.16A shows the Faradaic impedance spectra measured for two polymer films generated on Au-wire electrodes upon application of 60 and 30 cycles (curves a and d, respectively). The electron-transfer resistance values, R_{et}, extracted from the spectra are equal to 138 and 54 kΩ, respectively. The addition of 25 mM glucose resulted in a decrease in the electron-transfer resistance, R_{et}, for both polymer films to 93 and 23 kΩ, curves b and e, respectively. After the glucose solution was washed off, the impedance spectra of both polymer-modified electrodes returned to the initial values of R_{et}, curves c and f, respectively. It is evident from the impedance spectroscopy and SPR measurements (Fig. 7.16A and Fig.7.16B, respectively) that the extent of the polymer shrinking–swelling upon addition–depletion of glucose is almost proportional to the polymer film thickness.

The swelling of the copolymer **2** functionalized with *m*-acrylamidophenylboronic acid units in the presence of various concentrations of glucose is further supported by microgravimetric QCM measurements. An Au-quartz crystal was modified with the copolymer **2** by the electropolymerization under conditions

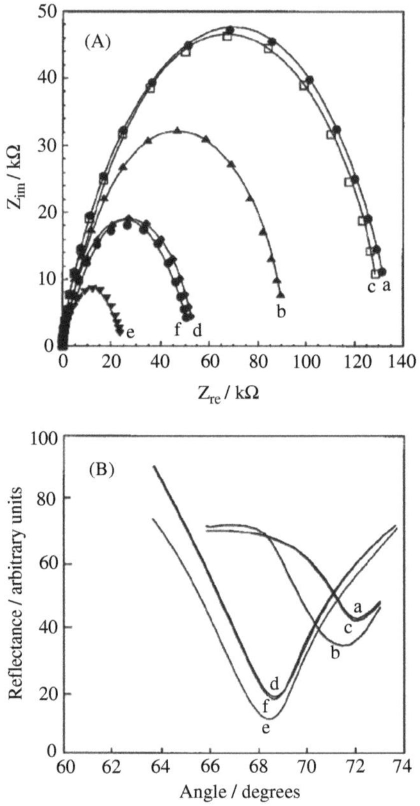

Figure 7.16 Swelling and shrinking processes induced on the copolymer **2** films of different thicknesses upon addition and removal of glucose solution: (A) Faradaic impedance spectra corresponding to (a) copolymer **2** generated by 60 cycles of electropolymerization prior to the addition of glucose; (b) the copolymer film after addition of 25 mM glucose; (c) the copolymer film after removal of glucose; (d) copolymer **2** produced by 30 cycles of electropolymerization prior to the addition of glucose; (e) the copolymer film after addition of 25 mM glucose; (f) the copolymer film after removal of glucose. The impedance measurements were performed at an Au-wire electrode in the presence of 0.05 M HEPES buffer, pH 7.4, containing $K_3[Fe(CN)_6]/K_4[Fe(CN)_6]$ (1:1)-mixture, 10 mM upon application of the biasing potential of 0.175 V. (B) SPR spectra corresponding to (a) copolymer **2** produced by five cycles of electropolymerization prior to the addition of glucose; (b) the copolymer film after addition of 25 mM glucose; (c) the copolymer film after removal of glucose; (d) copolymer **2** produced by three cycles of electropolymerization prior to the glucose addition; (e) the copolymer film after addition of 25 mM glucose; (f) the copolymer film after removal of glucose. The SPR measurements were performed in the presence of 0.05 M HEPES buffer, pH 7.4. Adapted with permission from Ref. [71]. Copyright (2001) American Chemical Society.

similar to those described for all other measurements (five cycles of electropolymerization). Figure 7.17, curve a, shows the frequency changes in the copolymer-**2**-modified Au-quartz crystal upon interaction with different concentrations of glucose. Prior to the treatment of the copolymer **2** with the different glucose concentrations, the copolymer was washed to exclude previously bound glucose. It is evident that in the presence of glucose, the crystal frequency decreases, indicating an increase in the mass associated with the crystal. The changes in the crystal frequency increase as the concentrations of glucose are elevated. Also, it should be noted that upon washing of the polymer, the crystal retains resonance frequency, which is almost similar to the value prior to the interaction with glucose. Control experiments reveal that an Au-quartz crystal that includes a pure acrylamide polymer film that does not include *m*-acrylamidophenylboronic acid units is unaffected upon the addition of glucose within the entire concentration range of glucose. Thus, the addition of glucose results in the swelling of the polymer film, and the extent of the swelling is increased upon the increase in the glucose concentration. The changes in the mass associated with the crystal may be roughly approximated by the Sauerbrey equation, Eq. (7.3), where Δf is the measured frequency shift, f_0 is the frequency of the quartz crystal prior to the mass change, Δm is the mass change, A is the piezoelectrically active area, ρ_q is the density of quartz (2.648 g/cm^3), and μ_q is the shear modulus (2.947 × 10^{11} dynes/cm^2 for AT-cut quartz):

$$\Delta f = -2f_0^2\left[\Delta m/A(\mu_q\rho_q)^{1/2}\right]. \tag{7.3}$$

For example, at a glucose concentration of 25 mM, the swelling process yields a frequency change of $\Delta f = -650$ Hz, which translates to a mass change of 2.5 × 10^{-6} g. Realizing that most of the mass change originates from the absorption of water (*vide infra*), and taking into account the area of the electrode, the thickness change upon swelling corresponds to 125 nm, a value that is in good agreement with the value derived from the surface plasmon resonance experiments. Figure 7.17, curve b, shows the changes in the thickness of the polymer as a result of swelling at different glucose concentrations as derived from the SPR experiments. The two calibration plots (curve a derived from QCM experiments and curve b derived from the SPR studies, respectively) are, as expected, very similar.

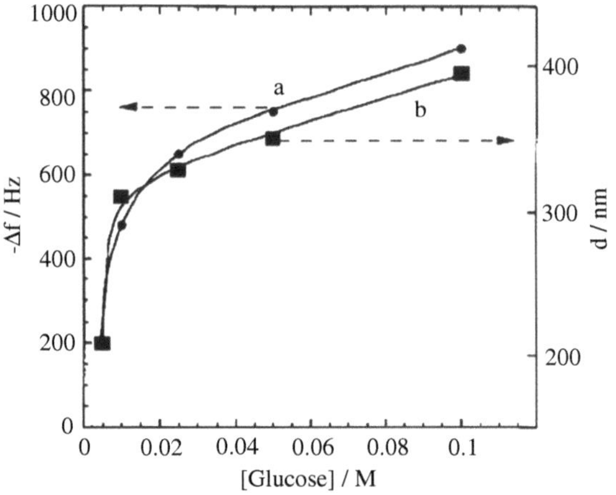

Figure 7.17 Calibration curves of the frequency changes Δf (curve a) and the film thickness d (curve b) extracted from QCM and SPR measurements, respectively, upon interaction of the copolymer **2** film with different concentrations of glucose. The QCM measurements were performed in air after the interaction of the functionalized crystal with the glucose solution of the corresponding concentration. The SPR measurements were performed in the presence of 0.05 M HEPES buffer, pH 7.4. The copolymer **2** was produced by five cycles of electropolymerization. Adapted with permission from Ref. [71]. Copyright (2001) American Chemical Society.

7.5 Conclusion

Specific studies described in this chapter demonstrate the possibilities of SPR method for the characterization and monitoring of conformational changes in different molecular systems. These include the characterization of interaction between macromolecules of different nature (polymers, proteins, receptors, etc.) with both small and large ligand molecules (odorants, sugars, proteins, etc.), allowing elucidation of the changes in geometrical and optical parameters of such systems. It has been shown that the SPR technique is a highly effective method for monitoring conformational changes in gel-like or soft polymeric materials due to the possibility of easy registration of changes in the refractive index during the measurement of the investigated object in real-time mode.

In one of the studies, it was shown that the structural changes in BSA/PAA complex can be monitored by SPR spectroscopy in real-time mode. The dynamics of conformational changes in the protein–polymer complex expressed the strong dependence on pH of liquid environment and proportion of components of PCs. Soluble PCs corresponded with high pH and a swelled structure, whereas insoluble PCs were responsible for more compact, shrunk complexes, which showed that at low pH, the ability of aggregation and sedimentation depended on the proportion of components. The rate constants of formation of insoluble PCs at high pH appeared higher than the ones for the transformation into soluble transparent structure at low pH. The fluorescence spectroscopy should be considered a useful additional method for the verification of results of SPR experiments. The combination of these two methods appeared to be a promising approach for the investigation of protein–polymer complexes and even for the development of synthetic vaccines.

The binding of odorant molecules to artificially created sensitive structures has been investigated using SPR and complementary methods in another study. Two different architectures of biological films carrying OR 17-40 receptors have been developed, which differ in the order and orientation of the biomolecules in relation to the gold film, carrier of the surface plasmon waves. The low thickness of the disordered biofilm and its high porosity lead to an improvement in the availability of the Gα subunit for the GTP-γ-S protein and can explain the better sensitivity of the A2 biofilm to the helional odorant. It has been demonstrated that the density of adsorption of nanosomes and the multilayer bulk thickness are crucial points in the design of the olfactory biofilms for the SPR-based screening of odorants. Sensitivity of the OR 17-40–bearing film to helional was specific and could be ascribed to the two main molecular events inducing the SPR signal change: desorption of Gα protein from membrane bilayer and intrinsic conformational changes in activated OR 17-40 receptor itself.

The last considered study has applied a battery of physical methods to characterize the conformational changes in a glucose-responsive copolymer, which are exhibited as its swelling and shrinking. The study has revealed different methods to characterize the structure of the hydrogels (thickness, water content) and the kinetics of glucose-induced swelling and polymer shrinking processes. The characterization of the swelling and shrinking processes of the boronic acid-acrylamide copolymer has important consequences on the future application of such polymers as glucose-sensing matrixes or as glucose-activated release matrixes. Besides the detailed analyses of the glucose-induced swelling and shrinking processes of the boronic acid copolymer by the different techniques, it was suggested that similar methods may be applied to characterize stimuli-triggered phase transitions and structural changes in other polymers.

References

1. Tsuchida E. and Abe K., Eds. (1982) *Interactions Between Macromolecules in Solution and Intermacromolecular Complexes. Advances in Polymer Science* **45** (Springer, Berlin, Heidelberg).
2. Chen, H. L. and Morawetz, H. (1983). Fluorometric study of the equilibrium and kinetics of poly (acrylic acid) association with polyoxyethylene or poly (vinyl pyrrolidone), *Eur. Polym. J.*, **19**, pp. 923–928.
3. Boussaad, S., Pean, J., and Tao, N. J. (2000). High-resolution multiwavelength surface plasmon resonance spectroscopy for probing conformational and electronic changes in redox proteins, *Anal. Chem.*, **72**, pp. 222–226.
4. Jachimska, B., Tokarczyk, K., Łapczyńska, M., Puciul-Malinowska, A., and Zapotoczny, S. (2016). Structure of bovine serum albumin adsorbed on silica investigated by quartz crystal microbalance, *Colloids Surf. A*, **489**, pp. 163–172.
5. Mervinetsky, E., Alshanski, I., Hamo, Y., Sandonas, L. M., Dianat, A., Buchwald, J., Gutierrez, R., Cuniberti, G., Hurevich, M., and Yitzchaik, S. (2017). Copper induced conformational changes of tripeptide monolayer based impedimetric biosensor, *Sci. Rep.*, **7**, 9498.
6. Ngo, K. X., Kodera, N., Katayama, E., Ando, T., and Uyeda, T. Q. (2015). Cofilin-induced unidirectional cooperative conformational changes in actin filaments revealed by high-speed atomic force microscopy, *eLife*, **4**, e04806.
7. Choi, B. Y., Kahng, S. J., Kim, S., Kim, H., Kim, H. W., Song, Y. J., Ihm, J., and Kuk, Y. (2006). Conformational molecular switch of the azobenzene molecule: A scanning tunneling microscopy study, *Phys. Rev. Lett.*, **96**, 156106.

8. Saibil, H. R. (2000). Conformational changes studied by cryo-electron microscopy, *Nat. Struct. Mol. Biol.*, **7**, pp. 711–714.
9. Mukherjee, S., Griffin, D. H., Horn, J. R., Rizk, S. S., Nocula-Lugowska, M., Malmqvist, M., Kim, S. S., and Kossiakoff, A. A. (2018). Engineered synthetic antibodies as probes to quantify the energetic contributions of ligand binding to conformational changes in proteins, *J. Biol. Chem.*, **293**, pp. 2815–2828.
10. Wei, M., Gao, Y., Li, X., and Serpe, M. J. (2017). Stimuli-responsive polymers and their applications, *Polym. Chem.*, **8**, pp. 127–143.
11. Brennan-Fournet, M. E., Huerta, M., Zhang, Y., Malliaras, G., and Owens, R. M. (2015). Detection of fibronectin conformational changes in the extracellular matrix of live cells using plasmonic nanoplates, *J. Mater. Chem. B*, **3**, pp. 9140–9147.
12. Petrov, R. V., Mustafaev, M. I., and Norimov, A. S. (1992). Physicochemical criteria for the construction of artificial immunomodulators and immunogens on the basis of polyelectrolyte complexes, *Sov. Med. Rev. D. Immunol. UK: Harwood Academic Publishers GmbH*, pp. 1–113.
13. Mustafaev, M., Cìrakoğlu, B., Sarac, S., Öztürk, S., Yücel, F., and Bermek, E. (1996). Soluble and insoluble ternary complexes of serum proteins with polyanions in the presence of Cu^{2+} in water, *J. Appl. Polym. Sci.*, **62**, pp. 99–109.
14. Laguecir, A., Ulrich, S., Labille, J., Fatin-Rouge, N., Stoll, S., and Buffle, J. (2006). Size and pH effect on electrical and conformational behavior of poly (acrylic acid): Simulation and experiment, *Eur. Polym. J.*, **42**, pp. 1135–1144.
15. Wang, P., Matsukawa, S., Kameda, T., Kurosu, H., and Ando, I. (1997). Conformational behaviour of poly (aspartic acid) and its sodium salt in the solid state as studied by high-resolution solid-state ^{13}C-NMR spectroscopy, *J. Mol. Struct.*, **435**, pp. 229–234.
16. Wang, J. and Somasundaran, P. (2006). Reversible conformational behavior of poly (acrylic acid) LB film with changes in pH, ionic strength and time, *Colloids Surf. A*, **273**, pp. 63–69.
17. Koenig, J. L., Angood, A. C., Semen, J., and Lando, J. B. (1969). Laser-excited Raman studies of the conformational transition of syndiotactic polymethacrylic acid in water, *J. Am. Chem. Soc.*, **91**, pp. 7250–7254.
18. Lando, J. B., Koenig, J. L., and Semen, J. (1973). Conformational studies of poly (methacrylic Acid). II. Laser-excited Raman studies of the conformational transition in aqueous solution, *J. Macromol. Sci. B*, **7**, pp. 319–343.

19. Ishiduki, K. and Esumi, K. (1997). The effect of pH on adsorption of poly (acrylic acid) and poly (vinylpyrrolidone) on alumina from their binary mixtures, *Langmuir*, **13**, pp. 1587–1591.

20. Fujiwara, M., Grubbs, R. H., and Baldeschwieler, J. D. (1997). Characterization of pH-dependent poly (acrylic acid) complexation with phospholipid vesicles, *J. Colloid Interface Sci.*, **185**, pp. 210–216.

21. Filenko, A., Demchenko, M., Mustafaeva, Z., Osada, Y., and Mustafaev, M. (2001). Fluorescence study of Cu^{2+}-induced interaction between albumin and anionic polyelectrolytes, *Biomacromolecules*, **2**, pp. 270–277.

22. Akkiliç, N., Mustafaeva, Z., and Mustafaev, M. (2007). High performance liquid chromatography study of water-soluble complexes and covalent conjugates of polyacrylic acid with bovine serum albumin, *J. Appl. Polym. Sci.*, **105**, pp. 3108–3120.

23. Chegel, V., Shirshov, Y., Avilov, S., Demchenko, M., and Mustafaev, M. (2002). A novel aldehyde dextran sulfonate matrix for affinity biosensors, *J. Biochem. Biophys. Methods*, **50**, pp. 201–216.

24. Akkilic, N., Mustafaev, M., and Chegel, V. (2008). Conformational dynamics of poly (acrylic acid)-bovine serum albumin polycomplexes at different pH conditions, *Macromol. Symp.*, **269**, pp. 138–144.

25. Burstein, E. A. (1968). Quenching of protein fluorescence. I. Principles of the method. Solutions of tryptophan, tyrosine and denatured proteins, *Biophysics (Moscow)*, **13**, pp. 433–442 (in Russian).

26. Lakowich, J. R. (1986). *Principles of Fluorescence Spectroscopy* (Plenum Press, New York and London).

27. Neurath, H. and Bailey, K., Eds. (1953) *The Proteins* (Academic Press Inc., New York).

28. Tollin, G., Salamon, Z., and Hruby, V. J. (2003). Techniques: Plasmon-waveguide resonance (PWR) spectroscopy as a tool to study ligand–GPCR interactions, *Trends Pharmacol. Sci.*, **24**, pp. 655–659.

29. Liu, Q., Cai, H., Xu, Y., Li, Y., Li, R., and Wang, P. (2006). Olfactory cell-based biosensor: A first step towards a neurochip of bioelectronic nose, *Biosens. Bioelectron.*, **22**, pp. 318–322.

30. Li, G., Ferrie, A. M., and Fang, Y. (2006). Label-free profiling of ligands for endogenous GPCRs using a cell-based high-throughput screening technology, *J. Lab. Autom.*, **11**, pp. 181–187.

31. Ko, H. J. and Park, T. H. (2005). Piezoelectric olfactory biosensor: Ligand specificity and dose-dependence of an olfactory receptor expressed in a heterologous cell system, *Biosens. Bioelectron.*, **20**, pp. 1327–1332.

32. Araneda, R. C., Peterlin, Z., Zhang, X., Chesler, A., and Firestein, S. (2004). A pharmacological profile of the aldehyde receptor repertoire in rat olfactory epithelium, *J. Physiol. (Lond.)*, **555**, pp. 743–756.

33. Araneda, R. C., Kini, A. D., and Firestein, S. (2000). The molecular receptive range of an odorant receptor, *Nat. Neurosci.*, **3**, pp. 1248–1255.

34. Benilova, I., Chegel, V. I., Ushenin, Y. V., Vidic, J., Soldatkin, A. P., Martelet, C., Pajot, E., and Jaffrezic-Renault, N. (2008). Stimulation of human olfactory receptor 17-40 with odorants probed by surface plasmon resonance, *Eur. Biophys. J.*, **37**, pp. 807–814.

35. Spinke, J., Liley, M., Guder, H. J., Angermaier, L., and Knoll, W. (1993). Molecular recognition at self-assembled monolayers: The construction of multicomponent multilayers, *Langmuir*, **9**, pp. 1821–1825.

36. Vidic, J. M., Grosclaude, J., Persuy, M. A., Aioun, J., Salesse, R., and Pajot-Augy, E. (2006). Quantitative assessment of olfactory receptors activity in immobilized nanosomes: A novel concept for bioelectronic nose, *Lab Chip*, **6**, pp. 1026–1032.

37. Vidic, J., Pla-Roca, M., Grosclaude, J., Persuy, M. A., Monnerie, R., Caballero, D., Errachid, A., Hou, Y., Jaffrezic-Renault, N., Salesse, R., Pajot-Augy, E., and Samitier, J. (2007). Gold surface functionalization and patterning for specific immobilization of olfactory receptors carried by nanosomes, *Anal. Chem.*, **79**, pp. 3280–3290.

38. Beketov, G. V., Shirshov, Y. M., Shynkarenko, O. V., and Chegel, V. I. (1998). Surface plasmon resonance spectroscopy: Prospects of superstrate refractive index variation for separate extraction of molecular layer parameters, *Sens. Actuators B Chem.*, **48**, pp. 432–438.

39. Johnsen, S. and Widder, E. A. (1999). The physical basis of transparency in biological tissue: Ultrastructure and the minimization of light scattering, *J. Theor. Biol.*, **199**, pp. 181–198.

40. Dannenberger, O., Weiss, K., Himmel, H. J., Jäger, B., Buck, M., and Wöll, C. (1997). An orientation analysis of differently endgroup-functionalised alkanethiols adsorbed on Au substrates, *Thin Solid Films*, **307**, pp. 183–191.

41. Bourne, H. R. (1997). How receptors talk to trimeric G proteins, *Curr. Opin. Cell Biol.*, **9**, pp. 134–142.

42. Wetzel, C. H., Oles, M., Wellerdieck, C., Kuczkowiak, M., Gisselmann, G., and Hatt, H. (1999). Specificity and sensitivity of a human olfactory receptor functionally expressed in human embryonic kidney 293 cells and *Xenopus Laevis* oocytes, *J. Neurosci.*, **19**, pp. 7426–7433.

43. Wilcox, M. D., Schey, K. L., Dingus, J., Mehta, N. D., Tatum, B. S., Halushka, M., Finch, J. W., and Hildebrandt, J. D. (1994). Analysis of G protein gamma subunit heterogeneity using mass spectrometry, *J. Biol. Chem.*, **269**, pp. 12508–12513.

44. Örnskov, E., Gottfries, J., Erickson, M., and Folestad, S. (2005). Experimental modelling of drug membrane permeability by capillary electrophoresis using liposomes, micelles and microemulsions, *J. Pharm. Pharmacol.*, **57**, pp. 435–442.

45. Pearce, T. C., Verschure, P. F., White, J., and Kauer, J. S. (2001). Stimulus encoding during the early stages of olfactory processing: A modeling study using an artificial olfactory system, *Neurocomputing*, **38**, pp. 299–306.

46. May, L. M. and Russell, D. A. (2002). The characterization of biomolecular secondary structures by surface plasmon resonance, *Analyst*, **127**, pp. 1589–1595.

47. Schild, H. G. (1992). Poly (*N*-isopropylacrylamide): Experiment, theory and application, *Prog. Polym. Sci.*, **17**, pp. 163–249.

48. Dušek, K. and Patterson, D. (1968). Transition in swollen polymer networks induced by intramolecular condensation, *J. Polym. Sci. A-2*, **6**, pp. 1209–1216.

49. Suzuki, A. and Tanaka, T. (1990). Phase transition in polymer gels induced by visible light, *Nature*, **346**, pp. 345–347.

50. Annaka, M. and Tanaka, T. (1992). Multiple phases of polymer gels, *Nature*, **355**, pp. 430–432.

51. Hirokawa, Y. and Tanaka, T. (1984). Volume phase transition in a non-ionic gel, *AIP Conf. Proc.*, **107**, pp. 203–208.

52. Hirotsu, S., Hirokawa, Y., and Tanaka, T. (1987). Volume-phase transitions of ionized N-isopropylacrylamide gels, *J. Chem. Phys.*, **87**, pp. 1392–1395.

53. Smith, T. A., Hotta, J. I., Sasaki, K., Masuhara, H., and Itoh, Y. (1999). Photon pressure-induced association of nanometer-sized polymer chains in solution, *J. Phys. Chem. B*, **103**, pp. 1660–1663.

54. Hofkens, J., Hotta, J., Sasaki, K., Masuhara, H., and Iwai, K. (1997). Molecular assembling by the radiation pressure of a focused laser beam: Poly (*N*-isopropylacrylamide) in aqueous solution, *Langmuir*, **13**, pp. 414–419.

55. Juodkazis, S., Mukai, N., Wakaki, R., Yamaguchi, A., Matsuo, S., and Misawa, H. (2000). Reversible phase transitions in polymer gels induced by radiation forces, *Nature*, **408**, pp. 178–181.

56. Otero, T. F. and Sansinena, J. M. (1995). Artificial muscles based on conducting polymers, *Bioelectrochem. Bioenerg.*, **38**, pp. 411–414.
57. Otero, T. F., Grande, H., and Rodriguez, J. (1996). Reversible electrochemical reactions in conducting polymers: A molecular approach to artificial muscles, *J. Phys. Org. Chem.*, **9**, pp. 381–386.
58. Okuzaki, H. and Funasaka, K. (2000). Electromechanical properties of a humido-sensitive conducting polymer film, *Macromolecules*, **33**, pp. 8307–8311.
59. Hoffman, A. S. (1987). Applications of thermally reversible polymers and hydrogels in therapeutics and diagnostics, *J. Control. Release*, **6**, pp. 297–305.
60. Willner, I., Rubin, S., Shatzmiller, R., and Zor, T. (1993). Reversible light-stimulated activation and deactivation of α-chymotrypsin by its immobilization in photoisomerizable copolymers, *J. Am. Chem. Soc.*, **115**, pp. 8690–8694.
61. James, T. D., Sandanayake, K. S., and Shinkai, S. (1996). Saccharide sensing with molecular receptors based on boronic acid, *Angew. Chem. Int. Ed. Engl.*, **35**, pp. 1910–1922.
62. James, T. D., Sandanayake, K. S., and Shinkai, S. (1995). Chiral discrimination of monosaccharides using a fluorescent molecular sensor, *Nature*, **374**, pp. 345–347.
63. Nakashima, K., Iguchi, R., and Shinkai, S. (2000). Diaza-18-crown-6-based saccharide receptor bearing two boronic acids. Possible communication between bound saccharides and metal cations, *Ind. Eng. Chem. Res.*, **39**, pp. 3479–3483.
64. James, T. D., Murata, K., Harada, T., Ueda, K., and Shinkai, S. (1994). Chiral discrimination of monosaccharides through gel formation, *Chem. Lett.*, **23**, pp. 273–276.
65. Kitano, S., Hisamitsu, I., Koyama, Y., Kataoka, K., Okano, T., and Sakurai, Y. (1991). Effect of the incorporation of amino groups in a glucose-responsive polymer complex having phenylboronic acid moieties, *Polym. Adv. Technol.*, **2**, pp. 261–264.
66. Miyazaki, H., Kikuchi, A., Koyama, Y., Okano, T., Sakurai, Y., and Kataoka, K. (1993). Boronate-containing polymer as novel mitogen for lymphocytes, *Biochem. Biophys. Res. Commun.*, **195**, pp. 829–836.
67. Kataoka, K., Miyazaki, H., Okano, T., and Sakurai, Y. (1994). Sensitive glucose-induced change of the lower critical solution temperature of poly [*N*, *N*-(dimethylacrylamide)-co-3-(acrylamido)-phenylboronic acid] in physiological saline, *Macromolecules*, **27**, pp. 1061–1062.

68. Kitano, S., Koyama, Y., Kataoka, K., Okano, T., and Sakurai, Y. (1992). A novel drug delivery system utilizing a glucose responsive polymer complex between poly (vinyl alcohol) and poly (*N*-vinyl-2-pyrrolidone) with a phenylboronic acid moiety, *J. Control. Release*, **19**, pp. 161–170.
69. Kanekiyo, Y., Sano, M., Iguchi, R., and Shinkai, S. (2000). Novel nucleotide-responsive hydrogels designed from copolymers of boronic acid and cationic units and their applications as a QCM resonator system to nucleotide sensing, *J. Polym. Sci. A*, **38**, pp. 1302–1310.
70. Kikuchi, A., Suzuki, K., Okabayashi, O., Hoshino, H., Kataoka, K., Sakurai, Y., and Okano, T. (1996). Glucose-sensing electrode coated with polymer complex gel containing phenylboronic acid, *Anal. Chem.*, **68**, pp. 823–828.
71. Gabai, R., Sallacan, N., Chegel, V., Bourenko, T., Katz, E., and Willner, I. (2001). Characterization of the swelling of acrylamidophenylboronic acid–acrylamide hydrogels upon interaction with glucose by faradaic impedance spectroscopy, chronopotentiometry, quartz-crystal microbalance (QCM), and surface plasmon resonance (SPR) experiments, *J. Phys. Chem. B*, **105**, pp. 8196–8202.
72. Katz, E., de Lacey, A. L., Fierro, J. L., Palacios, J. M., and Fernandez, V. M. (1993). Covalent binding of viologen to electrode surfaces coated with poly (acrylic acid) formed by electropolymerization of acrylate ions: I. Electrode preparation and characterization, *J. Electroanal. Chem.*, **358**, pp. 247–259.
73. Raether, H. (1988). *Surface Plasmons on Smooth and Rough Surfaces and on Gratings, Springer Tracts in Modern Physics* **111** (Springer-Verlag, Berlin Heidelberg).
74. Jordan, C. E. and Corn, R. M. (1997). Surface plasmon resonance imaging measurements of electrostatic biopolymer adsorption onto chemically modified gold surfaces, *Anal. Chem.*, **69**, pp. 1449–1456.
75. Drake, P. A. and Bohn, P. W. (1995). Anomalous diffusion. Surface plasmon resonance measurements as probes of nanometer-scale film-swelling dynamics for CH_3OH in poly (methyl methacrylate), *Anal. Chem.*, **67**, pp. 1766–1771.
76. Chen, X., Shakesheff, K. M., Davies, M. C., Heller, J., Roberts, C. J., Tendler, S. J. B., and Williams, P. M. (1995). Degradation of a thin polymer film studied by simultaneous in situ atomic force microscopy and surface plasmon resonance analysis, *J. Phys. Chem.*, **99**, pp. 11537–11542.
77. Peterlinz, K. A. and Georgiadis, R. (1996). In situ kinetics of self-assembly by surface plasmon resonance spectroscopy, *Langmuir*, **12**, pp. 4731–4740.

78. Jung, L. S., Campbell, C. T., Chinowsky, T. M., Mar, M. N., and Yee, S. S. (1998). Quantitative interpretation of the response of surface plasmon resonance sensors to adsorbed films, *Langmuir*, **14**, pp. 5636–5648.

79. Bond, A. M. (1980). *Modern Polarographic Methods in Analytical Chemistry* (Marcel Dekker, New York).

80. Alfonta, L., Bardea, A., Khersonsky, O., Katz, E., and Willner, I. (2001). Chronopotentiometry and Faradaic impedance spectroscopy as signal transduction methods for the biocatalytic precipitation of an insoluble product on electrode supports: Routes for enzyme sensors, immunosensors and DNA sensors, *Biosens. Bioelectron.*, **16**, pp. 675–687.

81. Bard, A. J. and Faulkner, L. R. (2001). *Electrochemical Methods: Fundamentals and Applications,* 2nd Ed. (John Wiley & Sons, Inc., New York).

82. Stoynov, Z. B., Grafov, B. M., Savova-Stoynova, B. S., and Elkin, V. V. (1991) *Electrochemical Impedance* (Nauka, Moscow) (in Russian).

Chapter 8

Gold Nanoparticles Modification and Aggregation: Applications from Bio- and Chemosensing to Drug Development

8.1 Introduction

The unique physical and chemical properties of noble metal nanoparticles such as gold nanoparticles (AuNPs) open up a substantial field for investigation of new science [1–6]. In comparison with bulk materials, nanoparticles possess a much higher concentration of electrons per surface unit. Therefore, attractive properties of AuNPs result from their coupling to an incident electromagnetic field (i.e., light) that appears as an enhancement of this field [7, 8]. Surface-enhanced Raman spectroscopy [9], fluorometry [10], and localized surface plasmon resonance (LSPR) spectroscopy [11] utilize this feature and have made important contributions to the analytical sciences. A number of practical applications of AuNPs, such as highly effective solar cells [12], cancer therapy [13], integrated optics [14], and chemical and biological sensing [15–17], have also been developed. The LSPR phenomenon depends not only on the wave frequency and structural parameters (shape, size, and chemical nature) of nanoparticles but also on the optical properties of adjacent medium as well as on the distance between nanoparticles. The two latter peculiarities are mainly used

Molecular Plasmonics: Theory and Applications
Volodymyr I. Chegel and Andrii M. Lopatynskyi

ISBN 978-981-4800-65-5 (Hardcover), 978-0-429-29511-9 (eBook)
www.jennystanford.com

to develop biological and chemical LSPR sensors based on colloidal AuNPs operating in the surface modification and aggregation modes [18, 19]. The processes of surface modification and aggregation of AuNPs also play a crucial role in novel drug development [20–22].

This chapter presents the results of specific practical studies of colloidal AuNPs interaction with various small molecules and biomolecules and their application for LSPR-based sensing and drug development. The first study deals with the characterization of colloidal AuNPs–biomolecules interaction in the view of a search for optimal LSPR sensor response modes, resulting in an experimental confirmation of theoretically predicted LSPR sensor response enhancement (see Section 2.3.3). The second study is focused on the experimental and theoretical characterization of small-molecules-induced aggregation of AuNPs, revealing how the differences between and combinations of functional groups in those molecules could promote the aggregation process to achieve a high-sensitive sensor platform. The third study reports on spectrophotometric characterization of interactions in the multicomponent doxorubicin–bovine serum albumin (BSA)–AuNPs system, providing insights on the design of potential prodrugs with regulated properties of antibiotic and protein complexation.

8.2 Optical Response of LSPR Sensor Based on Surface Modification of Colloidal Gold Nanoparticles

8.2.1 Mechanisms of LSPR Sensor Response Formation

The optical response of LSPR sensor depends on the structure of its sensitive element and on the sensory mechanism the device is based on. The most commonly used sensor elements of LSPR sensors are based on metal nanoparticles in the form of a colloidal solution or array located on the surface of a solid, and the mechanisms of the formation of the response include aggregation of nanoparticles and modification of their surface under the influence of the analyte. Colloidal solutions of noble metal nanoparticles were the first systems used in early biosensor experiments using the LSPR technique [23–25] and still remain a current platform for the development of new

LSPR sensors [26–29]. Therefore, the important task is to study the optical response of LSPR sensors based on a colloidal solution of gold nanoparticles upon the surface modification and aggregation of nanoparticles and to find the optimal mode for measuring the LSPR response.

This section presents a description of the LSPR sensor sensitive elements having a structure of a colloidal solution of AuNPs and the results of experimental studies of the influence of surface-modifying analytes on the optical response of this system. Also, the section compares the results of the application of different approaches for the LSPR sensor response estimation (see Section 2.3.3) to the experimental results obtained.

8.2.2 Morphological and Spectral Properties of Colloidal Gold Nanoparticles

One of the most exploited techniques for fabrication of colloidal AuNPs is the well-known Turkevich method [30, 31], which is based on the chemical reduction of gold ions in an aqueous solution under increased temperature. Gold nanoparticles resulting from this technology are almost monodisperse structures with a shape close to the spherical, and about 10 to 15 nm in size (Fig. 8.1a), which are stabilized in the solution by means of weakly bound citrate anions. From the histogram of the nanoparticles size distribution, presented in Fig. 8.1b, it is evident that the average diameter of the nanoparticles in the colloidal solution is about 13 nm. The concentration of AuNPs was estimated to be 8×10^{12} mL^{-1}, which corresponds to about 13 nM in terms of molar concentration. This value was obtained taking into account that the entire mass of gold in $HAuCl_4$ employed for colloidal dispersion preparation was fully transformed into nanoparticles.

Plasmonic properties of the AuNPs colloidal solution are manifested in the form of a peak in the light extinction spectrum, the maximum of which corresponds to a wavelength of 520 nm (Fig. 8.2). The fitting of the experimental extinction spectrum using a theoretical model based on the Mie scattering theory (see Section 2.3.1) showed that the least deviation from the experimental spectrum was observed with an AuNP diameter of 13 nm and a 1-nm-thick citrate-anion coating (Fig. 8.2). When fitted, the size effect of the electron mean free path reduction in AuNPs was taken into account; the citrate-anion layer was interpreted as a saturated monolayer of globular molecules according to Eq. (2.37), where the refractive index of the molecules was set to $n_{citrate}$ = 1.575 [33], and water was chosen as the ambient environment. The obtained values of the AuNP diameter and the thickness of the stabilizer layer are consistent with those indicated in the literature [31, 34].

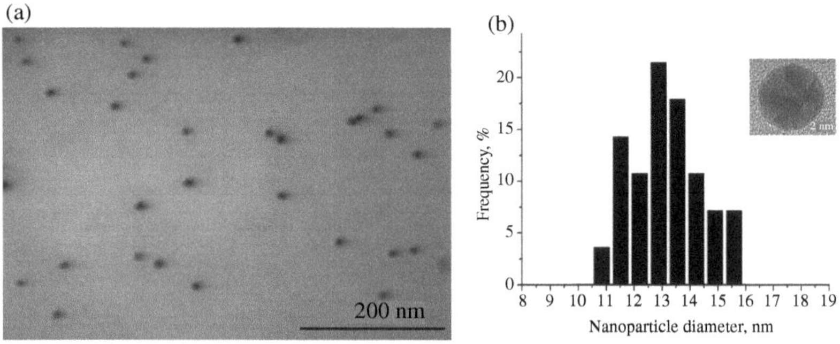

Figure 8.1 (a) Transmission electron microscopy (TEM) imaging of fabricated AuNPs. (b) Histogram of distribution of nanoparticles diameter measurements based on TEM. Inset: High-resolution TEM of single AuNP. Adapted with permission from Ref. [32]. Copyright (2012) American Chemical Society.

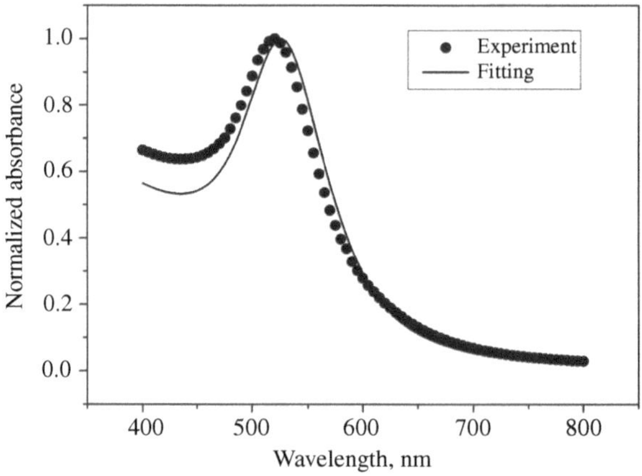

Figure 8.2 Measured light extinction spectrum of citrate-stabilized colloidal solution of AuNPs and the results of its fitting using a model based on the Mie theory.

8.2.3 Experimental Study of LSPR Response upon Colloidal Gold Nanoparticles Interaction with Different Analytes

During the interaction of an AuNP in a colloidal solution with analyte molecules, a change in the dielectric properties of the medium surrounding the nanoparticle may occur due to their adsorption and/or the replacement/neutralization of the stabilizer molecules on the surface of the nanoparticle by the molecules of the analyte. Moreover, the investigation of the zeta potential upon the interaction of colloidal AuNPs with various analytes showed that depending on the properties of the analyte molecules (in particular, on the magnitude and sign of their charge), the zeta potential may either increase or decrease [32]. This indicates a change in the charge state of the surface of nanoparticles, which can lead to surface modification and aggregation of nanoparticles with the preservation and loss of colloidal stability, respectively.

In order to study the peculiarities of the optical response of the LSPR sensor in the mode of surface modification of nanoparticles, measurements of the light extinction spectra were carried out when aqueous solutions of lipoic acid, glutathione, and BSA were added to colloidal AuNPs. Changes in the extinction spectra occurring after the addition of solutions of analytes (Fig. 8.3) indicate a gradual modification of the surface of nanoparticles by molecules of analytes. It should be noted that Fig. 8.3 does not show the initial extinction spectrum of colloidal AuNPs, since as a result of dilution with the addition of solutions of analytes, there is a decrease in the intensity of the extinction peak, which complicates the visual comparison of the initial spectrum with the presented ones. The shifts in the extinction peak position were measured relative to the initial position immediately after the addition of analyte solutions and equaled 1.7 nm, 1.1 nm, and 4.6 nm for lipoic acid, glutathione, and BSA, respectively. The values of the initial peak shifts in the extinction spectra are consistent with the concentrations of analyte solutions (1 mmol/L, 20 μmol/L, and 15 μmol/L) and their molecular weights (206 g/mol, 307 g/mol, and 66,463 g/mol) for lipoic acid, glutathione, and BSA, respectively.

Figure 8.4 shows the measured kinetic dependences of the peak position shift in the light extinction spectrum (LSPR response $\Delta\lambda_{max}$)

in comparison with the kinetic dependences of the LSPR response H_{top}^{+} corresponding to the spectra in Fig. 8.3, and the results of their fitting by a growing exponential function of type $R = R_0 + Ae^{\lambda/B}$. As can be seen from Fig. 8.4, the use of the proposed mode of measuring the LSPR response H_{top}^{+} on real samples of biomolecules improves the detection capability of LSPR sensor up to 4.5–48 times, depending on the type and concentration of biomolecules, which even exceeds the theoretically derived values (see Section 2.3.3).

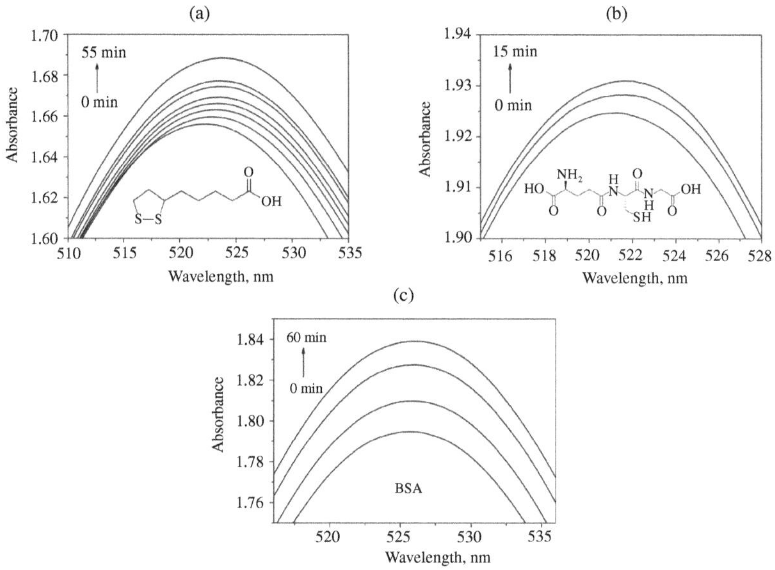

Figure 8.3 Measured extinction spectra of citrate-stabilized colloidal solutions of AuNPs after addition of aqueous solutions of (a) lipoic acid (up to a concentration of 1 mmol/L), (b) glutathione (up to a concentration of 20 μmol/L), and (c) BSA (up to a concentration of 15 μmol/L). The arrows indicate the chronological order of measuring the spectra.

Figure 8.5 shows the spectral dependences of the extinction shift in the case of the addition of BSA corresponding to the spectra in Fig. 8.3c. As a base spectrum for their calculation, the spectrum of light extinction measured immediately after the addition of an analyte was chosen. To determine the spectral position for measuring the LSPR response V_{right}, a derivative from the light extinction spectrum of a colloidal solution of AuNPs was calculated prior to the addition of an analyte, and its extremum was found on the right slope of the initial spectrum, whose position was 555 nm.

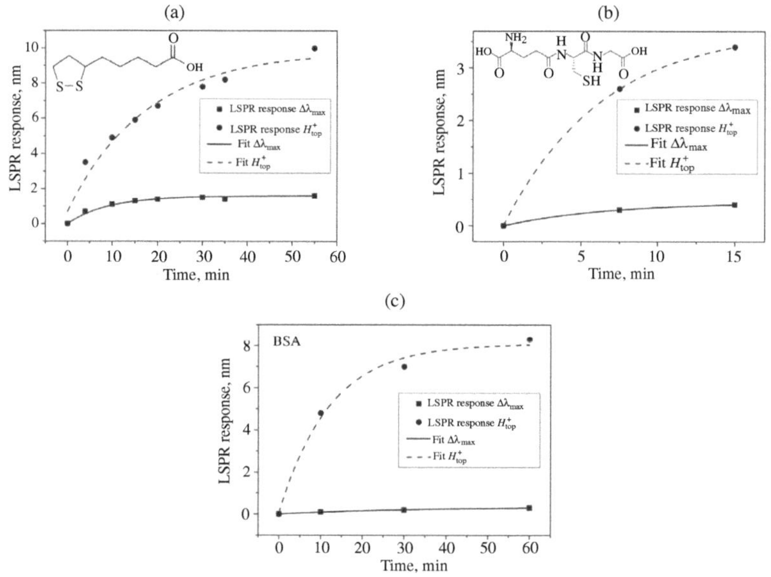

Figure 8.4 Measured kinetic dependences of LSPR response $\Delta\lambda_{max}$ compared with kinetic dependences of LSPR response H^{+}_{top}, obtained after the addition of aqueous solutions of (a) lipoic acid (up to a concentration of 1 mmol/L), (b) glutathione (up to a concentration of 20 μmol/L), and (c) BSA (up to a concentration of 15 μmol/L) to citrate-stabilized colloidal solutions of AuNPs.

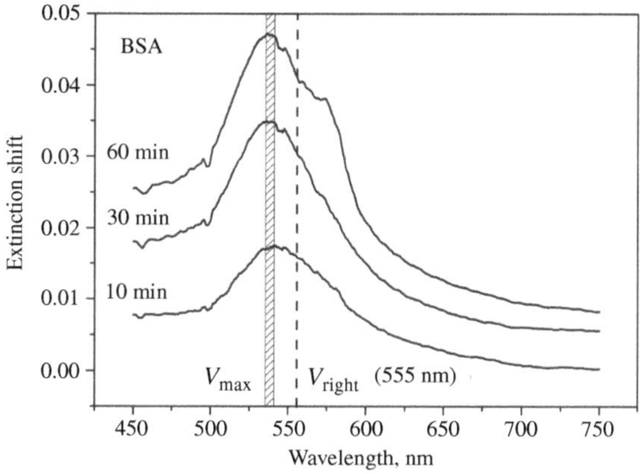

Figure 8.5 Measured spectral dependences of extinction shift, obtained after the addition of aqueous solution of BSA (up to a concentration of 15 μmol/L) to citrate-stabilized colloidal solution of AuNPs. Dashed line indicates the spectral position for measuring the LSPR response V_{right}; the shaded area indicates the spectral range in which the maximum extinction shift V_{max} is produced.

As can be seen from Fig. 8.5, the maximum extinction shift V_{max} is produced at slightly shorter wavelengths (535–540 nm), and the difference between V_{right} and V_{max} is 10–14%. Thus, the magnitude of response V_{right} on a real biomolecular sample is close to the maximum value of the "vertical" LSPR response V_{max}, which confirms the results of theoretical calculations.

8.3 Optical Response of LSPR Sensor Based on Aggregation of Colloidal Gold Nanoparticles

8.3.1 Gold Nanoparticles Aggregation as a Basis for Sensor Development

It is well known that aggregation of nanoparticles induces variations in the absorption spectra accompanied by significant color changes in solutions [35]. Similar color changes can be observed on the addition of an analyte, which initiates the aggregation of AuNPs, and this feature can be used for the development of chemical and biological sensors for the detection of specific compounds [27, 28, 36, 37]. Research work toward the development of LSPR biosensors based on the phenomenon of AuNPs aggregation has been performed [27, 38]. Interpretation of the LSPR phenomenon requires a clear understanding of physicochemical processes that occur during interactions between an analyte and AuNPs. Although a number of factors that influence AuNPs aggregation, e.g., molecular size of modifying alkanethiols, have been investigated individually [39, 40], the cooperative effect that takes place during multiple interactions between an analyte and AuNPs has not been fully discussed from a physical point of view. This makes understanding of aggregation mechanisms obscure and somewhat controversial, and it is not clear why aggregation occurs in some cases but not in others [41]. AuNPs have strong binding affinities for thiols and amines [39, 40, 42–44], where modification of the AuNP surfaces can lead to improved stability of AuNP dispersions permitting their industrial application in biosensing, immunological, and biochemical investigations [1, 9, 10]. The effect of amine functionality on binding to the surface of AuNPs has been investigated [45]. However, some authors have claimed that one type of amine group can readily bind to Au colloids, whereas others cannot [46]. Moreover, effects of the cooperative action of thiol and amine on aggregation of AuNPs have not been well investigated although they commonly coexist in biological samples.

This section covers the experimental and theoretical studies of cooperative functionalities of amine and thiol groups for aggregation of AuNPs [32]. Compounds containing various functional groups, including amine and thiol groups, sometimes connected within one molecule, were selected for the investigation of their influence on citrate-stabilized AuNPs. In addition, finite-difference time-domain (FDTD) modeling was applied for the first time to estimate the light extinction of nanoparticle aggregates involving single, chain-like, and globular structures, which can be a powerful method to overcome the drawbacks of conventional theoretical methods applicable only for one structure of AuNP. This research reveals physicochemical processes involved in the aggregation of citrate-stabilized AuNPs caused by interactions with a variety of model organic compounds and their influence on the optical response of a sensor based on aggregation of colloidal AuNPs.

8.3.2 Experimental Study of LSPR Response upon Colloidal Gold Nanoparticles Interaction with Different Analytes

To study the features of the optical response of the LSPR sensor operating in the mode of aggregation of nanoparticles, measurements of light extinction spectra of colloidal solutions of AuNPs were performed upon addition of aqueous solutions of thiourea, cysteamine hydrochloride, 6-mercapto-1-hexanol, tris(hydroxymethyl) aminomethane, and ethanol amine. Changes in the extinction spectra occurring after the addition of analyte solutions (Fig. 8.6) indicate a gradual loss of colloidal stability of AuNPs solution under the action of molecules of analytes. The peculiarities of the optical response, which reflect the AuNPs aggregation process, were discovered to be dependent on the structure and charge of the analyte molecules. Thus, the presence of sulfur atoms in the molecules of an analyte, which allows these compounds to covalently bind to gold atoms on the surface of nanoparticles, leads to the effective substitution of citrate ions that stabilize nanoparticles in a colloidal solution. If the charge of the analyte molecule is positive (for example, due to the presence of ionized amino groups, as in the molecules of thiourea and the cysteamine hydrochloride), then rapid aggregation and precipitation of AuNPs at small concentrations of the analyte (several μmol/L) are observed (Fig. 8.6a,b). If the sulfur compound has neutral functional groups (e.g., 6-mercapto-1-hexanol), then aggregation of AuNPs with further limited stability of the colloidal solution is evidenced (Fig. 8.6c).

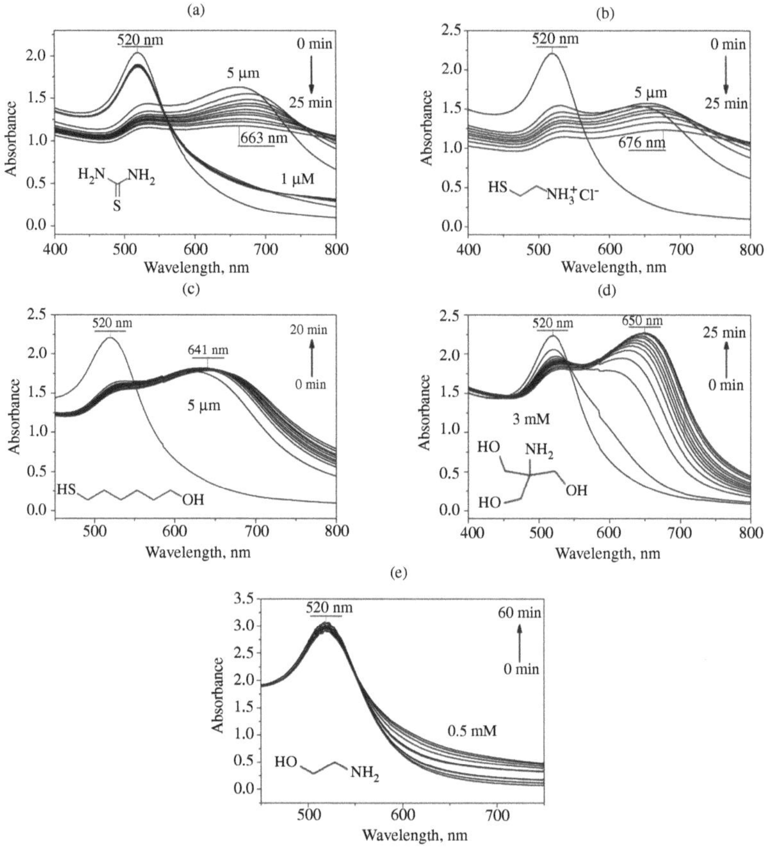

Figure 8.6 Measured extinction spectra of citrate-stabilized colloidal solutions of AuNPs after the addition of aqueous solutions of (a) thiourea (up to a concentration of 1 μmol/L, then 5 μmol/L), (b) cysteamine hydrochloride (up to a concentration of 5 μmol/L), (c) 6-mercapto-1-hexanol (up to a concentration of 5 μmol/L), (d) tris(hydroxymethyl)aminomethane (up to a concentration of 3 mmol/L), and (e) ethanol amine (up to a concentration of 0.5 mmol/L). The arrows indicate the chronological order of measuring the spectra. Adapted with permission from Ref. [32]. Copyright (2012) American Chemical Society.

In the absence of a sulfur atom in the analyte molecule, only positively charged molecules (for example, due to the presence of ionized amino groups, as in the molecules of tris(hydroxymethyl) aminomethane and ethanol amine) have a significant effect on the stability of the citrate-stabilized colloidal solutions of AuNPs, with their much higher concentration (up to several mmol/L) (Fig. 8.6d,e). Probably, in this case, the formation of a diffusive layer of analyte molecules near the surface of AuNP occurs, which leads to a partial neutralization of its negative charge.

8.3.3 Theoretical Study of LSPR Response upon Colloidal Gold Nanoparticles Aggregation

To clarify the experimental results and to find out the reason for changes in the extinction spectrum of the colloidal solution of AuNPs in the process of their aggregation, a theoretical study was conducted. As mentioned earlier, the LSPR phenomenon in noble metal nanoparticles, which is responsible for their properties of light extinction, is dependent on their size and shape. Thus, the absorption spectrum of the aggregated gold colloid should reflect both an increase in the average size of gold nanoparticles due to the agglomeration of nanoparticles, and changes in the shape of nanoparticles from a spherical to a variety of aggregate forms, which may include spherically symmetric and chain-shaped configurations. Therefore, the FDTD method of electromagnetic modeling was used for investigation of both effects and consideration of electromagnetic interactions between nanoparticles.

The aggregate geometries studied included close-packed spherically symmetric aggregates consisting of 13 nanoparticles and linear aggregates consisting of two to six nanoparticles, with each AuNP having a diameter of 13 nm [30, 31] and covered with 1-nm-thick citrate ion shell [34], immersed in water. The distances between the gold cores of citrate-capped nanoparticles in the linear aggregate and between central and surrounding nanoparticles in the spherical aggregate (interparticle gap) were fixed at 0.5, 1, and 2 nm, which is close to the size of a small organic molecule [39]. The developed model takes into account size-dependent gold optical constants and treats the citrate ion shell surrounding the nanoparticles by means of symmetrical Bruggeman effective medium theory [32].

Since the AuNP aggregates in the solution are randomly oriented and do not interact with each other due to their low concentration, one can assume that the extinction spectrum of the investigated system when illuminated by unpolarized light for a given geometry of an aggregate is directly proportional to the sum of the extinction spectra for one aggregate, calculated for different directions of incidence of light. In this case, the extinction spectrum of unpolarized light by an aggregate for each angle of incidence of light was determined as the arithmetic mean of the two extinction spectra obtained for the orthogonal polarizations of linearly polarized light [47]. To simulate the aggregated colloidal system response to unpolarized light, the total extinction spectrum for each of the aggregate geometries studied was estimated (Fig. 8.7) by summarizing the spectra calculated for a complete range of angles of incidence of linearly polarized light on the individual nanoparticle aggregate with a step of 10°.

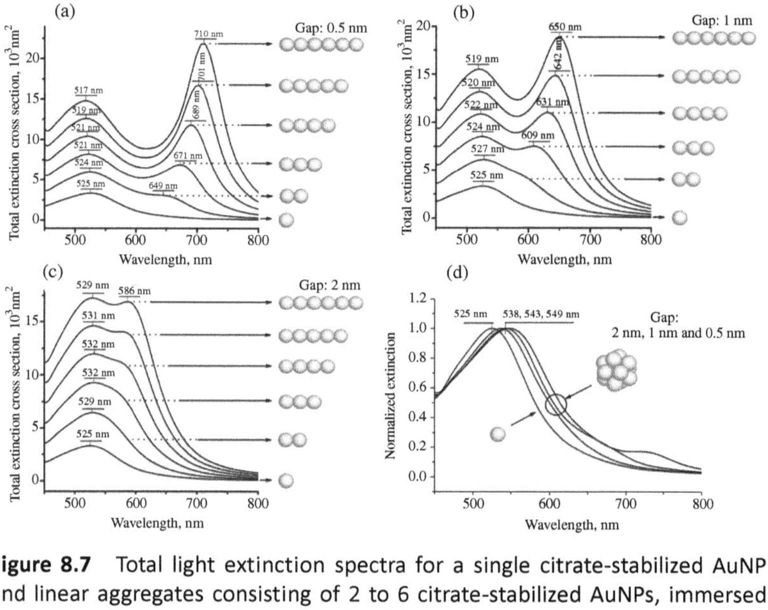

Figure 8.7 Total light extinction spectra for a single citrate-stabilized AuNP and linear aggregates consisting of 2 to 6 citrate-stabilized AuNPs, immersed in water, with interparticle gap fixed at (a) 0.5 nm, (b) 1 nm, and (c) 2 nm. (d) Normalized total extinction spectra for a single citrate-stabilized AuNP and a close-packed spherically symmetric aggregate consisting of 13 citrate-stabilized AuNPs, immersed in water, with interparticle gap fixed at 0.5 nm, 1 nm, and 2 nm. Adapted with permission from Ref. [32]. Copyright (2012) American Chemical Society.

It was found that the total extinction spectrum for a spherical aggregate (Fig. 8.7d) exhibits only one distinct peak (at 538, 543, and 549 nm for interparticle gap of 2 nm, 1 nm, and 0.5 nm, respectively) in the wavelength range of 450–800 nm, where experimental peaks are located. In contrast, the total extinction spectra for linear aggregates contain two peaks in the above-mentioned region, magnitudes and positions of which are dependent on the aggregate length and the interparticle gap (Fig. 8.7a–c). However, both experimental absorption peaks for an aggregated colloid are red-shifted with respect to the peak for an unaggregated colloid, while simulated absorption peaks behave differently: For a spherical aggregate, it is red-shifted, but for all of linear aggregates, one of the peaks is blue-shifted, and the other is red-shifted. This implies that the optical response of an aggregated Au colloid should be due to close to spherical and chain-like aggregates present in solution, which is consistent with the published data of microscopic studies [39, 40, 48, 49].

The simulated spectra for different aggregate geometries were also applied to fit an experimental spectrum in order to reveal the character of aggregate formation. The fitting procedure was carried out under the assumption that differently shaped nanoparticle aggregates act as independent light absorbers/scatterers and contribute to total extinction in a linear manner in proportion to their concentrations. The fitting algorithm was based on Monte Carlo enumeration of percentages of aggregates of all studied geometries and formation of a summed extinction spectrum, which was fitted to the experimental data by means of a least-squares technique.

Fitting results for the case of AuNP solutions after the addition of thiourea up to 3 μM are depicted in Fig. 8.8a. After enumeration of one billion aggregate concentration percentage sets, a root-mean-square deviation of 5.6% from the experimental extinction spectrum was achieved. As can be seen from the histogram (Fig. 8.8a, inset), the optical response is formed mainly by linear and small quantity of spherically symmetric nanoparticle aggregates. This proves the assumption about the origin of optical response, which is due to contributions from both chain-like and spherically shaped aggregates and is also confirmed by TEM results (Fig. 8.8b). The increased error of the fitting of the experimental spectrum in the range of wavelengths of 700–800 nm can be explained by the contribution of unaccounted chain-like and spherical aggregates consisting of more than 6 and 13 gold nanoparticles, respectively, as well as by the deviation of real aggregate shapes from those considered in a model.

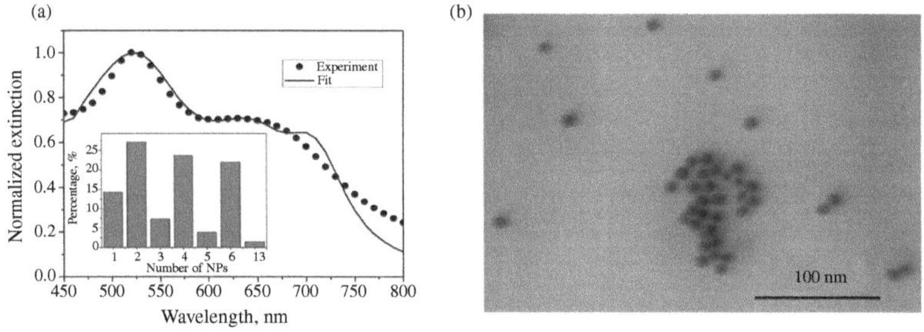

Figure 8.8 (a) Normalized extinction spectrum of citrate-stabilized colloidal solution of AuNPs after the addition of thiourea up to 3 μM (dots) and a theoretical best fit, which was obtained using the total extinction spectra of all studied aggregate geometries (linear (1–6 mer) and spherical aggregates (13 mer) shown in Fig. 8.7a–d) (solid line). Inset: histogram of distribution of nanoparticle number in an aggregate obtained as a result of fitting procedure. (b) TEM imaging of thiourea-covered AuNPs aggregate with a structure similar to that used in the simulation procedure. Adapted with permission from Ref. [32]. Copyright (2012) American Chemical Society.

8.4 Optical Characterization of Physicochemical Interactions in Multicomponent Doxorubicin–BSA–Gold Nanoparticle System

8.4.1 Gold Nanoparticles as a Factor of Influence on Doxorubicin–BSA Complex

The effectiveness of chemotherapy for cancer treatment when using traditional antitumor drugs is insufficient, as it is accompanied by a number of adverse side effects. In the first place, all drug products of this direction are characterized by high total toxicity. Doxorubicin (Dox) belongs to the anthracycline group of antibiotics due to its wide range of chemotherapeutic effects on malignant tumors and oncological diseases of the blood [50–52]. To overcome the mechanisms of development of Dox resistance, to increase the selectivity of its action, and to ensure the detoxification associated with radical forms of oxygen, a variety of methods are used for the creation of modified drugs (prodrugs) through the attachment of chemical fragments to antitumor agents, synthesis of conjugates with metallic, semiconductor, carbon nanoparticles, and metal ions [53–56].

A special place among doxorubicin-based prodrugs is taken by its conjugates with the noble metal nanoparticles, primarily Au. A high surface-to-volume ratio inherent to nanoparticles enables rapid response kinetics and provides improved drug-loading capabilities. In addition, AuNPs provide a high degree of biocompatibility and controlled synthesis. Since the LSPR spectra of AuNPs are very sensitive to changes in the dielectric permittivity of their environment, they can react to processes of electron-conformational transformations in the molecules of drugs, including their complexation with antioxidants. In addition, nanoparticles can heat up due to light absorption, providing thermal effects on tumor cells, and serve as containers for transport and release of drugs in malignant tumors [57–59]. An important factor in the expediency of using gold nanoparticles as nanocontainers for medicinal products is the stability of a nanoparticle–drug system, which eliminates the effects of aggregation with the subsequent withdrawal of aggregates by a macrophage system leading to a reduction or complete cancellation of the therapeutic effect of drugs.

Antioxidant drugs administered to the body to reduce the harmful effect of Dox include BSA, one of the transport proteins [60]. It is known that a free sulfide group, which occurs in 70% of albumin molecules, participates in disulfide exchange with the formation of intermolecular complexes and has a Dox effect [61]. It is known that BSA also forms complexes with Dox [61, 62]. It is obvious that physical interaction in BSA complexes with doxorubicin is accompanied by electron-conformational transformations in the molecules of the drug itself. Similar formation of complexes takes place in conjugates of BSA and Dox with AuNPs. At the same time, the influence of gold nanoparticles on the processes of complexation between molecules of antitumor and antioxidant drugs has not been sufficiently studied [63]. This section presents a relevant study of the AuNPs effect on Dox–BSA complexation [64].

8.4.2 Concentration-Dependent Evolution of Light Absorbance in Doxorubicin–BSA–Gold Nanoparticle System

The pharmaceutical form of doxorubicin containing lactose as the most often used form for such type of antibiotic was used to investigate the physicochemical interaction between doxorubicin and the BSA complex in the presence of gold nanoparticles. Deionized water titrated to a pH of 6.9 and chemically synthesized citrate-stabilized gold nanoparticles with an average size of 13 nm and an absorption peak in the vicinity of 520 nm were used to prepare the multicomponent doxorubicin–BSA–gold nanoparticles solutions with different content of protein molecules.

Figure 8.9 shows the optical density spectra for solutions of pure Dox and Dox–BSA conjugates with gold nanoparticles. Upon the addition of AuNPs, a new band at about 498 nm appears, which corresponds to the Dox–AuNPs conjugate. With the addition of the lowest concentration of BSA (1.33×10^{-5} M), upon the creation of gold nanoparticle conjugates with Dox and albumin solutions, the spectral reconstruction of optical absorption is observed with preserving a doxorubicin-specific band near $\lambda = 481$ nm and sharp increase in optical density. The band at 498 nm shifts with growing BSA concentration to a maximum at about $\lambda = 500$ nm followed by a slight optical density decrease (Fig. 8.9a, inset). Changes in this part of the spectrum are more likely to be the result of conformational changes in the Dox molecule. Creating conjugates of protein and antibiotic with gold nanoparticles leads to a substantial visible rearrangement of the right side of the optical density spectrum of Dox molecule, which may be due to both the change in the Dox conformation and the aggregation of gold nanoparticles. Since the stabilization of gold nanoparticles in an aqueous solution is provided by a layer of citrate, which leads to the appearance of negative charges on their surface, the coating of AuNPs with Dox molecules in monocationic prototropic form is considered a result of electrostatic interaction. The magnitude of such binding essentially depends on the pH of the medium that surrounds the resulting conjugates and is manifested as the red shift of the resonant band characteristic to AuNPs at about 520 nm and the emergence of new optical absorption bands in the range up to 700 nm, if aggregation of nanoparticles is observed [32].

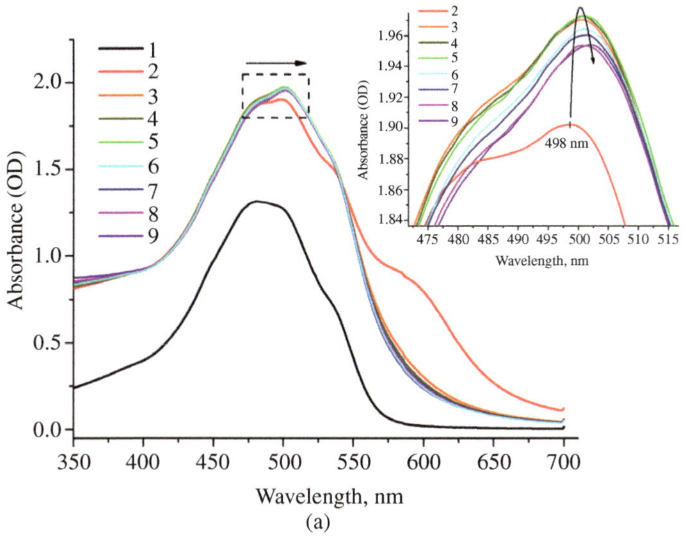

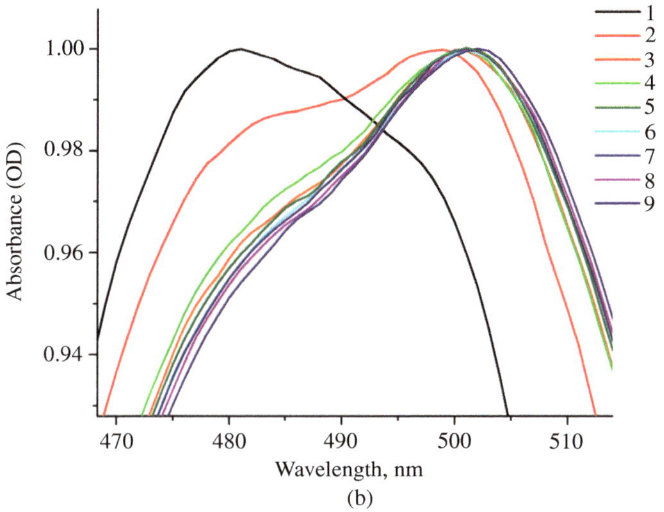

Figure 8.9 (a) Optical density spectra of solutions of Dox (1.33×10^{-4} M) (1), Dox conjugates with AuNPs (2) and Dox with 1.33×10^{-5} M (3), 2.66×10^{-5} M (4), 3.98×10^{-5} M (5), 5.31×10^{-5} M (6), 6.64×10^{-5} M (7), 9.29×10^{-5} M (8), 13.3×10^{-5} M bovine serum albumin (9) with gold nanoparticles at a concentration of 1.25 nM. Inset: parts of spectra in the wavelength range near $\lambda = 481$ nm. (b) Parts of the same normalized spectra in the wavelength range near $\lambda = 481$ nm. Adapted by permission from Springer Customer Service Centre GmbH: Springer *Applied Nanoscience,* Ref. [64], Copyright 2018.

Aggregation of gold nanoparticles with their subsequent precipitation is most likely due to the loss of stabilizing properties of citrate coating as a result of the transfer of protons with a change in the prototropic form of Dox. Indeed, in accordance with Ref. [32], only compounds with highly protonated amino groups and at high concentrations can neutralize the negative charge of citrate-stabilized AuNPs with subsequent aggregation. The Dox molecule has an amino group, but under conditions of neutral pH, its presence cannot be a key factor in destabilizing the colloidal solution and initiating the aggregation process. However, the addition of colloidal gold with pH = 5 changes the acidity of a neutral solution of Dox, which leads to the protonation of the Dox molecule and the aggregation of gold nanoparticles. The indicated protonation is also evidenced as the displacement of the peak of the vibrational mode of the amino group at a frequency of 3777 cm^{-1} to the 3373 cm^{-1} position in the infrared absorption spectrum [65]. That is, the aggregation process in the present case may be due to the charge transfer process in the Dox–AuNPs conjugate, which is accompanied by spectral changes in the Dox molecule, characteristic of the anionic prototropic form of doxorubicin. The calculation of the difference between the optical densities obtained for Dox solutions with AuNPs ($A_{Dox+AuNPs}$) and the sum of the optical densities of constituents $A_{Dox} + A_{AuNPs}$ reveals an intense absorption band located near $\lambda = 600$ nm (Fig. 8.10). It can be assumed that this band is characteristic to the anionic prototropic form of doxorubicin [66], the occurrence of which is due to the transfer of charges in the conjugate with subsequent aggregation of nanoparticles and their precipitation, which depends on the concentration of BSA in solutions.

In the presence of BSA in solution, this band gradually disappears and a new band near 550 nm appears (Fig. 8.10), which belongs to the cationic double-charged form of doxorubicin, that is, when the initial concentration of BSA increases, an anionic prototropic form of doxorubicin is transformed into its cationic double-charged form. At significant concentrations of protein in the presence of AuNPs, the most probable is the transition of Dox from a single-cationic prototropic form to a double-cationic form. Thus, protein molecules suspend the proton transfer process and play a stabilizing role in conjugates with AuNPs, forming not only C, but also D forms of antibiotic molecules, which contributes to the preservation of negative citrate-induced charges on their surfaces, and thus keeps AuNPs from aggregation. With increasing concentration of BSA, the precipitate of aggregated gold nanoparticles gradually disappears and, at the highest content of 1.33×10^{-4} M, is not observed (Fig. 8.10, inset).

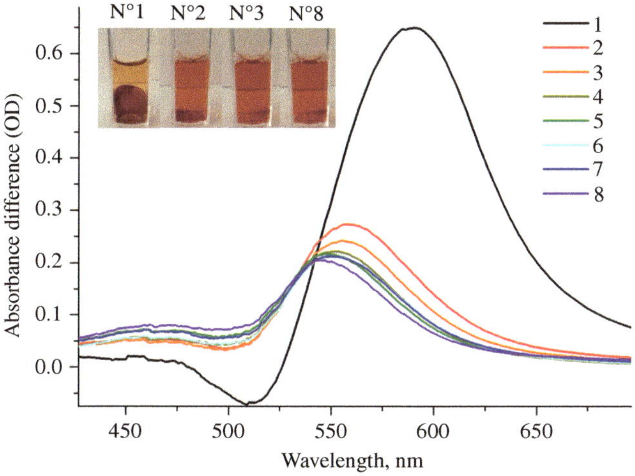

Figure 8.10 Spectra of the optical densities difference $A_{Dox+AuNPs}-(A_{Dox}+A_{AuNPs})$ (1) and $A_{Dox+BSA+AuNPs}-(A_{Dox+BSA}+A_{AuNPs})$ at BSA concentrations of 1.33×10^{-5} M (2), 2.66×10^{-5} M (3), 3.98×10^{-5} M (4), 5.31×10^{-5} M (5), 6.64×10^{-5} M (6), 9.29×10^{-5} M (7), 13.3×10^{-5} M (8), 1.33×10^{-4} M Dox and 1.25 nM AuNPs. Inset: photograph of samples 1, 2, 3, and 8, where differences in precipitation of AuNP–Dox–BSA aggregates for samples with different BSA concentration are observed. Reprinted by permission from Springer Customer Service Centre GmbH: Springer *Applied Nanoscience,* Ref. [64], Copyright 2018.

The results presented not only evidence the substantial binding between BSA and Dox with the formation of the complex, but also imply the significant effect of gold nanoparticles on the mechanism of complex formation and conformational changes in the protein–antibiotic system at LSPR conditions. That is, the presence of AuNPs enhances the regulatory capabilities of the Dox–BSA complex and can be used to create prodrugs.

8.5 Conclusion

Specific studies described in this chapter demonstrate the versatility of plasmonic nanoparticles, namely AuNPs, for a range of applications from molecular sensing to drug development. These include the high-sensitive LSPR-based sensing of small molecules and biomolecules based on AuNPs light extinction spectral analysis in different optical response formation modes and AuNPs-assisted characterization of interactions in antibiotic–protein system with tunable complexation properties.

In one of the studies, the validity of the proposed methods of measuring the LSPR response H_{top}^{+} and V_{right} in the LSPR biosensor based on colloidal gold in the mode of surface modification of AuNPs was confirmed. Using the H_{top}^{+} LSPR response measurement mode on real samples of biomolecules improved the detectivity of the LSPR biosensor by 4.5–48 times, depending on the type and concentration of biomolecules, which even exceeds the theoretically derived values (see Section 2.3.3). The magnitude of response V_{right} on a real biomolecular sample was shown to be close to the maximum value of the "vertical" LSPR response V_{max}, which confirms the results of theoretical calculations.

Drastic effect of cooperative functionalities in a single molecular conjugate on the AuNP aggregation has been investigated in another study, explaining the optimal route for building a high-sensitive LSPR sensor based on the aggregation of colloidal AuNPs. Namely, the degree of aggregation of AuNPs was found to be dependent on the chemical structure and charge of the analyte molecule. Thus, the presence in the molecules of the analyte of atoms that interact with gold atoms on the surface of nanoparticles in a covalent manner, in conjunction with a positive charge of the molecule, caused rapid aggregation and precipitation of AuNPs detectable by a naked eye at an analyte concentration starting from 5 μmol/L. A theoretical model based on the FDTD method was proposed for consideration of evolution of the light extinction properties of the AuNP system during the aggregation process. It was shown that the optical response of an aggregated Au colloid should be formed both by close to spherical and chain-like nanoparticle aggregates present in solution. The theoretical approach applied was demonstrated to be useful for the estimation of the distribution of aggregate shapes corresponding to the experimental light extinction spectrum.

The last considered study applied light extinction spectroscopy to study the interactions within the multicomponent Dox–BSA–AuNPs system. Dox conjugate with citrate-stabilized AuNPs was characterized by the appearance of a new intense absorption band located near λ = 600 nm, which was assumed to belong to the anionic prototropic form of Dox as a result of the observed AuNPs aggregation. In the presence of BSA in solution and with an increase in its concentration, this band disappeared with a transformation in a band near 550 nm, which may belong to the cationic double-charged form of Dox due to the absence of AuNP aggregation. When adding AuNPs to Dox solutions with different albumin content, the optical absorption spectra underwent significant changes due to a decrease in the interaction of an antibiotic with nanoparticles, a significant association of albumin molecules with AuNPs, and, as a consequence, the rearrangement of the Dox–BSA–AuNPs system to a more stable state, which is important in the drug-delivery process. The results of the study indicate the possibility of creating effective prodrugs comprising Dox–BSA–AuNPs with regulated properties of antibiotic and protein complexation due to the presence of AuNPs.

References

1. Thaxton, C. S., Georganopoulou, D. G., and Mirkin, C. A. (2006). Gold nanoparticle probes for the detection of nucleic acid targets, *Clin. Chim. Acta*, **363**, pp. 120–126.
2. Caruthers, S. D., Wickline, S. A., and Lanza, G. M. (2007). Nanotechnological applications in medicine, *Curr. Opin. Biotechnol.*, **18**, pp. 26–30.
3. Jain, K. K. (2007). Applications of nanobiotechnology in clinical diagnostics, *Clin. Chem.*, **53**, pp. 2002–2009.
4. Baptista, P., Pereira, E., Eaton, P., Doria, G., Miranda, A., Gomes, I., Quaresma, P., and Franco, R. (2008). Gold nanoparticles for the development of clinical diagnosis methods, *Anal. Bioanal. Chem.*, **391**, pp. 943–950.

5. Sengupta, S. and Sasisekharan, R. (2007). Exploiting nanotechnology to target cancer, *Br. J. Cancer*, **96**, pp. 1315–1319.
6. Souza, G. R., Christianson, D. R., Staquicini, F. I., Ozawa, M. G., Snyder, E. Y., Sidman, R. L., Arap, W., and Pasqualini, R. (2006). Networks of gold nanoparticles and bacteriophage as biological sensors and cell-targeting agents, *Proc. Natl. Acad. Sci. USA*, **103**, pp. 1215–1220.
7. Haes, A. J., Haynes, C. L., McFarland, A. D., Schatz, G. C., Van Duyne, R. P., and Zou, S. (2005). Plasmonic materials for surface-enhanced sensing and spectroscopy, *MRS Bull.*, **30**, pp. 368–375.
8. Hao, E. and Schatz, G. C. (2004). Electromagnetic fields around silver nanoparticles and dimers, *J. Chem. Phys.*, **120**, pp. 357–366.
9. Witlicki, E. H., Johnsen, C., Hansen, S. W., Silverstein, D. W., Bottomley, V. J., Jeppesen, J. O., Wong, E. W., Jensen, L., and Flood, A. H. (2011). Molecular logic gates using surface-enhanced Raman-scattered light, *J. Am. Chem. Soc.*, **133**, pp. 7288–7291.
10. Nerambourg, N., Praho, R., Werts, M. H. V., Thomas, D., and Blanchard-Desce, M. (2008). Hydrophilic monolayer-protected gold nanoparticles and their functionalisation with fluorescent chromophores, *Int. J. Nanotechnol.*, **5**, pp. 722–740.
11. Willets, K. A. and Van Duyne, R. P. (2007). Localized surface plasmon resonance spectroscopy and sensing, *Annu. Rev. Phys. Chem.*, **58**, pp. 267–297.
12. Derkacs, D., Lim, S. H., Matheu, P., Mar, W., and Yu, E. T. (2006). Improved performance of amorphous silicon solar cells via scattering from surface plasmon polaritons in nearby metallic nanoparticles, *Appl. Phys. Lett.*, **89**, 093103.
13. Kang, B., Mackey, M. A., and El-Sayed, M. A. (2010). Nuclear targeting of gold nanoparticles in cancer cells induces DNA damage, causing cytokinesis arrest and apoptosis, *J. Am. Chem. Soc.*, **132**, pp. 1517–1519.
14. Wang, T. J., Tu, C. W., and Liu, F. K. (2005). Integrated-optic surface-plasmon-resonance biosensor using gold nanoparticles by bipolarization detection, *IEEE J. Sel. Top. Quantum Electron.*, **11**, pp. 493–499.
15. Hunt, H. K. and Armani, A. M. (2010). Label-free biological and chemical sensors, *Nanoscale*, **2**, pp. 1544–1559.
16. Haes, A. J. and Van Duyne, R. P. (2004). A unified view of propagating and localized surface plasmon resonance biosensors, *Anal. Bioanal. Chem.*, **379**, pp. 920–930.

17. Anker, J. N., Hall, W. P., Lyandres, O., Shah, N. C., Zhao, J., and Van Duyne, R. P. (2008). Biosensing with plasmonic nanosensors, *Nat. Mater.*, **7**, pp. 442–453.

18. Lopez, G. A., Estevez, M. C., Soler, M., and Lechuga, L. M. (2017). Recent advances in nanoplasmonic biosensors: Applications and lab-on-a-chip integration, *Nanophotonics*, **6**, pp. 123–136.

19. Oliveira, E., Nunez, C., Santos, H. M., Fernandez-Lodeiro, J., Fernandez-Lodeiro, A., Capelo, J. L., and Lodeiro, C. (2015). Revisiting the use of gold and silver functionalised nanoparticles as colorimetric and fluorometric chemosensors for metal ions, *Sens. Actuators B Chem.*, **212**, pp. 297–328.

20. Li, X., Robinson, S. M., Gupta, A., Saha, K., Jiang, Z., Moyano, D. F., Sahar, A., Riley, M. A., and Rotello, V. M. (2014). Functional gold nanoparticles as potent antimicrobial agents against multi-drug-resistant bacteria, *ACS Nano*, **8**, pp. 10682–10686.

21. Daraee, H., Eatemadi, A., Abbasi, E., Fekri Aval, S., Kouhi, M., and Akbarzadeh, A. (2016). Application of gold nanoparticles in biomedical and drug delivery, *Artif. Cells Nanomed. Biotechnol.*, **44**, pp. 410–422.

22. Cui, T., Liang, J. J., Chen, H., Geng, D. D., Jiao, L., Yang, J. Y., Qian, H., Zhang, C., and Ding, Y. (2017). Performance of doxorubicin-conjugated gold nanoparticles: Regulation of drug location, *ACS Appl. Mater. Interfaces*, **9**, pp. 8569–8580.

23. Alivisatos, A. P., Johnsson, K. P., Peng, X., Wilson, T. E., Loweth, C. J., Bruchez Jr, M. P., and Schultz, P. G. (1996). Organization of 'nanocrystal molecules' using DNA, *Nature*, **382**, pp. 609–611.

24. Mirkin, C. A., Letsinger, R. L., Mucic, R. C., and Storhoff, J. J. (1996). A DNA-based method for rationally assembling nanoparticles into macroscopic materials, *Nature*, **382**, pp. 607–609.

25. Englebienne, P. (1998). Use of colloidal gold surface plasmon resonance peak shift to infer affinity constants from the interactions between protein antigens and antibodies specific for single or multiple epitopes, *Analyst*, **123**, pp. 1599–1603.

26. Li, X., Jiang, L., Zhan, Q., Qian, J., and He, S. (2009). Localized surface plasmon resonance (LSPR) of polyelectrolyte-functionalized gold-nanoparticles for bio-sensing, *Colloids Surf. A*, **332**, pp. 172–179.

27. Wei, F., Lam, R., Cheng, S., Lu, S., Ho, D., and Li, N. (2010). Rapid detection of melamine in whole milk mediated by unmodified gold nanoparticles, *Appl. Phys. Lett.*, **96**, 133702.

28. Qi, W. J., Wu, D., Ling, J., and Huang, C. Z. (2010). Visual and light scattering spectrometric detections of melamine with polythymine-stabilized gold nanoparticles through specific triple hydrogen-bonding recognition, *ChemComm*, **46**, pp. 4893–4895.

29. Castellana, E. T., Gamez, R. C., and Russell, D. H. (2011). Label-free biosensing with lipid-functionalized gold nanorods, *J. Am. Chem. Soc.*, **133**, pp. 4182–4185.

30. Turkevich, J., Stevenson, P. C., and Hillier, J. (1951). A study of the nucleation and growth processes in the synthesis of colloidal gold, *Discuss. Faraday Soc.*, **11**, pp. 55–75.

31. McFarland, A. D., Haynes, C. L., Mirkin, C. A., Van Duyne, R. P., and Godwin, H. A. (2004). Color my nanoworld, *J. Chem. Educ.*, **81**, 544.

32. Chegel, V., Rachkov, O., Lopatynskyi, A., Ishihara, S., Yanchuk, I., Nemoto, Y., Hill, J. P., and Ariga, K. (2012). Gold nanoparticles aggregation: Drastic effect of cooperative functionalities in a single molecular conjugate, *J. Phys. Chem. C*, **116**, pp. 2683–2690.

33. http://www.chemspider.com/Chemical-Structure.305.html

34. Wanner, M., Gerthsen, D., Jester, S. S., Sarkar, B., and Schwederski, B. (2005). Treatment of citrate-capped Au colloids with NaCl, NaBr and Na_2SO_4: A TEM, EAS and EPR study of the accompanying changes, *Colloid Polym. Sci.*, **283**, pp. 783–792.

35. Faraday, M. (1857). X. The Bakerian Lecture. Experimental relations of gold (and other metals) to light, *Philos. Trans. R. Soc. Lond.*, **147**, pp. 145–181.

36. Jain, P. K., Lee, K. S., El-Sayed, I. H., and El-Sayed, M. A. (2006). Calculated absorption and scattering properties of gold nanoparticles of different size, shape, and composition: Applications in biological imaging and biomedicine, *J. Phys. Chem. B*, **110**, pp. 7238–7248.

37. De La Escosura-Muniz, A., Parolo, C., and Merkoçi, A. (2010). Immunosensing using nanoparticles, *Mater. Today*, **13**, pp. 24–34.

38. Li, H. and Rothberg, L. (2004). Colorimetric detection of DNA sequences based on electrostatic interactions with unmodified gold nanoparticles, *Proc. Natl. Acad. Sci. U.S.A.*, **101**, pp. 14036–14039.

39. Zhong, Z., Patskovskyy, S., Bouvrette, P., Luong, J. H., and Gedanken, A. (2004). The surface chemistry of Au colloids and their interactions with functional amino acids, *J. Phys. Chem. B*, **108**, pp. 4046–4052.

40. Bellino, M. G., Calvo, E. J., and Gordillo, G. (2004). Adsorption kinetics of charged thiols on gold nanoparticles, *Phys. Chem. Chem. Phys.*, **6**, pp. 424–428.

41. Kanaras, A. G., Wang, Z., Bates, A. D., Cosstick, R., and Brust, M. (2003). Towards multistep nanostructure synthesis: Programmed enzymatic self-assembly of DNA/gold systems, *Angew. Chem. Int. Ed. Engl.*, **42**, pp. 191–194.

42. Weisbecker, C. S., Merritt, M. V., and Whitesides, G. M. (1996). Molecular self-assembly of aliphatic thiols on gold colloids, *Langmuir*, **12**, pp. 3763–3772.

43. Katz, E. and Willner, I. (2004). Integrated nanoparticle–biomolecule hybrid systems: Synthesis, properties, and applications, *Angew. Chem. Int. Ed. Engl.*, **43**, pp. 6042–6108.

44. Aryal, S., Remant, B. K. C., Dharmaraj, N., Bhattarai, N., Kim, C. H., and Kim, H. Y. (2006). Spectroscopic identification of S–Au interaction in cysteine capped gold nanoparticles, *Spectrochim. Acta A*, **63**, pp. 160–163.

45. Mandal, S., Phadtare, S., and Sastry, M. (2005). Interfacing biology with nanoparticles, *Curr. Appl. Phys.*, **5**, pp. 118–127.

46. Selvakannan, P. R., Mandal, S., Phadtare, S., Pasricha, R., and Sastry, M. (2003). Capping of gold nanoparticles by the amino acid lysine renders them water-dispersible, *Langmuir*, **19**, pp. 3545–3549.

47. Bohren, C. F. and Huffman, D. R. (1983). *Absorption and Scattering of Light by Small Particles* (Wiley-Interscience, New York).

48. Li, M., Johnson, S., Guo, H., Dujardin, E., and Mann, S. (2011). A generalized mechanism for ligand-induced dipolar assembly of plasmonic gold nanoparticle chain networks, *Adv. Funct. Mater.*, **21**, pp. 851–859.

49. Zhang, F. X., Han, L., Israel, L. B., Daras, J. G., Maye, M. M., Ly, N. K., and Zhong, C. J. (2002). Colorimetric detection of thiol-containing amino acids using gold nanoparticles, *Analyst*, **127**, pp. 462–465.

50. Arcamone, F. (1981). *Doxorubicin: Anticancer Antibiotics* (Academic Press, New York).

51. Zhong, Y. J., Shao, L. H., and Li, Y. A. N. (2013). Cathepsin B-cleavable doxorubicin prodrugs for targeted cancer therapy, *Int. J. Oncol.*, **42**, pp. 373–383.

52. Wang, Q., Zhong, Y. J., Yuan, J. P., Shao, L. H., Zhang, J., Tang, L., Liu, S. P., Hong, Y. P., Firestone, R. A., and Li, Y. (2013). Targeting therapy of hepatocellular carcinoma with doxorubicin prodrug PDOX increases anti-metastatic effect and reduces toxicity: A preclinical study, *J. Transl. Med.*, **11**, 192.

53. Dykman, L. A., Bogatyrev, V. A., Shchyogolev, S. U., and Khlebtsov, N. G. (2008). *Gold Nanoparticles: Synthesis, Properties, Biomedical Application* (Nauka, Moscow) (in Russian).

54. Orel, V., Romanov, A., Rykhalskyi, O., Shevchenko, A., Orel, I., Burlaka, A., and Lukin, S. (2016). Antitumor effect of superparamagnetic iron oxide nanoparticles conjugated with doxorubicin during magnetic nanotherapy of Lewis Lung carcinoma, *Materwiss. Werksttech.*, **47**, pp. 165–171.

55. Dellinger, A., Zhou, Z., Connor, J., Madhankumar, A. B., Pamujula, S., Sayes, C. M., and Kepley, C. L. (2013). Application of fullerenes in nanomedicine: An update, *Nanomedicine*, **8**, pp. 1191–1208.

56. Mosunov, A. A., Pashkova, I. S., Sidorova, M., Pronozin, A., Lantushenko, A. O., Prylutskyy, Y. I., Parkinson, J. A., and Evstigneev, M. P. (2017). Determination of the equilibrium constant of C_{60} fullerene binding with drug molecules, *Phys. Chem. Chem. Phys.*, **19**, pp. 6777–6784.

57. You, J., Shao, R., Wei, X., Gupta, S., and Li, C. (2010). Near-infrared light triggers release of paclitaxel from biodegradable microspheres: Photothermal effect and enhanced antitumor activity, *Small*, **6**, pp. 1022–1031.

58. Yeshchenko, O. A., Kutsevol, N. V., and Naumenko, A. P. (2016). Light-induced heating of gold nanoparticles in colloidal solution: Dependence on detuning from surface plasmon resonance, *Plasmonics*, **11**, pp. 345–350.

59. Daraee, H., Eatemadi, A., Abbasi, E., Fekri Aval, S., Kouhi, M., and Akbarzadeh, A. (2016). Application of gold nanoparticles in biomedical and drug delivery, *Artif. Cells Nanomed. Biotechnol.*, **44**, pp. 410–422.

60. Peters Jr, T. (1985). Serum albumin, *Adv. Protein. Chem.*, **37**, pp. 161–245.

61. Gorchakova, N. O., Chekman, I. S., Vlasova, N. M., Golovkova, L. P., Gerashenko, I. I., and Maximchuk, O. O. (2011). Doxorubicin complexation with bovine serum albumin, *Rep. Natl. Acad. Sci. Ukr.*, **4**, pp. 177–181 (in Ukrainian).

62. Honary, S., Jahanshahi, M., Golbayani, P., Ebrahimi, P., and Ghajar, K. (2010). Doxorubicin-loaded albumin nanoparticles: Formulation and characterization, *J. Nanosci. Nanotechnol.*, **10**, pp. 7752–7757.

63. Gautier, J., Allard-Vannier, E., Munnier, E., Soucé, M., and Chourpa, I. (2013). Recent advances in theranostic nanocarriers of doxorubicin based on iron oxide and gold nanoparticles, *J. Control. Release*, **169**, pp. 48–61.

64. Goncharenko, N. A., Pavlenko, O. L., Dmytrenko, O. P., Kulish, M. P., Lopatynskyi, A. M., and Chegel, V. I. (2019). Gold nanoparticles as a factor of influence on doxorubicin–bovine serum albumin complex, *Appl. Nanosci.*, **9**, pp. 825–833.

65. Mirza, A. Z. and Shamshad, H. (2011). Preparation and characterization of doxorubicin functionalized gold nanoparticles, *Eur. J. Med. Chem.*, **46**, pp. 1857–1860.

66. Sturgeon, R. J. and Schulman, S. G. (1977). Electronic absorption spectra and protolytic equilibria of doxorubicin: Direct spectrophotometric determination of microconstants, *J. Pharm. Sci.*, **66**, pp. 958–961.

Chapter 9

Metamaterials with Reversible Optoelectronic and Physicochemical Properties

9.1 Introduction

The development of nanostructured materials with reversible properties [1–3] is one of the key research areas that yield various innovative functions. In particular, metamaterials with unique optical properties have attracted great interest over the last decades because of the wide-ranging possibilities for practical applications in laser optics, optoelectronics, chemical and biosensing, etc. [4–9]. Important contributions to the creation of such metamaterials have been obtained by using plasmonic nanostructures, such as gold and silver nanoparticles, nanorods, nanowires, nanodiscs, and nanoholes [10–15]. Most works describe metamaterials developed having static physical characteristics, and only a few articles have been devoted to preparations of metamaterials that permit tuning of physical properties or, in other words, exhibit the so-called "dynamic plasmonic" behavior [16–21]. This discrepancy is due to difficulties related to physical and technological limitations especially when different types of thin composite films are considered. It is

Molecular Plasmonics: Theory and Applications
Volodymyr I. Chegel and Andrii M. Lopatynskyi

ISBN 978-981-4800-65-5 (Hardcover), 978-0-429-29511-9 (eBook)
www.jennystanford.com

obvious that tuning of the physical characteristics of plasmonic colloidal solutions can be achieved more easily than for plasmonic nanostructures embedded in polymer matrices or, moreover, within solid materials due to the limited possible variation in the distance between nanoparticles. Unfortunately, only nanostructured thin-film matrices or ordered nanostructure arrays can be used for the preparation of chip-based formats that exhibit reversible properties and are suitable for practical applications. The physical aspect of this problem might be overcome if ordered arrays of shape-anisotropic nanoparticles that are sensitive to polarization of light are used. However, to fabricate such nanochips, only time-consuming and expensive nanolithographic methods are generally used, e.g., nanoimprint lithography based on specially constructed two-dimensional nanostructure array (nanoblock) molds derived from one-dimensional gratings [22].

This chapter presents the results of two specific studies related to the development of plasmonic nanocomposite matrix films exhibiting features of the "dynamic interface," which can be prepared using quick and simple methods. The first study shows that it is possible to vary the distance between Ag nanoparticles embedded in poly(acrylic acid) (PAA) matrices in three dimensions, while the PAA network fixes their relative positions within the matrix. The Ag nanocomposite matrix developed showed reversible spectral properties both in surface plasmon resonance (SPR) and localized surface plasmon resonance (LSPR) measurements after immersion in liquids of different pH. pH-induced swelling and shrinking activity of the composite metamaterial was proposed and theoretically simulated to explain this phenomenon. The second study highlights the promising potential of electrochemical SPR method by the examples of redox switching of optical and electrical properties of nanocomposite polymeric films. Specifically, it deals with the SPR transduction of the redox transformations in electropolymerized thin Cu^{2+}/polyacrylic acid films, which exhibit properties of a metamaterial with redox switching of electrorefractive, electrochromic, and conductivity functions.

9.2 Nanocomposite Polymer Matrix Containing Ag Nanoparticles with Dynamic Plasmonic Properties

9.2.1 Preparation of Nanocomposite Matrix

To prepare the nanocomposite matrix, known properties of some polymers to reduce the ions of plasmonic metals (Ag, Au) were exploited. For example, Gradess et al. [23] used polyvinyl alcohol to reduce Ag^+ ions at high temperature. In the considered study [24], poly(acrylic acid) was used since polyacrylate films exhibit stronger pH-dependent swelling and shrinking properties [25]. For the reduction of metal ions, UV irradiation was used instead of heating as a more effective and rapid process [26].

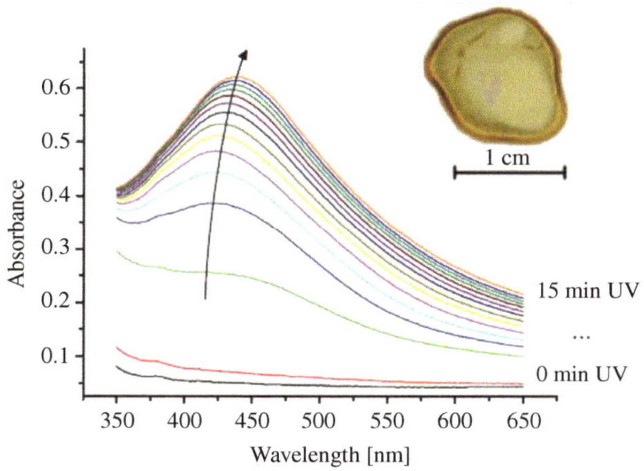

Figure 9.1 Electronic absorption spectra of a PAA film during treatment with UV light that leads to formation of embedded Ag nanoparticles. Inset: Image of PAA/AgNPs matrix (in swelled state at the center of the spot and in shrunk state around). Reproduced from Ref. [24] with permission from CSIRO Publishing.

The synthesis of Ag nanoparticles (AgNPs) in PAA-based hydrogel was carried out by spincoating the water–ethanol solution of PAA containing 0.2 M $AgNO_3$ onto clean microscope glass substrates (for spectrophotometric measurements) and SPR chips based on thin Au film (for SPR measurements) with subsequent drying and

UV irradiation using a mercury–xenon lamp. The colorless polymer films changed color to yellow (which was evident on the transparent glass substrates) after exposure to UV light due to the formation of light-absorbing Ag nanoparticles by photoreduction.

Typical absorbance spectra of composite PAA film with embedded Ag nanoparticles, deposited on a glass substrate, acquired during the UV treatment are shown in Fig. 9.1. The gradual increase in absorbance and redshift of absorbance peak (from 420 to 440 nm) originated from the intensifying LSPR response of nanoparticles due to the increase in size and concentration of Ag nanoparticles over the course of the UV photoreduction process. These results (Fig. 9.1) demonstrate that it is possible to control the process of Ag nanoparticle formation by adjustment of UV exposure time.

9.2.2 Reversible pH-Induced Changes in Optical Properties of Nanocomposite Matrix: LSPR Study

After exposure to UV, the sample was immersed in deionized water and a blue shift of the absorbance peak to ~425 nm was observed. This change in optical properties of the composite film was explicitly related to the alteration of the Ag nanoparticle's environment. It is known that the LSPR of noble metal nanoparticles depends on both dielectric properties of the surrounding medium [27] and intensity of the electromagnetic interaction between adjacent nanoparticles [28]. According to these specific plasmonic properties, a blue shift in absorbance for an ensemble of Ag nanoparticles can typically be induced by a decrease in the refractive index of environment and a decrease in electromagnetic interaction between nanoparticles (e.g., due to increase in interparticle distance). Because of the strong adsorption of water to PAA leading to hydrogel formation [25], optical evidence for the interaction of the PAA film with embedded Ag nanoparticles with deionized water can be treated as a simultaneous influence of the aforementioned mechanisms. Namely, adsorption of water by the PAA polymer network leads to reduction in its average refractive index due to the difference in their refractive indices—1.527 for PAA [29] and 1.333 for water. Additionally, water adsorption by PAA induces swelling of the polymer film due to hydrogel formation and subsequently results in an increase in the average distance between Ag nanoparticles contained in the film.

To study swelling and shrinking behavior of the composite films, they were successively immersed in deionized water, and then in aqueous 0.1 M sulfuric acid for 1 min, which led to a cyclic reversible shift in the absorbance peak (Fig. 9.2). Namely, the absorbance maximum switched between ~425 and ~411 nm for deionized water and 0.1 M aqueous sulfuric acid, respectively. It should be noted that the refractive index of aqueous 0.1 M sulfuric acid is almost identical to that of pure water [30], and the absorption spectrum should not undergo a blue shift due to change in the ambient refractive index. This implies that a blue shift in absorbance under the influence of sulfuric acid solution was induced by a further swelling of the polymer due to the change in ambient medium pH. However, it is known that swelling of pure PAA usually occurs at high pH [31], so it was supposed that the presence of AgNPs in the PAA matrix drastically changes the conformation of this polymer most probably either during reduction of Ag^+ ions or by chelation of Ag atom/ions by carboxylic acid in PAA.

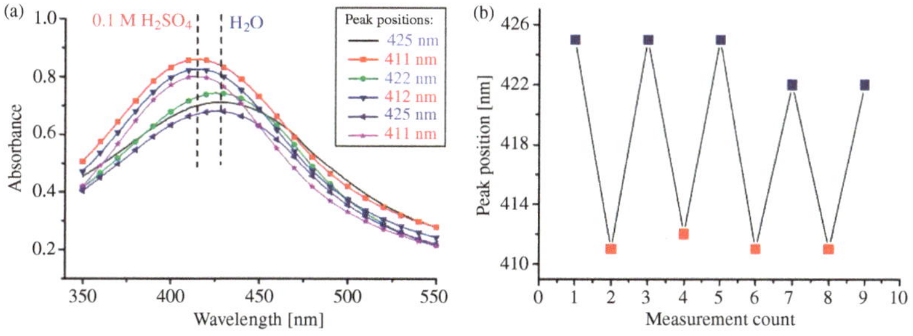

Figure 9.2 (a) Electronic absorption spectra of PAA hydrogel film containing Ag nanoparticles after alternating immersion in deionized water and 0.1 M sulfuric acid aqueous solution. (b) Respective absorbance spectrum peak positions. Reproduced from Ref. [24] with permission from CSIRO Publishing.

9.2.3 Reversible pH-Induced Changes in Optical Properties of Nanocomposite Matrix: SPR Study

To verify the results of spectrophotometric measurements, SPR experiments on the same composite Ag nanoparticle–PAA film deposited on an SPR chip were carried out. To study swelling and shrinking behavior of the composite films, they were successively immersed in deionized water and then in aqueous 0.1 M sulfuric acid until SPR response stabilization. As the main mode of SPR measurements, the time dependence of the reflected light intensity from SPR chip at a fixed angle of incidence was recorded. The resulting observation of cyclic shifts in the SPR reflectance upon successive immersion of the sample into deionized water and aqueous 0.1 M sulfuric acid confirmed the swell/shrink process (Fig. 9.3).

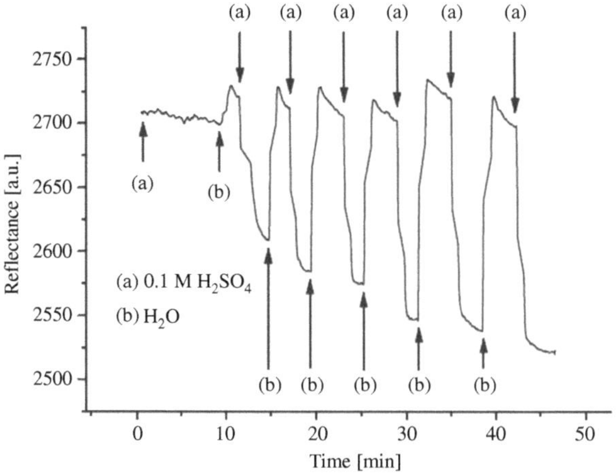

Figure 9.3 SPR sensogram corresponding to the shrinking and swelling of the PAA hydrogel film containing embedded Ag nanoparticles, during successive immersions in deionized water and aqueous 0.1 M sulfuric acid, respectively. Reproduced from Ref. [24] with permission from CSIRO Publishing.

9.2.4 Theoretical Study of Nanocomposite Polymer Matrix with Dynamic Plasmonic Properties by Means of FDTD Simulations

Simulation of nanocomposite matrix optical properties was performed using the finite-difference time-domain (FDTD) method in the FDTD Solutions package (Lumerical Solutions, Inc.) by means of a model consisting of nine spherical Ag nanoparticles located in a 3×3 square grid embedded into a polymer medium resting on a glass substrate (Fig. 9.4). The diameter of nanoparticles was fixed at 12 nm taking into consideration the experimentally measured electronic absorption spectrum of a PAA/AgNPs matrix in air

(Fig. 9.1), which peaks at 440 nm. Shrunken and swollen composite films were designed by adjusting the polymer medium refractive index and interparticle distance. It should be noted that the chosen parameters differ significantly for shrunken and swollen films, and their values as well as a value of nanoparticle diameter most probably do not reflect the exact experimental parameters. These values were introduced intentionally in order to obtain more pronounced differences in the simulated optical properties and to determine the direction of absorption peak shift. However, the estimated interparticle distance in a shrunken state based on the composite film preparation procedure was equal to ~22 nm, which is near the value of 15 nm used in modelling. After obtaining the absorption spectra, two-dimensional electric field intensity distributions at light wavelengths corresponding to the absorbance peak maxima were simulated.

The swell/shrink process was simulated numerically as the change in ambient refractive index and interparticle distance in a layer of Ag nanoparticles (Fig. 9.4). The absorbance spectra (Fig. 9.5) and electric field intensity distributions at light wavelengths corresponding to absorbance peak maxima (Fig. 9.6) were calculated. According to modelling results, the absorbance peak undergoes a blue shift due to a decrease in the ambient medium refractive index and an increase in the interparticle distance (i.e., decrease in interparticle interaction, which is evident by a decrease in electric field intensity) during the swelling process.

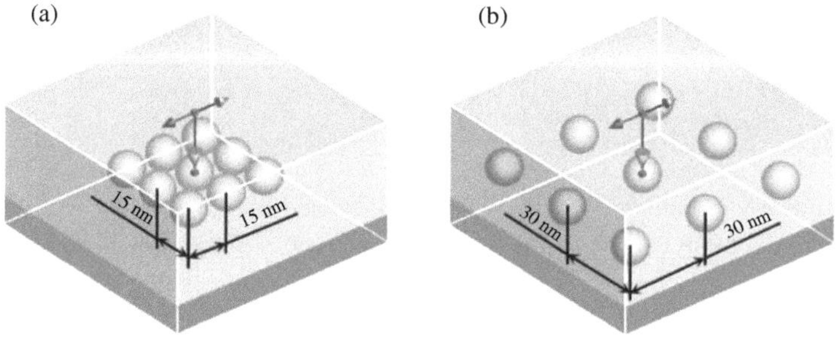

Figure 9.4 Models of the (a) shrunken and (b) swollen PAA hydrogel containing Ag nanoparticles. Reproduced from Ref. [24] with permission from CSIRO Publishing.

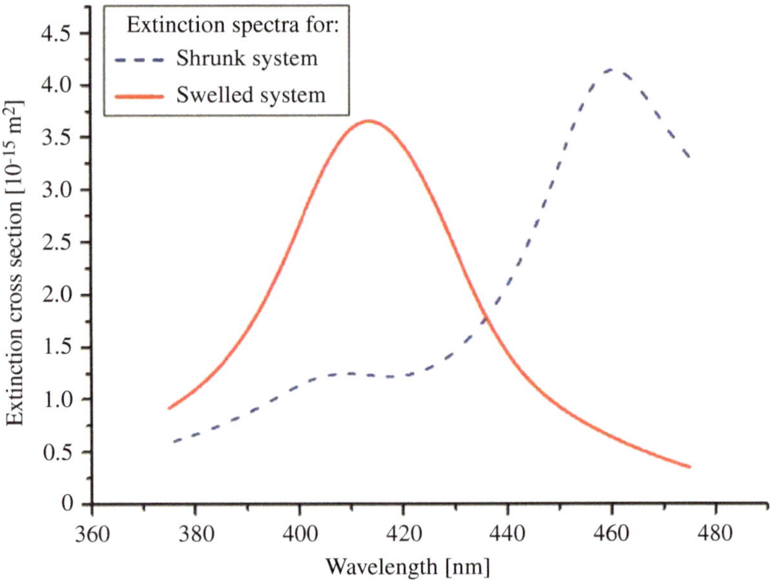

Figure 9.5 Simulated absorbance spectra for shrunken and swollen PAA hydrogel film containing Ag nanoparticles. Reproduced from Ref. [24] with permission from CSIRO Publishing.

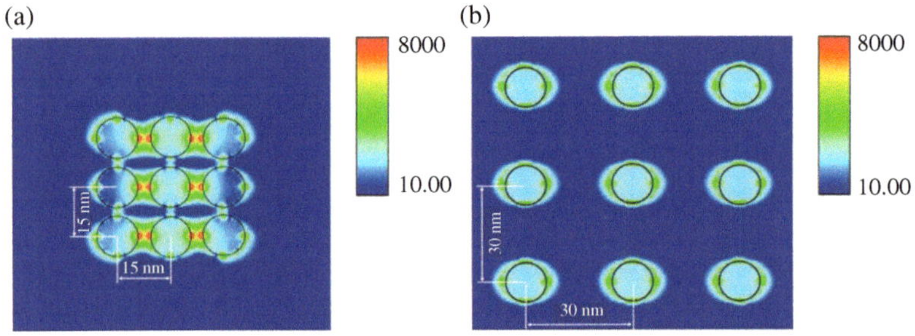

Figure 9.6 Simulated electric field intensity distributions for (a) shrunken and (b) swollen PAA hydrogel film containing Ag nanoparticles at light wavelengths corresponding to absorbance peak maxima in Fig. 9.5. Reproduced from Ref. [24] with permission from CSIRO Publishing.

9.3 Redox Switching of Electrorefractive, Electrochromic, and Conductivity Functions of Cu^{2+}/Polyacrylic Acid Films on the SPR Electrode Surface

9.3.1 Functional Polymers with Controlled Optoelectronic Functions

Functional polymers of controlled refractivity [32, 33], photochromic [34, 35], electrochromic [36–38], and optoelectronic [39, 40] functions have attracted substantial research activities in the last decades. The use of polymers with tunable refractive properties as optical modulators, optical filters, or electro-optic guided wave devices has been reported [41, 42]. Similarly, polymers exhibiting photorefractive properties have been extensively used for optical storage [43, 44]. Photochromic and electrochromic polymers have been suggested as active materials for the development of smart windows [45, 46], and conductive polymers have been extensively employed as active components in electrically induced light-emitting devices (LEDs) [39, 40, 47].

A huge effort has been directed toward the development of functional metal or semiconductor nanoparticle–polymer hybrid systems exhibiting tailored sensoric [48–50], electronic [51, 52], and photoelectrochemical functions [53, 54]. In this section, the possibilities of molecular plasmonics are considered for development and investigation of a redox-switchable Cu^{2+}-ion/PAA composite that undergoes electrochemical reduction to a Cu^0-nanoparticle/PAA composite. SPR spectroscopy, absorption spectroscopy, and conductivity measurements were applied to examine the reversible redox switching of the electrorefractive, electrochromic, and conductivity functions of the hybrid polymer. Previous studies have reported on the chemical preparation of Cu-nanoparticle/polymer hybrid systems [55]. Also, the reversible electrochemical deposition of metals [56], e.g., Pb or Cu, on conductive glass as a photochromic layer has been examined. The considered study represents a Cu^{2+}/PAA hybrid system demonstrating redox-switchable optoelectronic features [57].

9.3.2 Investigation of Functional Properties of the Cu^{2+}/PAA Composite

A polyacrylic acid film was electropolymerized on an Au-coated glass plate [58–60], and the film was saturated with Cu^{2+} ions (Fig. 9.7). Values of the polymer film thickness, *d*, and complex refractive index, *n*, obtained by fitting the experimental SPR curve were $d \sim 280$ nm and $n = 1.372 + 0.02i$.

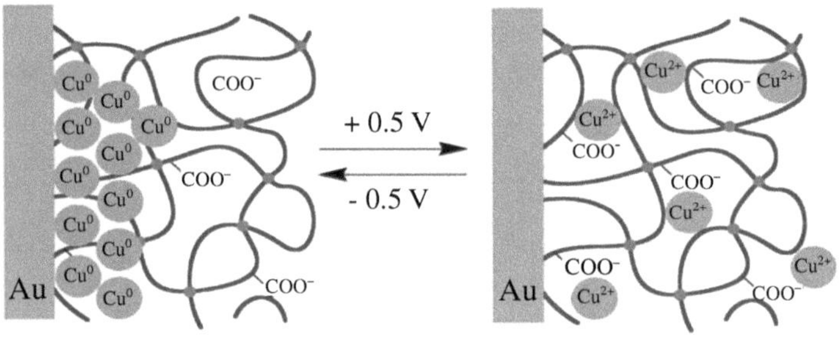

Figure 9.7 Simplified scheme of the redox-switchable Cu^{2+}/PAA metamaterial-modified Au electrode.

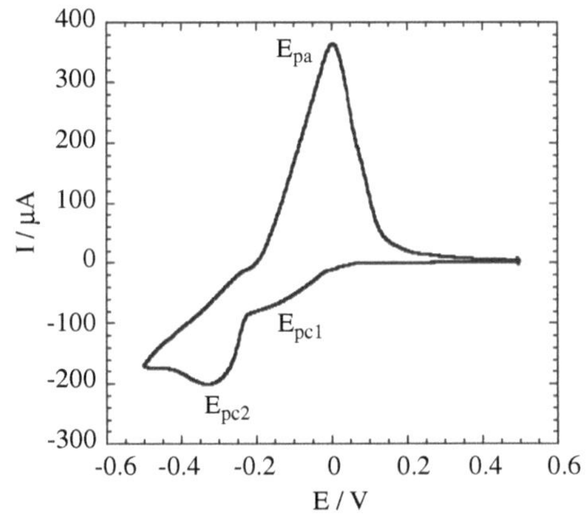

Figure 9.8 Cyclic voltammogram of the Cu^{2+}/PAA-modified Au electrode. The data were recorded in 0.1 M tris-buffer, pH 5.5, under argon, potential scan rate 3 mV/s. Reprinted with permission from Ref. [57]. Copyright 2002, WILEY-VCH Verlag GmbH & Co. KGaA, Weinheim.

Figure 9.8 shows the cyclic voltammogram of the resulting Cu^{2+}/PAA film using 0.1 M tris-buffer, pH 5.5, as the electrolyte solution.

This cyclic voltammogram follows the known mechanism for the electrodeposition of copper on an Au electrode [61]. Upon sweeping the potential from +0.5 V to −0.5 V, a poorly resolved cathodic wave, corresponding to the reduction of Cu^{2+} to Cu^{+}, was observed at E_{pc1} = −0.12 V, followed by the reduction wave of Cu^{+} to Cu^{0} at E_{pc2} = −0.31 V. Upon sweeping the potential from −0.5 V to +0.5 V, the reverse anodic wave was observed at E_{pa} = 0.0 V, corresponding to the oxidation of Cu^{0} to Cu^{2+}. By coulometric assay of the waves corresponding to the oxidation of Cu^{0} to Cu^{2+}, and assuming that all Cu^{0} is oxidized to Cu^{2+}, the Cu^{0} content in the film was estimated to be ca. 6.3×10^{-8} mol/cm^{2}.

The redox transitions Cu^{2+}/PAA to Cu^{0}/PAA and back were followed by in situ electrochemical/SPR measurements. Figure 9.9 shows the time-dependent SPR curves of the film upon switching the electrode potential from +0.5 V to −0.5 V. The reduction of Cu^{2+} to Cu^{0} metallic nanoparticles was accompanied by a minor change in the position of the SPR minimum reflectivity angle (from 67.3° to 67.7°), but by a substantial increase in the film reflectance and the formation of shallow SPR curves, which became shallower as the Cu^{0} nanoparticles accumulate in the film. A fitting of the SPR reflectance curve corresponding to the Cu^{2+}/PAA film, Fig. 9.9A, curve a, was performed. The results indicated no significant difference in film thickness in comparison with the PAA film without Cu^{2+}, and a complex refractive index of the film, n = 1.374 + 0.007i. Fitting of the data obtained upon electrochemical formation of the Cu^{0} nanoparticles in the film revealed that the real component of the refractive index of the film, n_{Re}, was almost unaltered, while the imaginary part of the refractive index, n_{Im}, was significantly increased. For example, fitting of the SPR curve obtained after 210 s of Cu^{0} accumulation, Fig. 9.9A, curve h, revealed no significant change in film thickness and a complex refractive index of the film, n = 1.360 + 0.126i. The complex dielectric constant of the film, ε, could be calculated from the obtained values of the complex refractive index [62], using Eqs. (9.1–9.2), where Re and Im are notations for the real and imaginary parts of the corresponding value, respectively:

$$\varepsilon_{Re} = n_{Re}^2 - n_{Im}^2, \tag{9.1}$$

$$\varepsilon_{Im} = 2n_{Re}n_{Im}, \tag{9.2}$$

where $n = n_{Re} + in_{Im}$ and $\varepsilon = \varepsilon_{Re} + i\varepsilon_{Im}$.

The electrochemically induced change in the refractive index implies that the complex dielectric constant ε of the film also alters upon reduction of the Cu^{2+} ions to metallic particles. It was found that the dielectric constant of the Cu^{2+}/PAA film is $\varepsilon = 1.89 + 0.02i$, whereas the dielectric constant of Cu^0/PAA is $\varepsilon = 1.83 + 0.34i$. Clearly, the real part of the dielectric constant was almost unaltered, while the imaginary one changed significantly. At optical frequencies, the complex dielectric constant, ε, is related directly to the metallic conductivity of the material. For light of wavelength λ incident on a material with conductivity σ, the imaginary part of the dielectric constant can be expressed by Eq. (9.3), where c is the speed of light [62]:

$$\varepsilon_{Im} = 2\sigma\lambda/c. \quad (9.3)$$

From the optical constants [63], and using Eq. (9.3), an effective in-plane conductivity of Cu^0/PAA corresponding to 8.5×10^3 S/m was calculated. Thus, the major changes in the complex refractive index of the film may be attributed to the enhanced conductivity of the film containing Cu^0 particles.

Figure 9.9A, inset, shows the time-dependent SPR reflectance changes at a fixed incidence angle of $\theta = 66°$ upon transformation of the Cu^{2+}/PAA film to the Cu^0/PAA state. The reduction of Cu^{2+} to Cu^0 particles was a relatively slow process and proceeded for ca. 300 s for complete metallization of the film. This was attributed to the fact that the ions have to migrate through the polymer film and reach the electrode surface in order to be reduced. The Cu^0 metallic particles generated electrochemically at the electrode surface provide further conductive sites for the formation of the Cu^0 particles across the polymeric film. Figure 9.9B shows the SPR reflectance curves of the Cu^0/PAA film upon application of a potential step from −0.5 V to +0.5 V. The Cu^{2+}/PAA was regenerated as evident by the reflectance curve with a minimum reflectance angle at 67.3°. Figure 9.9B, inset, shows the time-dependent reflectance changes upon oxidation of the Cu^0/PAA film to the Cu^{2+}/PAA film. In contrast to the slow reduction process, the oxidation of the film was fast and proceeded within ca. 30 s. This is consistent with the fact that the conductive Cu^0 particles were aggregated in a metallic network associated with the electrode, and their oxidation was fast.

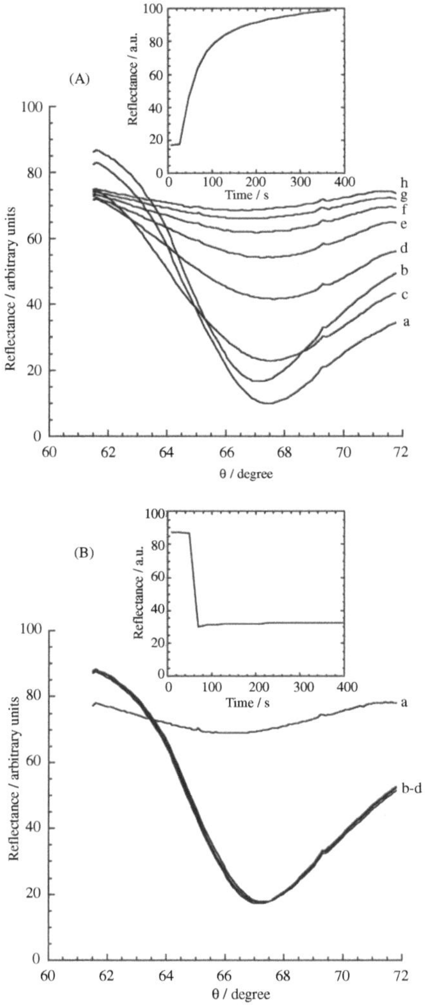

Figure 9.9 (A) SPR reflectance curves of the Cu^{2+}/PAA-modified Au electrode recorded after the application of a potential step from 0.5 V to −0.5 V: (a) 0 s, (b) 30 s, (c) 60 s, (d) 90 s, (e) 120 s, (f) 150 s, (g) 180 s, (h) 210 s. Inset: Time-dependent reflectance changes recorded at a fixed angle of incidence ($\vartheta = 66°$) immediately after application of the potential step from 0.5 V to −0.5 V. (B) SPR reflectance curves of the Cu^0/PAA-modified Au electrode recorded after the application of a potential step from −0.5 V to 0.5 V: (a) 0 s, (b) 30 s, (c) 60 s, (d) 90 s. Inset: Time-dependent reflectance changes recorded at a fixed angle of incidence ($\vartheta = 66°$) immediately after the application of the potential step from −0.5 V to 0.5 V. All SPR measurements were performed in 0.1 M tris-buffer, pH 5.5. Reprinted with permission from Ref. [57]. Copyright 2002, WILEY-VCH Verlag GmbH & Co. KGaA, Weinheim.

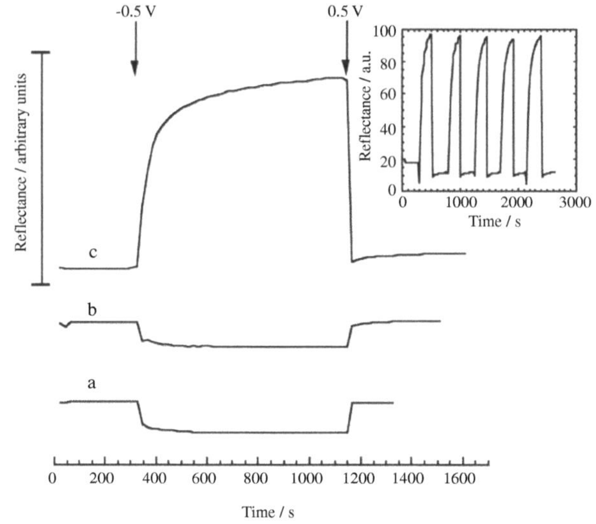

Figure 9.10 Time-dependent reflectance changes recorded at a fixed angle of incidence ($\vartheta = 66°$) upon reversible application of a potential step from 0.5 V to −0.5 V and back from −0.5 V to 0.5 V: (a) the bare Au electrode, (b) the PAA-modified Au electrode, (c) the Cu^{2+}/PAA-modified Au electrode. Arrows show the time when the respective potential was applied. Inset: Time-dependent reversible reflectance changes in the Cu^{2+}/PAA-modified Au electrode recorded at a fixed angle of incidence ($\vartheta = 66°$) upon multi-potential step switching between 0.5 V and −0.5 V. All SPR measurements were performed in 0.1 M tris-buffer, pH 5.5. Reprinted with permission from Ref. [57]. Copyright 2002, WILEY-VCH Verlag GmbH & Co. KGaA, Weinheim.

Figure 9.10 shows the time-dependent reflectance changes at a fixed angle of $\theta = 66°$ of the bare Au electrode, curve a, the PAA-modified Au surface, curve b, and the Cu^{2+}/PAA-modified electrode, curve c, upon the application of a potential step from −0.5 V to +0.5 V and back. The potential step had a small effect on the reflectance of the bare Au surface and the PAA-modified electrode. However, the changes in the reflectance intensities were of opposite direction to that observed for the Cu^{2+}/PAA film. That is, the reflectance of the bare Au electrode and the PAA-modified electrode showed a small decrease upon the application of a potential of −0.5 V, whereas the reflectance of the Cu^{2+}/PAA-modified electrode displayed a large increase upon the application of a potential of −0.5 V, and the formation of the Cu^0/PAA film. The redox switching of the reflectance features of the Cu^{2+}/PAA film was fully reversible

(Fig. 9.10, inset). While the increase in reflectance corresponding to the formation of the Cu^0/PAA film was slow, the transformation of the film to the Cu^{2+}/PAA state was fast. Interestingly, no degradation was observed in the reflectance signals, indicating that there was no leakage of Cu^{2+} ions from the polymer film. Presumably, oxidation of the Cu^0 metallic particles associated with the electrode yielded Cu^{2+} ions that are trapped in the PAA network. Thus, SPR analysis of the optical properties of the Cu^{2+}/PAA film revealed the redox-switchable electrorefractive properties of the polymer.

The Cu^{2+}/PAA film associated with the Au-coated glass electrode also revealed redox-switchable electrochromic properties. Figure 9.11A shows the time-dependent evolution of the absorption features of the film upon the application of a potential step from +0.5 V to −0.5 V on the Cu^{2+}/PAA film. It can be seen that transformation of the Cu^{2+}/PAA film to the Cu^0/PAA state resulted in an increase in absorbance and in the appearance of an absorbance band with a maximum at ca. 630 nm. This band was attributed to the surface plasmon excitation upon build-up of the Cu^0 nanoparticle network at the electrode support. An enlarged width and red shift of this band relative to the published values [64] of 570–590 nm are probably related to the non-spherical form, coalescence, and interparticle coupling of the Cu^0 particles formed.

Figure 9.11B shows the absorbance changes in the Cu^0/PAA electrode upon switching the potential from −0.5 V to +0.5 V. A fast decrease in the absorbance was observed due to electrochemical dissolution of the Cu^0 nanoparticles. The film transmittance changed from 85% (Cu^{2+}/PAA state) to 70% (Cu^0/PAA state) at 670 nm. The redox-switchable absorbance changes in the film were reversible, revealing that the Cu^{2+}/Cu^0/PAA film represents a novel electrochromic system.

Analysis of the optical properties of the redox-switchable Cu^{2+}/PAA film also revealed that the conductivity features of the film were altered upon redox switching of Cu^{2+}/PAA to the Cu^0/PAA state. The transverse (perpendicular) resistivity of the film has been characterized in its different states by measuring the voltammetric response between the Au-conductive support and a conductive tip (Au wire 0.5 mm diameter) introduced into the polymeric film. A small range (50 mV) potential sweep (2 mV/s) was applied between these two electrodes to measure the conductive properties of the

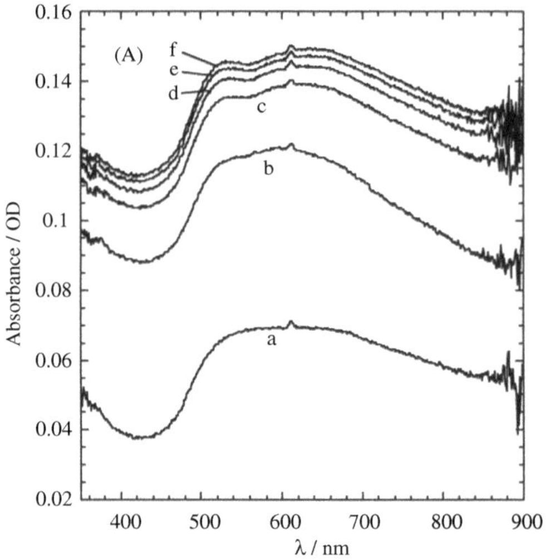

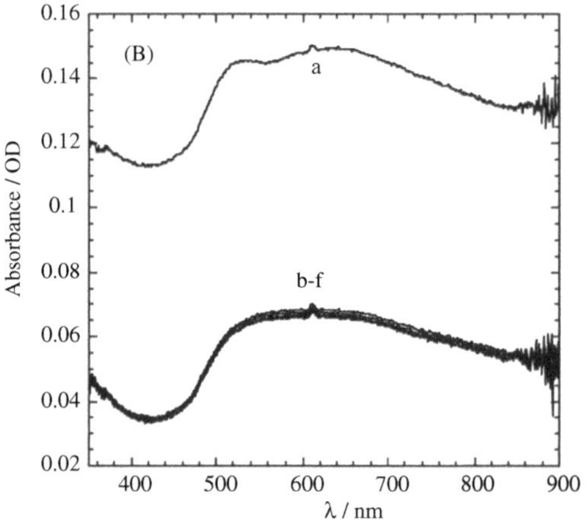

Figure 9.11 (A) Absorbance spectra of the Cu^{2+}/PAA-modified Au electrode measured after the application of a potential step from 0.5 V to −0.5 V: (a) 0 min, (b) 1 min, (c) 3 min, (d) 5 min, (e) 7 min, (f) 9 min. (B) Absorbance spectra of the Cu^0/PAA-modified Au electrode measured after the application of a potential step from −0.5 V to 0.5 V: (a) 0 min, (b) 1 min, (c) 3 min, (d) 5 min, (e) 7 min, (f) 9 min. The spectra were recorded in 0.1 M tris-buffer, pH 5.5, and the spectrum of the bare Au electrode was subtracted from the spectra of the modified electrode. Reprinted with permission from Ref. [57]. Copyright 2002, WILEY-VCH Verlag GmbH & Co. KGaA, Weinheim.

medium under conditions where no electrochemical Faradaic process occurs in the film. Figure 9.12, curve a, shows the I–V curve that corresponds to Cu^{2+}/PAA. From the slope, the film resistance was estimated to be ca. 300 kΩ. Reduction of the film to the Cu^0/PAA state followed by measuring the resistance of the film resulted in the *I–V* curve shown in Fig. 9.12, curve b. The resistance of the film that included the Cu^0 metallic nanoparticles was substantially lower, R = 2.2 kΩ, implying that the film exhibited a significantly higher conductivity. By reversible switching of the film between the Cu^0/PAA and Cu^{2+}/PAA states, the polymer film was cycled between high- and low-conductive states, respectively (Fig. 9.12, inset).

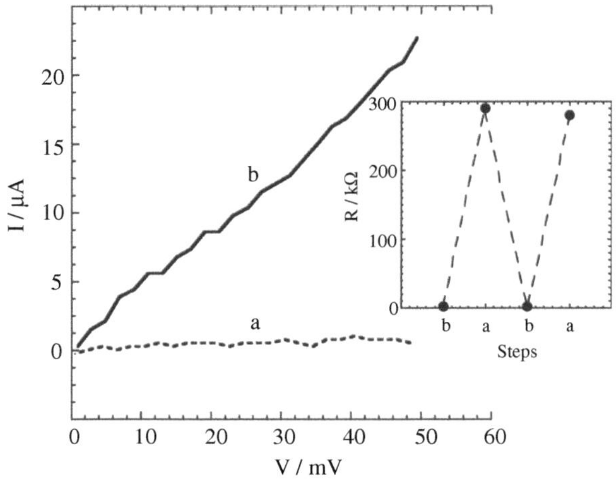

Figure 9.12 *I–V* curves of the redox-switched polymer states: (a) The Cu^{2+}/PAA-modified Au electrode after application of a potential of 0.5 V. (b) The Cu^0/PAA-modified Au electrode after application of a potential of −0.5 V. Inset: Reversible changes in the film resistance extracted from the *I–V* curves upon reversible application of a potential of 0.5 V (steps "a") or −0.5 V (steps "b"). The *I–V* curves were measured between the Au conductive support and an Au wire (0.5 mm diameter) introduced into the polymeric film. Reprinted with permission from Ref. [57]. Copyright 2002, WILEY-VCH Verlag GmbH & Co. KGaA, Weinheim.

9.4 Conclusion

Specific studies described in this chapter demonstrate the possibilities of molecular plasmonics techniques for development and characterization of nanocomposite metamaterials exemplified by nanocomposite thin films possessing a range of reversible properties. These include pH-induced swelling and shrinking of Ag/PAA films, as well as redox-switchable electrorefractive, electrochromic, and conductivity functions of Cu^{2+}/PAA films.

The first study is related to the fabrication and investigation of plasmonic metamaterial with reversible optical and physicochemical properties based on a polymer–metal nanoparticles composite. Specifically, PAA films containing Ag nanoparticles have been shown to exhibit optical absorbance properties controllable by the pH of the surrounding aqueous medium. Swelling and shrinking of the studied composite film inducing absorbance peak shifts were confirmed by theoretical simulations. However, optimization of preparation protocol (i.e., reagent concentrations and photoreduction process duration) is required to achieve the maximum sensitivity of optical properties due to pH changes. Hydrogels containing metal nanoparticles might be employed as pH sensors or as pH-driven dynamic systems, as well as sensor elements for substances and factors that influence LSPR properties of nanoparticles (e.g., adsorption of molecules, swelling/shrinking of polymer, changes in ambient medium, etc.).

The next considered study revealed a redox-switchable functional polymer consisting of a Cu^{2+}/polyacrylic acid film associated with an electrode support. By redox transformation of the polymer between the Cu^{2+}/PAA and Cu^{0}/PAA states, the electrorefractive, electrochromic, and conductivity properties of the film were shown as those that could be reversibly switched multiple times. These electroswitchable properties of the polymer matrix may be used to assemble new types of optical filters [65], optical modulators [66], optoelectronic devices [67], smart windows [68], and interfaces of controlled conductivity. Preliminary studies indicate that other metal-ion/PAA films such as Pb^{2+}/PAA or Zn^{2+}/PAA show similar redox-switchable optoelectronic properties.

References

1. Ariga, K., Ishihara, S., Labuta, J., and Hill, J. P. (2011). Supramolecular approaches to nanotechnology: Switching properties and dynamic functions, *Curr. Org. Chem.*, **15**, pp. 3719–3733.

2. Wales, D. J., Cao, Q., Kastner, K., Karjalainen, E., Newton, G. N., and Sans, V. (2018). 3D-printable photochromic molecular materials for reversible information storage, *Adv. Mater.*, **30**, 1800159.

3. Lai, W. C., Hsueh, C. Y., and Chang, C. W. (2019). Nanostructured polymers with embedded self-assembled networks: Reversibly tunable phase behaviors and physical properties, *Soft Matter*, https://doi.org/10.1039/C9SM00997C.
4. Austin, D., Mullin, N., Bismuto, A., Luxmoore, I., Adawi, A. M., Revin, D. G., Soulby, M., Cockburn, J. W., Jiang, Q., Krysa, A. B., Cullis, A. G., Faist, J., Hobbs, J. K., and Wilson, L. R. (2010). Transmission properties of plasmonic metamaterial quantum cascade lasers, *IEEE Photonics Technol. Lett*, **22**, pp. 1217–1219.
5. Ren, M., Jia, B., Ou, J. Y., Plum, E., Zhang, J., MacDonald, K. F., Nikolaenko, A. E., Xu, J., Gu, M., and Zheludev, N. I. (2011). Nanostructured plasmonic medium for terahertz bandwidth all-optical switching, *Adv. Mater.*, **23**, pp. 5540–5544.
6. Gu, Y., Li, Q., Xiao, J., Wu, K., and Wang, G. P. (2011). Plasmonic metamaterials for ultrasensitive refractive index sensing at near infrared, *J. Appl. Phys.*, **109**, 023104.
7. Kabashin, A. V., Evans, P., Pastkovsky, S., Hendren, W., Wurtz, G. A., Atkinson, R., Pollard, R., Podolskiy, V. A., and Zayats, A. V. (2009). Plasmonic nanorod metamaterials for biosensing, *Nat. Mater.*, **8**, pp. 867–871.
8. Neira, A. D., Olivier, N., Nasir, M. E., Dickson, W., Wurtz, G. A., and Zayats, A. V. (2015). Eliminating material constraints for nonlinearity with plasmonic metamaterials, *Nat. Commun.*, **6**, 7757.
9. Zhang, X., Zheng, Y., Liu, X., Lu, W., Dai, J., Lei, D. Y., and MacFarlane, D. R. (2015). Hierarchical porous plasmonic metamaterials for reproducible ultrasensitive surface-enhanced Raman spectroscopy, *Adv. Mater.*, **27**, pp. 1090–1096.
10. Zhang, X., Sun, B., Friend, R. H., Guo, H., Nau, D., and Giessen, H. (2006). Metallic photonic crystals based on solution-processible gold nanoparticles, *Nano Lett.*, **6**, pp. 651–655.
11. Malynych, S. and Chumanov, G. (2003). Light-induced coherent interactions between silver nanoparticles in two-dimensional arrays, *J. Am. Chem. Soc.*, **125**, pp. 2896–2898.
12. Aslan, K., Leonenko, Z., Lakowicz, J. R., and Geddes, C. D. (2005). Fast and slow deposition of silver nanorods on planar surfaces: Application to metal-enhanced fluorescence, *J. Phys. Chem. B*, **109**, pp. 3157–3162.
13. Zheng, Y. B., Yang, Y. W., Jensen, L., Fang, L., Juluri, B. K., Flood, A. H., Weiss, P. S., Stoddart, J. F., and Huang, T. J. (2009). Active molecular plasmonics: Controlling plasmon resonances with molecular switches, *Nano Lett.*, **9**, pp. 819–825.

14. Jonsson, M. P., Dahlin, A. B., Feuz, L., Petronis, S., and Höök, F. (2010). Locally functionalized short-range ordered nanoplasmonic pores for bioanalytical sensing, *Anal. Chem.*, **82**, pp. 2087–2094.
15. Zheludev, N. I. and Plum, E. (2016). Reconfigurable nanomechanical photonic metamaterials, *Nat. Nanotechnol.*, **11**, pp. 16–22.
16. Shim, J. Y. and Gupta, V. K. (2007). Reversible aggregation of gold nanoparticles induced by pH dependent conformational transitions of a self-assembled polypeptide, *J. Colloid Interface Sci.*, **316**, pp. 977–983.
17. Iyer, K. S., Zdyrko, B., Malynych, S., Chumanov, G., and Luzinov, I. (2011). Reversible submergence of nanoparticles into ultrathin block copolymer films, *Soft Matter*, **7**, pp. 2538–2542.
18. Otsuka, H., Akiyama, Y., Nagasaki, Y., and Kataoka, K. (2001). Quantitative and reversible lectin-induced association of gold nanoparticles modified with α-lactosyl-ω-mercapto-poly (ethylene glycol), *J. Am. Chem. Soc.*, **123**, pp. 8226–8230.
19. Kristensen, A., Yang, J. K., Bozhevolnyi, S. I., Link, S., Nordlander, P., Halas, N. J., and Mortensen, N. A. (2017). Plasmonic colour generation, *Nat. Rev. Mater.*, **2**, 16088.
20. Wang, G., Chen, X., Liu, S., Wong, C., and Chu, S. (2016). Mechanical chameleon through dynamic real-time plasmonic tuning, *ACS Nano*, **10**, pp. 1788–1794.
21. Papaioannou, M., Plum, E., Rogers, E. T., and Zheludev, N. I. (2018). All-optical dynamic focusing of light via coherent absorption in a plasmonic metasurface, *Light Sci. Appl.*, **7**, 17157.
22. Lucas, B. D., Kim, J. S., Chin, C., and Guo, L. J. (2008). Nanoimprint lithography-based approach for the fabrication of large-area, uniformly-oriented plasmonic arrays, *Adv. Mater.*, **20**, pp. 1129–1134.
23. Gradess, R., Abargues, R., Habbou, A., Canet-Ferrer, J., Pedrueza, E., Russell, A., Valdés, J. L., and Martínez-Pastor, J. P. (2009). Localized surface plasmon resonance sensor based on Ag-PVA nanocomposite thin films, *J. Mater. Chem.*, **19**, pp. 9233–9240.
24. Chegel, V., Lopatynskyi, A., Ishihara, S., Hill, J. P., and Ariga, K. (2012). Ag nanoparticle-poly (acrylic acid) composite film with dynamic plasmonic properties, *Aust. J. Chem.*, **65**, pp. 1223–1227.
25. Elliott, J. E., Macdonald, M., Nie, J., and Bowman, C. N. (2004). Structure and swelling of poly (acrylic acid) hydrogels: Effect of pH, ionic strength, and dilution on the crosslinked polymer structure, *Polymer*, **45**, pp. 1503–1510.

26. Zhou, Y., Yu, S. H., Wang, C. Y., Li, X. G., Zhu, Y. R., and Chen, Z. Y. (1999). A novel ultraviolet irradiation photoreduction technique for the preparation of single-crystal Ag nanorods and Ag dendrites, *Adv. Mater.*, **11**, pp. 850–852.
27. Murray, W. A., Auguié, B., and Barnes, W. L. (2009). Sensitivity of localized surface plasmon resonances to bulk and local changes in the optical environment, *J. Phys. Chem. C*, **113**, pp. 5120–5125.
28. Rechberger, W., Hohenau, A., Leitner, A., Krenn, J. R., Lamprecht, B., and Aussenegg, F. R. (2003). Optical properties of two interacting gold nanoparticles, *Opt. Commun.*, **220**, pp. 137–141.
29. Gooch, J. W. (2010) *Encyclopedic Dictionary of Polymers*, 2nd Ed. (Springer, New York).
30. https://www.mt.com/us/en/home/supportive_content/concentration-tables-ana/Sulfuric_Acid_re_e.html
31. Akkilic, N., Mustafaev, M., and Chegel, V. (2008). Conformational dynamics of poly (acrylic acid)–bovine serum albumin polycomplexes at different pH conditions, *Macromol. Symp.*, **269**, pp. 138–144.
32. Burland, D. M., Miller, R. D., and Walsh, C. A. (1994). Second-order nonlinearity in poled-polymer systems, *Chem. Rev.*, **94**, pp. 31–75.
33. Fouda, I. M. (2001). Opto-thermal properties of fibers XVIII—Skin-core structures variations in nylon 6 fibers due to different annealing conditions, *J. Appl. Polym. Sci.*, **81**, pp. 3349–3360.
34. Walheim, S., Schäffer, E., Mlynek, J., and Steiner, U. (1999). Nanophase-separated polymer films as high-performance antireflection coatings, *Science*, **283**, pp. 520–522.
35. Feng, W., Zhang, T. R., Liu, Y., Wei, L., Lu, R., Li, T. J., Zha, Y. Y., and Yao, J. N. (2002). Novel hybrid inorganic–organic film based on the tungstophosphate acid–polyacrylamide system: Photochromic behavior and mechanism, *J. Mater. Res.*, **17**, pp. 133–136.
36. Rosseinsky, D. R. and Mortimer, R. J. (2001). Electrochromic systems and the prospects for devices, *Adv. Mater.*, **13**, pp. 783–793.
37. Arman, S. (2001). Electrochromic materials for display applications: An introduction, *J. New Mat. Electrochem. Syst.*, **4**, pp. 173–180.
38. Monk, P. M. S., Mortimer, R. J., and Rosseinsky, D. R. (1995). *Electrochromism: Fundamentals and Applications* (VCH, Weinheim).
39. Grote, J. G., Zetts, J. S., Nelson, R. L., Hopkins, F. K., Dalton, L. R., Zhang, C. H., and Steier, W. H. (2001). Effect of conductivity and dielectric constant on the modulation voltage for nonlinear optic polymer based opto-electronic devices, *Opt. Eng.*, **40**, pp. 2464–2473.

40. Friend, R. H. (2001). Conjugated polymers. New materials for optoelectronic devices, *Pure Appl. Chem.*, **73**, pp. 425–430.
41. Steier, W. H., Chen, A., Lee, S. S., Garner, S., Zhang, H., Chuyanov, V., Dalton, L. R., Wang, F., Ren, A. S., Zhang, C., Todorova, G., Harper, A., Fetterman, H. R., Chen, D., Udupa, A., Bhattacharya, D., and Tsap, B. (1999). Polymer electro-optic devices for integrated optics, *Chem. Phys.*, **245**, pp. 487–506.
42. Donaldson, A. (1991). Candidate materials and technologies for integrated optics: Fast and efficient electro-optic modulation, *J. Phys. D*, **24**, 785.
43. Lucchetti, L. and Simoni, F. (2000). Soft material for optical storage, *Riv. Nuovo Cimento*, **23**, pp. 1–27.
44. Moerner, W. E. and Silence, S. M. (1994). Polymeric photorefractive materials, *Chem. Rev.*, **94**, pp. 127–155.
45. Lampert, C. M. (1993). Optical switching technology for glazings, *Thin Solid Films*, **236**, pp. 6–13.
46. Granqvist, C. G. (1990). Chromogenic materials for transmittance control of large-area windows, *Crit. Rev. Solid State*, **16**, pp. 291–308.
47. Zhan, C., Yu, G., Lu, Y., Wang, L., Wujcik, E., and Wei, S. (2017). Conductive polymer nanocomposites: A critical review of modern advanced devices, *J. Mater. Chem. C*, **5**, pp. 1569–1585.
48. Godovsky, D. Y. (2000). Device applications of polymer-nanocomposites, In *Biopolymers–PVA Hydrogels, Anionic Polymerisation Nanocomposites*, (Springer, Berlin, Heidelberg) pp. 163–205.
49. Bauer, G., Pittner, F., and Schalkhammer, T. (1999). Metal nano-cluster biosensors, *Microchim. Acta*, **131**, pp. 107–114.
50. Schalkhammer, T. (1998). Metal nano-clusters as transducers for bioaffinity interactions, *Monatsh. Chem.*, **129**, pp. 1067–1092.
51. Sheeney-Haj-Ichia, L., Sharabi, G., and Willner, I. (2002). Control of the electronic properties of thermosensitive poly (*N*-isopropylacrylamide) and Au-nanoparticle/poly (*N*-isopropylacrylamide) composite hydrogels upon phase transition, *Adv. Funct. Mater.*, **12**, pp. 27–32.
52. Selvan, S. T., Hayakawa, T., Nogami, M., and Möller, M. (1999). Block copolymer mediated synthesis of gold quantum dots and novel gold–polypyrrole nanocomposites, *J. Phys. Chem. B*, **103**, pp. 7441–7448.
53. Pardo-Yissar, V., Bourenko, T., Wasserman, J., and Willner, I. (2002). Solvent-switchable photoelectrochemistry in the presence of CdS–nanoparticle/acrylamide hydrogels, *Adv. Mater.*, **14**, pp. 670–673.

54. Khazraji, A. C., Hotchandani, S., Das, S., and Kamat, P. V. (1999). Controlling dye (Merocyanine-540) aggregation on nanostructured TiO_2 films. An organized assembly approach for enhancing the efficiency of photosensitization, *J. Phys. Chem. B*, **103**, pp. 4693–4700.

55. Gotoh, Y., Igarashi, R., Ohkoshi, Y., Nagura, M., Akamatsu, K., and Deki, S. (2000). Preparation and structure of copper nanoparticle/poly (acrylic acid) composite films, *J. Mater. Chem.*, **10**, pp. 2548–2552.

56. Mascaro, L. H., Kaibara, E. K., and Bulhoes, L. O. (1997). An electrochromic system based on the reversible electrodeposition of lead, *J. Electrochem. Soc.*, **144**, pp. L273–L274.

57. Chegel, V. I., Raitman, O. A., Lioubashevski, O., Shirshov, Y., Katz, E., and Willner, I. (2002). Redox-switching of electrorefractive, electrochromic, and conductivity functions of Cu^{2+}/polyacrylic acid films associated with electrodes, *Adv. Mater.*, **14**, pp. 1549–1553.

58. Collins, G. L. and Thomas, N. W. (1977). Mechanism of coating by electropolymerization on metal cathodes from zinc chloride solutions of acrylamide, *J. Polymer Sci. Polymer Chem. Ed.*, **15**, pp. 1819–1831.

59. Katz, E., de Lacey, A. L., Fierro, J. L., Palacios, J. M., and Fernandez, V. M. (1993). Covalent binding of viologen to electrode surfaces coated with poly (acrylic acid) formed by electropolymerization of acrylate ions: I. Electrode preparation and characterization, *J. Electroanal. Chem.*, **358**, pp. 247–259.

60. Gabai, R., Sallacan, N., Chegel, V., Bourenko, T., Katz, E., and Willner, I. (2001). Characterization of the swelling of acrylamidophenylboronic acid–acrylamide hydrogels upon interaction with glucose by faradaic impedance spectroscopy, chronopotentiometry, quartz-crystal microbalance (QCM), and surface plasmon resonance (SPR) experiments, *J. Phys. Chem. B*, **105**, pp. 8196–8202.

61. Gileadi, E. and Tsionsky, V. (2000). Studies of electroplating using an EQCM. I. Copper and silver on gold, *J. Electrochem. Soc.*, **147**, pp. 567–574.

62. Jackson, J. D. (1998). *Classical Electrodynamics*, 3rd Ed. (John Wiley & Sons, New York).

63. Palik, E. D., Ed. (1998). *Handbook of Optical Constants of Solids* (Academic Press, San Diego, CA).

64. Sosebee, T., Giersig, M., Holzwarth, A., and Mulvaney, P. (1995). The nucleation of colloidal copper in the presence of poly (ethyleneimine), *Ber. Bunsenges. Phys. Chem.*, **99**, pp. 40–49.

65. Lenz, G., Eggleton, B. J., Madsen, C. K., and Slusher, R. E. (2001). Optical delay lines based on optical filters, *IEEE J. Quantum Electron.*, **37**, pp. 525–532.

66. Gorecki, C. (2001). Recent advances in silicon guided-wave MOEMS: From technology to applications, *Opto-Electron. Rev.*, **9**, pp. 248–256.

67. Cacialli, F. (2000). Organic semiconductors for the new millennium, *Philos. Trans. Royal Soc. A*, **358**, pp. 173–192.

68. Granqvist, C. G., Azens, A., Hjelm, A., Kullman, L., Niklasson, G. A., Rönnow, D., Strømme Mattsson, M., Veszelei, M., and Vaivars, G. (1998). Recent advances in electrochromics for smart windows applications, *Sol. Energy*, **63**, pp. 199–216.

Index